North Fork Mountain, Grant County, West Virginia.

Land Snails

of

West Virginia

Daniel C. Dourson

**With Important Contributions
from
Craig Stihler and Judy Dourson**

**Developed in cooperation with the West Virginia
Division of Natural Resources**

All illustrations by Rachael Grabowski unless otherwise stated

All photographs by the author unless otherwise stated

~2015~

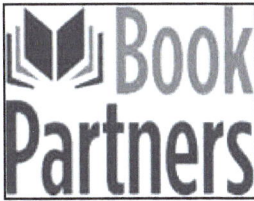

Book Partners

A division of The HF Group
North Manchester, IN 46962

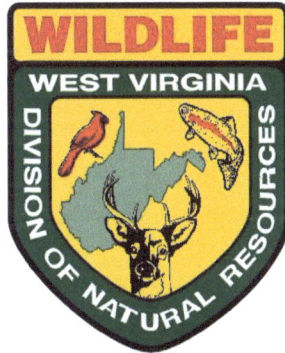

WILDLIFE
WEST VIRGINIA
DIVISION OF NATURAL RESOURCES

LINCOLN MEMORIAL
U N I V E R S I T Y

CUMBERLAND MOUNTAIN
CMRC
RESEARCH CENTER

Goatslug Publications
Bakersville, NC 28705

Contents

Acknowledgments

This manuscript would not have been possible without the support and assistance of numerous individuals and organizations.

I would like to acknowledge the printing and binding company, Book Partners, a division of The HF Group, located in North Manchester, Indiana for its high quality printing and binding, attention to detail and its commitment to print the book entirely in the USA. This is the fourth book that Book Partners has produced for me and is proof that quality and value are obtainable in the USA. I am pleased to continue to work with Book Partners.

I also thank the Cumberland Mountain Research Center at Lincoln Memorial University and Dr. Ron Caldwell, Director, for supporting this project by generously providing access to its Wild M420 Apozoom macroscope with a Jenoptik ProgRes C5 camera and i-Solution Lite software that allowed the creation of perfectly focused images of the micro-snails (under 5 mm) as well as accommodations during several visits to photograph snails for the book. The informative introduction was written by Jim Vanderhorst, state ecologist with WVDNR. Many thanks to Walt Kordek, WVDNR, for his initial and sustaining support of the statewide snail survey work and the resulting information presented here. I am grateful to Jeff Nekola, University of New Mexico, for DNA sequencing conducted in his lab to compare an undetermined species of *Vertigo* and information regarding land snail records from West Virginia. Thanks to Ken Hotopp and Tim Pearce for their earlier publication which included database records from West Virginia from all the major museum collections in the USA, county maps and information about each species. This document proved invaluable and a starting point for this manuscript.

Most of the data presented in this book were compiled under two projects overseen by the WVDNR's Wildlife Diversity Unit and funded through the US Fish and Wildlife Service's State Wildlife Grants Program with matching funds provided by the State of West Virginia as well as Dourson Biological Consulting. One project, a cooperative project with Hotopp and Pearce, compiled existing museum records from all the major institutions with collections from West Virginia. The second project, coordinated by Craig Stihler (WVDNR), involved the collection of additional specimens statewide to fill data gaps, provide recent distribution data, and collect specimens for taxonomic studies. Production of this book was funded by the West Virginia Nongame Wildlife Fund.

Specimen Loans: Jochen Gerber, Collections Manager at The Field Museum of Natural History, Chicago, Illinois assisted in obtaining multiple specimen loans of land snail and access to specimens in the Hubricht Collection for review and photography. Timothy Pearce, Curator of Mollusks at the Carnegie Museum of Natural History in Pittsburgh, Pennsylvania and Paul Robb, collection assistant, also of the Carnegic Museum, for specimen loans of land snails from West Virginia as well as the updated database of land snail records for the state of West Virginia.

Collectors: The data presented in this book represents both historical records and results from collections that began in the 1980's with a more concerted effort beginning in 2006 and continuing to the present. Past collectors include Henry A. Pilsbry, Leslie Hubricht, Gordon MacMillan, Ralph Taylor of Marshall University, Wayne F. Grimm, Amy and Wayne Van Devender, John MacGregor, James Kiser, G. Thomas Watters,

John Slapcinsky, Ronald Caldwell, Mark Gumbert, Daniel Douglas, Ken Hotopp, and Timothy Pearce.

Collectors from 2006 to Present
First and foremost, many thanks to Craig Stihler of the WVDNR, for his tireless efforts to document the land snail fauna of the state of West Virginia as well as acknowledging the important role that land snails play in the ecosystem. I would like to give special recognition to Jeff Hajenga and Kieran O'Malley, wildlife biologists with WVDNR, who went above and beyond the call of duty to pursue specimens of special interest including some new species described here in this book as well as thoroughly surveying the areas under their jurisdiction. Thanks also to the following individuals who contributed significantly to the overall knowledge of the land snail fauna of West Virginia by collecting land snails across the state: Thomas Allen, Richard Bailey, Whitney Bailey, Spencer Bell, Ron Blankenship, Eric Boehm, C. J. Bowman, Kelly Boyce, Barbara Breshock, Terry Bronson, Tony Bronson, Albert "Jay" Buckelew, Alton Byers, Elizabeth Byers, Roger Channell, Amy Cimarolli, Janet Clayton, Marquette Crockett, Jim Crum, Traci Cummings, Zachary Dillard, Barbara Douglas, Dan Dourson, Judy Dourson, Rick Doyle, Kevin Eliason, Dick Esker, Thomas Fox, Jim Fregonara, Isaac Gibson, Celeste Good, Idun Guenther, Dave Haar, Jason Hajenga, Stacey Hajenga, Alyssa Hanna, Jeff Hansbarger, Michael Hatten, Rosalie Haizlett, Cheryl Jennings, Andrew Jones, Paul Lenza, Dan Lukich, D. W. Mahan, Brian McDonald, Joe Miller, Maryann Millet, Donna Mitchell, Patty Morrison, Sam Norris, Kieran O'Malley, Susan Olcott, Nate Owens, Nate Parish, Thomas Pauley, Maggi Perl, Judith Polan, William Roody, Nicole Sadecky, Luke Sadecky, Barbara Sargent, Katie Scott, Mark Scott, Ben Shamblin, Kem Shaw, John Slapcinsky, Frank Slider, Christine Stihler, Elizabeth Stout, Lee Strawn, Brian Streets, Ken Sturm, Alana Sucke, Dave Summerfield, Rob Tallman, Nate Taylor, Eric Tennant, Eric Tidmore, Rosa Tolin, William Tolin, Robert Vagnetti, Jim Vanderhorst, Anne Wakeford, Al Waldron, Jack Wallace, Dean Walton, Lila Warren, Annie Williams, Josh Winn, and Douglas Wood.

Illustrations, Photo Contributions, and Maps: Thanks to Rachael Grabowski for outstanding artwork, some redrawn from Pilsbry's two volume treatises and Burch's *How to Know the Eastern Land Snails*; Jeffrey C. Nekola, Dept. of Biology, University of New Mexico, and Brian F. Coles, Mollusca Section, Department of Biodiversity, National Museum of Wales, UK for photos and illustrations from *Pupillid Land Snails of Eastern North America*; Jochen Gerber, photos of *Anguispira* species; Aleta Karstad, watercolor images of slugs from *Identifying Land Snails and Slugs of Canada* by F. Wayne Grimm *et al.* (2010); Wayne Van Devender, Appalachian State University, for SEM images; Jodi White-McLean and Joris Koene for images of love darts; George Boorujy for illustration of the great tit, Andrea Badgley for the images of Breaks Interstate Parkand Larry Watrous for various images throughout the book. Thanks to Michael McCumber, www.mdmpix.com, for his incredible landscape images that highlight West Virginia's varied landscapes. Thanks to Piper Roby, Copperhead Consulting, for her excellent maps of the Great Appalachian Valley, and other West Virginia maps.

Reviewers: Many thanks to the following individuals for their candid and thorough reviews: Ron Caldwell, who reviewed the entire text, John Slapcinsky, who reviewed the *Anguispira* and *Mesomphix* sections, and Jeff Nekola for his review of the Pupillidae segment. The following persons from the WVDNR spent over 100 hours as reviewers: Craig Stihler, Jack Wallace, Rick Doyle, and Barbara Sargent reviewed the entire document and Jeff Hajenga reviewed portions of it. Finally, thanks to my wife, Judy Dourson, for her help as a field assistant, her work on county distribution maps, literature searches, and editing.

Historic & Recent Land Snail Studies in West Virginia

Early land snail investigators in West Virginia included the prominent Henry R. Pilsbry, who compiled previous literature and described new taxa with his comprehensive two volume treatise. These fine works were followed by Gordon Kutchka MacMillan, who in 1949, published "The Land Snails of West Virginia." MacMillan's publication was the result of extensive collecting throughout the state aided by a network of citizen scientists. Another early survey effort that would build on the work of Pilsbry and MacMillan was by Leslie Hubricht, who in 1985 published, "The Distribution Records of the Native Land Mollusks of the Eastern United States." In addition, important contributions came from Ralph Taylor of Marshall University as well as his student, Clement Counts (1977) who focused on the Polygyrids of the state for his master's thesis. Kenneth Emberton studied the Polygyrids of the eastern United States and included land snails collected from West Virginia in his subsequent publications in 1988 and 1991. A recent land snail survey of the Cheat River Gorge in Preston and Monongalia Counties, West Virginia found 66 species in a 26 km segment of the canyon (Hotopp *et al.* 2008). This is the richest single site reported in the state.

In 2008, Ken Hotopp and Tim Pearce conducted a thorough review and compilation of West Virginia land snail distributions from past literature and museum records. They compiled a digital database of West Virginia land snails found in nine regional collections based on specimen records, standardized the records, and mapped by county the locations of each species. As Hotopp and Pearce stated in their 2008 report, "while literature records are important contributions to the overall knowledge, specimens upon which the records are based are even more valuable as they can be studied to verify reports or make new taxonomic, ecological or biogeographical discoveries." Records that were deemed questionable by Hotopp and Pearce were not verified, however, by specimen examination, leaving many doubtful records unsettled. In total, nineteen land snail species had "highly-questionable" occurrences in West Virginia, falling substantially outside Hubricht's 1985 distribution maps. Most of these ambiguous taxa are housed in the mollusk collection at the Carnegie Museum of Natural History and were provided to the author for evaluation. Fifteen species from Carnegie were studied, ending with a judgment as to their identity, while another eight taxa (which could not be located), remain unconfirmed with notes as to their likely identification. In addition, a dozen or so land snails with curious range extensions in West Virginia were highly scrutinized, the results of which are found throughout the text under species accounts.

Most recently, a statewide land snail survey led by the WVDNR and initiated in 2006, has had no less than amazing results. This land snail inventory has been one of the most comprehensive in the state, and quite possibly, North America. Biologists, Master Naturalists, and volunteers across West Virginia have collected thousands of specimens, resulting in 459 new county records, 5 new state records and 7 species previously unknown to science. Over 17,000 specimens have been examined by the author alone as a result of this statewide endeavor. There are currently 168 native land snail species confirmed from West Virginia, 11 of which are endemic (more than any neighboring state). Eighteen species are restricted to the borders of West Virginia and Virginia along the Ridge and Valley region. While the land snails in West Virginia are fairly well documented, there remains a dearth of basic biological and life history information for nearly all 168 species. One of the few snails that has received more than casual observation is the federally threatened Cheat three-tooth, *Triodopsis platysayoides* (Hotopp 2003; Dourson 2008). Below is a summary of results from the statewide survey and several predictions on additional discoveries.

WVDNR Statewide Snail Survey Results from West Virginia Beginning in 2006 & Ending in 2014

Number of Native Land Snail Species In West Virginia Confirmed By The Author	168
Endemic Species To West Virginia	11
Globally Rare Species (G1/G2)	19
New County Records	459
New State Records	5
New Species To Science-Documented in WV	7
New Species Described in Book From West Virginia	4
New Species Needing Additional Investigation (DNA)	3
New Species Not Yet Discovered (an estimate)	5
Additional Species Close To West Virginia Borders But Not Yet Confirmed (see Part 2 of book)	26
New State Records Waiting Discovery (an estimate)	13
Exotic Species (including slugs) In WV	10
Additional Exotic Species Not Yet Detected In WV	5
Total Number of Specimens Examined & Identified By The Author During Statewide Survey	17,130
Total Number of Specimens Examined From Museums By The Author (2012-2014)	576
Total Number of Native Land Snail Species Expected to Occur in West Virginia	186

Land Snails of West Virginia and Bordering States

The book is presented in two parts; **Part I** specifically covers 180 land snails (including both native & exotic) known within state borders. **Part II** expands beyond state boundaries (illustrated below between the dashed-lines) to include an additional 26 species found in counties of adjacent states (Ohio, Pennsylvania, Maryland, Virginia, and Kentucky) that border West Virginia. The book is fully illustrated with hundreds of color photographs and drawings. A pictorial key for both parts is included and was designed for the beginner and advanced malacologist alike. Also included is a comprehensive section devoted to exotic (invasive) shelled snails and slugs.

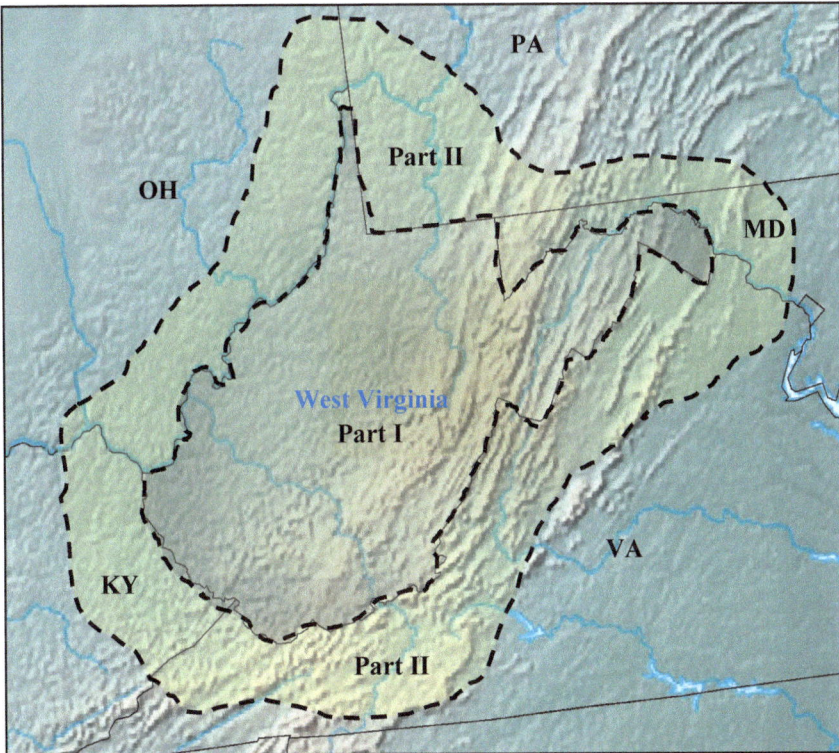

For its comparatively small size, West Virginia has an exceptionally high number of land snail species, eleven of which are endemic. The richest counties occur along the state's border with Virginia; Pendleton County being the most speciose. This is largely tied to the great span of elevation, soils, geology and the age of the mountains. These conditions, which are less dramatic to the north and northeast, have fueled the engines of species diversity within state boundaries. Many snails that were described as forms in early works by Pilsbry, Hubricht, and others have been elevated to species by later investigators including Emberton and Fairbanks. There are several interesting subspecies (i.e. *Mesomphix rugeli oxycoccus*) found in West Virginia, although currently synonymized with *M. rugeli*, may in fact represent speciation in its infancy. DNA and genital dissection await these gastropods. Furthermore, these diverging snail forms may be key to bridging our better understanding of evolutionary biology.

West Virginia County
Map Locations

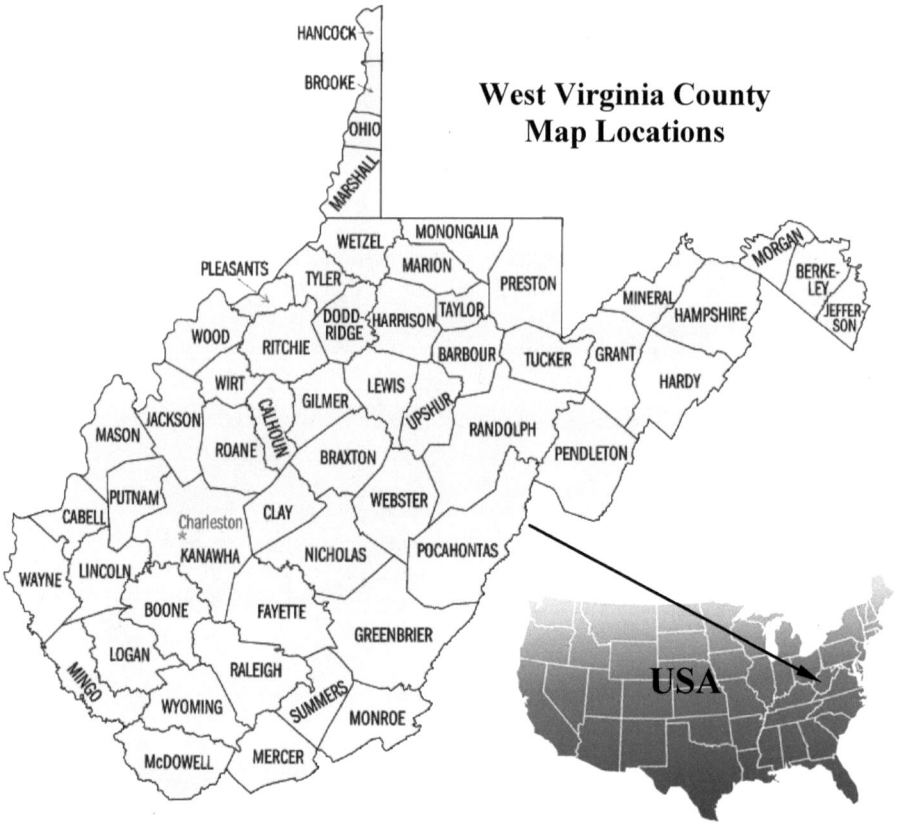

Number of Land Snails
reported from each
West Virginia County

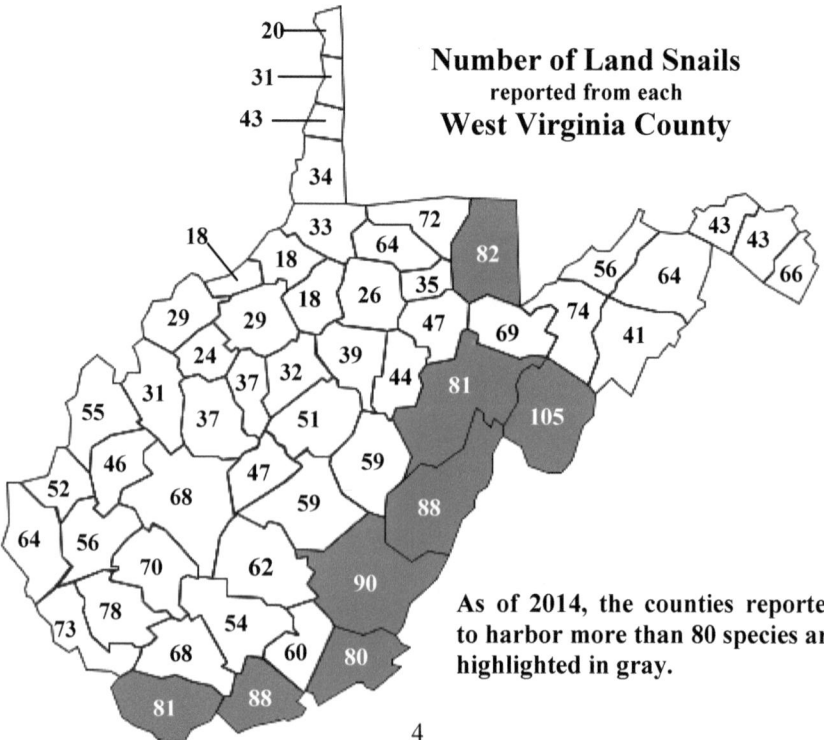

As of 2014, the counties reported
to harbor more than 80 species are
highlighted in gray.

4

Species Accounts

As stated earlier, the following **Species Accounts**, under PART I, list every land snail reported within state boundaries. This includes all native species as well as reported and potential exotic snails and slugs. PART II of the book includes land snails not yet reported from West Virginia, but known to occur close to state borders and therefore have a reasonable probability of occurring.

For each species illustrated, the **Common Name**, **Scientific Name**, **Description**, **Similar Species**, **Habitat**, **Status** in West Virginia (including rankings) and **Specimen** (information as to the location of the particular specimen photographed) are included, making the text useful and comprehensive for Natural Resources agencies and other land managers whose lands harbor terrestrial gastropods.

Maps

Three colors were used to indicate when the species was recorded in the state and are as follows:

Blue- records from museums, previous collectors and other sources

Yellow-records from the WVDNR Snail Atlas Project

Green-records from both the WVDNR Snail Atlas Project and other sources

For maps showing overall species distribution in the eastern USA, refer to Hubricht (1985), "The Distributions of the Native Land Mollusks of the Eastern United States", Fieldiana Zoology New Series, No. 24, Publication 1359: Field Museum of Natural History, Chicago, Illinois.

Abundance Categories of West Virginia's Land Snails

The abundance of land snails within West Virginia was determined by investigating records for each species and assigning a category based on that examination. The categories used in the book are as follows:

1) Common: Species is found in some quantities in its habitat throughout the state.

2) Relatively Common: Species is usually encountered in its habitat in smaller numbers, but with regularity.

3) Uncommon: Species not encountered frequently within its habitat, but widespread in small numbers.

4) Rare: Rarely encountered even in suitable habitat.

Global Ranks (from NatureServe): Each land snail species has two ranks, the **Global Rank** followed by the **State Rank** (example-G1S3).

G1-Critically Imperiled—At very high risk of extinction or elimination due to very restricted range, very few populations or occurrences, very steep declines, very severe threats, or other factors.
G2-Imperiled—At high risk of extinction or elimination due to restricted range, few populations or occurrences, steep declines, severe threats, or other factors.
G3-Vulnerable—At moderate risk of extinction or elimination due to a fairly restricted range, relatively few populations or occurrences, recent and widespread declines, threats, or other factors.
G4-Apparently Secure—At fairly low risk of extinction or elimination due to an extensive range and/or many populations or occurrences, but with possible cause for some concern as a result of local recent declines, threats, or other factors.
G5-Secure—At very low risk or extinction or elimination due to a very extensive range, abundant populations or occurrences, and little to no concern from declines or threats.

State Ranks: State ranks are listed below and in most cases, parallel the Global rankings in terms of their connotation. Ranks listed are considered "provisional" until each species can be thoroughly reviewed and the state ranks updated.

SX-Extirpated
SH-Possibly Extirpated
S1-Critically Imperiled
S2-Imperiled
S3-Vulnerable
S4-Apparently Secure
S5-Secure
SU-Unranked
SNR-Not Ranked (Data not sufficient to rank at this time)
SNA-Not Applicable (Used for Exotic Species)

As inventories continue across West Virginia, no doubt new records will be added to the already diverse list. This is especially true if inventories target adult land snails under 5 mm in size which, in general, make up no less than 40 percent of the total land snail fauna in ordinary sites (hillsides and ravines) and as much as 75 percent in rare habitats (e.g., glades). In some cases, pupillids (*Gastrocopta* and *Vertigo*) represent 80-100 percent of the total molluscan diversity within sites (Nekola and Coles 2010). In West Virginia, the most promising genera to harbor species unknown to science are *Paravitrea*, *Glyphyalinia* and *Helicodiscus*.

The Mountain State
of
West Virginia

From mountain tops shrouded in mist to raging rivers below, West Virginia is a land of varied expressions. In a seemingly endless sea of green, few states have a greater proportion of forested land. Although these vast woodlands are dotted and crisscrossed by the workings of humans, it is easy to imagine that much of West Virginia is still wild, where nature continues to thrive and prosper. Extensive forests, small openings, wetlands, streams, cliffs, and caves are all natural communities. These biological communities are clusters of organisms including plants, animals, fungi, and microbes which live together in a particular environment.

For a small state, West Virginia has a great diversity of natural community types. This can be attributed to its geographic spread and geologic and topographic variation. Prevailing winds traveling eastward across the state first cross a section of low hills eroded from nearly horizontal layers of rock. These lowlands support extensive deciduous forests of oaks, hickories, maples, beech, tulip poplar and other trees. The species composition of these forests varies subtly depending on which direction the slope faces, slope position (whether the forest is high or low on the slope), geologic substrate, and disturbance by humans and natural forces.

Chimney Top, North Fork Mountain, Grant County, WV

7

Blackwater Falls State Park in fall colors, Tucker County, West Virginia.

Landscapes of West Virginia

Bald Knob Overlook, Canaan Valley is below, Tucker County, West Virginia

Blackwater Falls State Park, Tucker County, West Virginia

Flora and Fauna of West Virginia

Black bear

Painted trillium

Yellow clintonia

Virginia bluebells

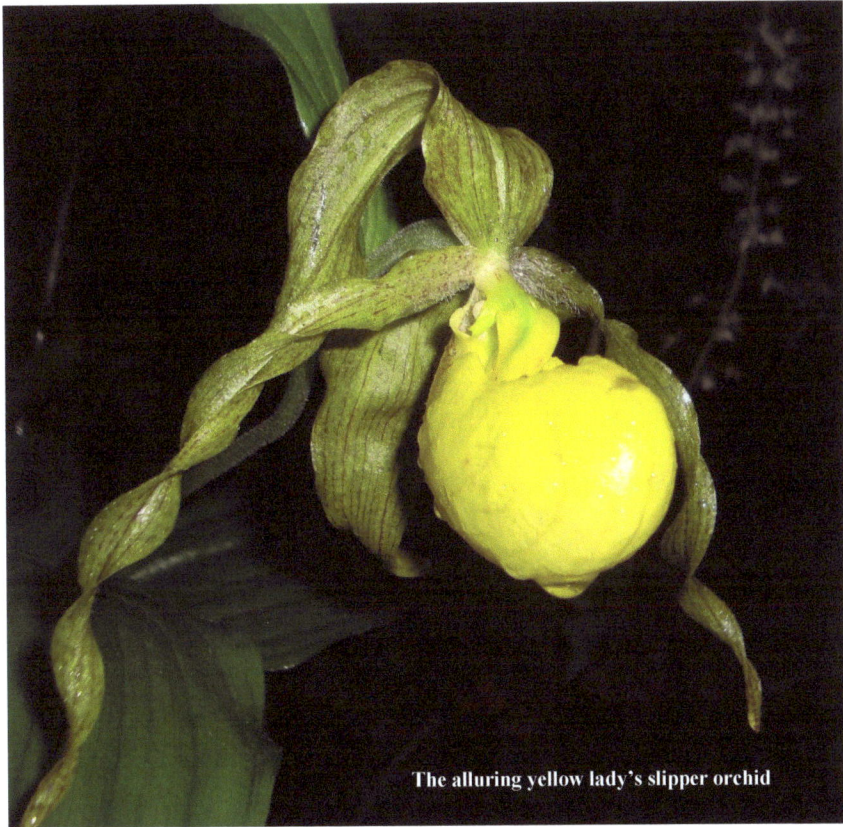
The alluring yellow lady's slipper orchid

Moving east, the hills become bigger. As the uplifted Allegheny Mountains are approached, climatic effects of elevation become dominant influences on vegetation composition and structure. The west sides of the mountains receive the highest rainfall in the state and support some of our most productive forests. Great rhododendron, the state flower, often forms impenetrable thickets in the forest understory. Trilliums, wild orchids, and a large variety of ferns are common plants of the forest floor. Vegetation at the highest elevations often resembles that found at lower elevations further north. Here we find forests dominated by yellow birch, cherry, hemlock, and red spruce. Open communities in the highlands, including rock outcrops, heath grasslands, and bog-like wetlands, are some of the state's most visited scenic attractions.

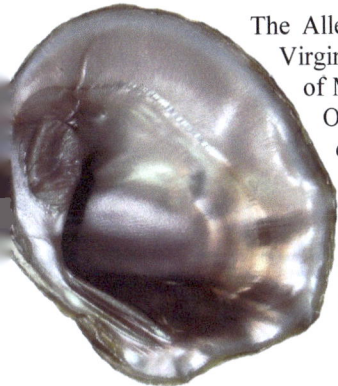

The Allegheny Front is the eastern continental divide. West Virginia waters to the west of the divide flow into the Gulf of Mexico, while those on the east flow into the Atlantic Ocean. These rivers and streams long ago created the dramatic landscape of gorges and valleys that we see today. Waters on each side of the divide support distinct aquatic communities, including many native species of fish, crayfish, and exquisite freshwater mussels like the Purple wartyback (pictured left) found on one side but not the other. Riparian vegeta-

11

tion, plants found along waterways, is some of the most diverse in the state. The timing, duration, and energy of floodwaters determine what kind of vegetation can develop and persist. These types range from tall grass prairies and battered woodlands along high-energy rapids to floodplain forests and backwater swamps along slower reaches.

East of the Allegheny Front, the landscape and vegetation patterns shift again. The climate is drier and hotter. Pine trees are more abundant compared to most other parts of the state. Folded and faulted geologic layers have eroded to form parallel ridges and valleys, exposing different rock types that support contrasting vegetation. Open shale barrens and limestone and sandstone glades occur on hot, dry southern slopes that cannot support forests. These sunny habitats host many rare and distinctly Appalachian plant species and a high diversity of butterflies.

Exposures of limestone provide the conditions necessary for development of underground caves that support communities of bats, rare troglobitic cave snails like the hollow glyph, *Glyphyalinia specus* (below image), and other strange animals and microbes adapted to life in the dark. Above ground, limestone geology is associated with gentle topography and soil fertility, which has resulted in a large proportion of natural vegetation being converted to agriculture. (Introduction by Jim Vanderhorst, ecologist with WVDNR).

Hollow glyph, *Glyphyalinia specus*

12

The copious Hercules beetle

Northern copperhead

Green salamander

14

Introduction to the

Land Snails

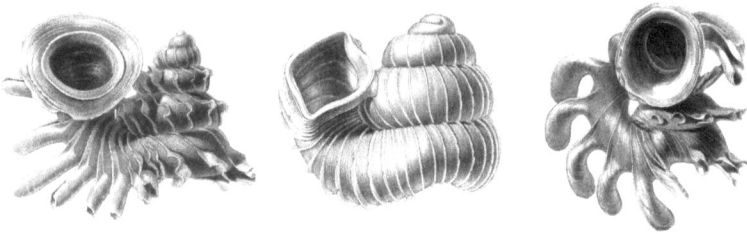

Land snails come in every color, form, and size imaginable; many beyond belief. Take, for example, terrestrial gastropods in the genus *Opisthostoma* (three species illustrated above). These eye-catching gems have taken calcium carbonate sculpturing to extremes. No less complicated than a Roman Cathedral, their architecture represents the most spectacular achievements in convoluted evolution ever seen in a land snail. The genus is known only from limestone outcrops and caves of Borneo (Illustrations by Jaap Vermeulen).

As part of a large and diverse group of organisms known as Gastropods, land snails fall within the prodigious and immense Phylum Mollusca. Containing around 85,000 described and living species world-wide, mollusks include aquatic snails (both marine and freshwater), land snails, terrestrial slugs, sea-slugs, sea-hairs, limpets, bivalves (clams, oysters, and mussels), squids, octopuses, and the famous nautiluses. Octopuses are considered the most neurologically advanced invertebrates on earth, having feet divided into a number of prehensile and skillful tentacles capable of twisting the lid off a glass jar to access the food within. Although snails are slow moving, squids are among the fastest animals known, exceeding underwater speeds of more than 70 mph in short bursts, a result of their water-propelled jet propulsion. At 18 meters long and weighing in at around two tons, the colossal squid is the largest living invertebrate having 15-inch eyes, the biggest of any animal. Mollusks are also the longest lived, multi-celled animals. One species of bivalve, the ocean quahog, *Arctica islandica* can live more than 500 years.

There is scarcely a place on the dry surface of the world, outside the polar regions, where one cannot find at least a few examples of gastropods (Abbott 1989). From hot, nearly waterless deserts to cold mountain tops, land snails are thriving. Despite their biological and physiological limitations, land snails have developed efficient mechanisms for coping with freezing, starvation, and desiccation. For example, when conditions become increasingly dry, snails cover the aperture of their shells with an epiphragm, a mucous sheet that hardens, sealing in critical moisture and slowing desiccation. Some snails can remain

15

dormant for years, resuming activity during wet weather. Land snails in North America are typically nocturnal and are most active in wet weather when temperatures are between 50-75 degrees Fahrenheit.

Even though mollusks rank as one of the most numerous and speciose groups of organisms on Earth, they remain largely unstudied. As a result, little is known of their importance in ecosystems. Land snails, like most invertebrates, suffer from being in a conservation "blind-spot". As snail research moves forward, however, our understanding of the value of these organisms is increasing. Research has shown, for example, that land snails play an important role in micronutrient cycling in terrestrial ecosystems (Dallinger *et al.* 2000) and in dispersing plant seeds and fungal spores (Richter 1980). Land snail diversity has been shown to predict vertebrate conservation priorities (Moritz *et al.* 2001). Further, live snails and their vacant shells provide a food and calcium carbonate source to animals in many systematic groups. These include ants, firefly larva, snail-killing flies (Foote 1959), *Cychrine* beetles (which feed chiefly on land snails) (Symondson 2004), Arachnids including harvestmen, carnivorous snails, numerous species of salamanders (Petranka 1998), turtles and frogs (Burch 1962), a variety of small mammals including shrews, mice and moles (Reid 2006), snakes (Lee 1994, Dourson 2012), a variety of passerine birds (Graveland *et al.* 1994; Graveland 1996), thrushes, ruffed grouse and wild turkey (Martin *et al.* 1951), bats, (Bonato *et al.* 2004; Thabah *et al.* 2007; Dourson 2012), and primates, including humans.

While a building body of evidence suggests the importance of mollusks in present-day ecosystems, their historical value is less well known; namely their contributions made to existing plant communities, animals and, in particular, caves. In the past, the colossal accumulation of deceased mollusks, corals, and tiny creatures known as Foraminifera (which have calcareous skeletons), provided the necessary building material to create the limestone where caves are formed. Many species uniquely adapted to caves roost or otherwise live in these vast underworlds, a number of species occurring nowhere else. These ancient shells have also provided the necessary limestone (in cement) to form the foundations of our cities and homes.

Declining land snail populations can have ripple effects to surrounding ecosystems. The great tit, *Parus major,* in the Netherlands, for example, has declined precipitously with declining land snails as a result of acid rain (Graveland *et al.* 1994; Graveland 1996). A lack of snail shells in the bird's diet causes the bird to lay eggs with thin shells which break, and reduces the reproductive success rates. In North America, Hames *et al.* (2002) documented a correlation between a reduced number of wood thrushes and acid rain, hypothesizing a connection to reduced land snail populations.

Great tit by George Boorujy

Sensitive to changes in the environment, native land snails could provide an early warning to impending habitat deterioration, similar to the way that fresh-water mussels found in streams and rivers are used to determine the quality of waterways. For example, research has shown that when snails feed on various foods such as mushrooms, green vegetation, or forest litter (detritus), environmental contaminants present are ingested and sequestered in their tissues (Dallinger and Wieser 1984a), the midgut gland being the main accumulation site of these trace elements (Dallinger 1993). Further laboratory experiments by Dallinger and Wieser (1984) have shown that land snails that ate lettuce laced with zinc, cadmium, lead, and copper readily sequestered these elements. More concerning though, snails quickly become poisoned when simply raised on soils contaminated with cadmium, raising fear that toxins in polluted soils may be more bio-active than previously believed (Scheifler *et al.* 2003). The environmental concerns to ecosystems, and the consequence to snails and higher organisms which feed on contaminated gastropods are valid. Rimmer *et al.* (2005) found elevated levels of mercury in the blood of Bicknell's thrush on mountains in New England, and thrushes are reported to eat snails (Martin *et al.* 1951), presumably to obtain calcium for egg laying. Native land snails could, therefore, be used to forecast impending problems created by anthropogenic pollutants reported to be accumulating in forests of West Virginia. This hypothesis remains untested.

Parasites in Land Snails

Although nearly every kind of mollusk is inhabited by some form of parasite, only a few gastropods are of medical or veterinary importance (Burch 1962). Of these, almost all live in fresh water environments. Snails are required hosts in the life cycles of parasitic trematode worms. A few land snails, such as *Cochlicopa lubrica,* are vectors of lancet liver flukes in sheep, cattle, deer, and groundhogs (Burch 1962). *Zonitoides arboreus* and *Anguispira alternata*, both native land snails in West Virginia, are implicated in the spread of lungworms in domestic sheep (Burch 1962). Multiple species of land snails and white-tailed deer play a significant role in the transmission of a parasitic nematode known as brainworm, *Parelaphostrongylus tenuis*. The initial host of the nematode is the white-tailed deer and the intermediate hosts are several species of snails and slugs. Interestingly, brainworm nematodes do not appear to significantly affect white-tailed deer populations yet may cause debilitation or even fatality in elk, moose, goats, and sheep. The life cycle of the nematode begins with adult worms which are normally located between the membranes (meninges) that cover the brain and in the spinal cord of the white-tailed deer. Eggs are deposited either on these membranes or directly into blood vessels. Those deposited on the membranes hatch, and the larvae enter small blood vessels to be carried to the lungs where they enter the alveoli. Eggs deposited into blood vessels are carried to the lungs and eventually hatch with larval penetration of the alveoli. Activity in the lung tissue produces an interstitial pneumonia. The larvae pass up the respiratory tract from the alveoli, are swallowed, and then are eliminated in the feces of the deer. Larvae appear in the feces about three months after the host becomes infected. The larvae then enters into a snail or slug (the intermediate host) through the gastropods foot while crawling over infected deer feces or direct ingestion of deer feces by the gastropod. Development of the larvae in the gastropod to a stage when they are infective to the vertebrate host takes about three weeks. Cervids become parasitized by ingesting infected gastropods that are clinging to grass or other browse foods.

In the final host, cervids, development of the larva to the adult worm takes place in tissues of the central nervous system, particularly the spinal cord. Parasites leave the tissues of the spinal cord after about 20-40 days and locate between the spinal membranes where they mature. Subsequently, they tend to accumulate in the cranial region. The adult worms are about 50 mm in length and may be seen fairly readily when free in the cranial cavity. From one to 20 worms have been found in the crania of infected deer, yet as previously stated, *P. tenuis* seldom causes damage in white-tailed deer. In other cervids, there is often extensive damage to tissues of the brain and spinal cord. The resulting neurologic disease is characterized by weakness, fearlessness, lack of coordination of movement, circling, deafness, impaired vision, paralysis, and death. When found in moose, this disease is often called "moose sickness" or "moose disease." A correlation to prevalence of the disease in elk, moose, and other cervids has been linked to an increase in population of the white-tailed deer. In Canada, studies by Anderson and Prestwood (1981) have substantiated the importance of this problem in management of big game and have given some indication of the dynamics of the host/parasite relationships among wild populations. Land snail species that have been implicated in the spread of brainworm infestation include: *Anguispira alternata, Deroceras leave* (a slug), *Deroceras reticulatum* (a slug), *Discus catskillensis, Euchemotrema fraternum, Philomycus carolinianus* (a slug), *Neohelix albolabris, Pallifera dorsalis* (a slug), *Striatura exigua, Triodopsis tridentata, Ventridens intertextus,* and *Zonitoides arboreus.* Any species of land snail that occurs where there is a high prevalence of white-tailed deer could potentially be a host and contribute to the spread of this disease (Anderson and Prestwood 1981).

Land Snails as Pests
Land snails, including slugs, can be agricultural pests. By and large, snails and slugs that are problematic in gardens are species that are non-native introductions. Most have been accidently released into North America by way of plants, potting soils, or shipping crates, but a few species, like *Cepaea nemoralis,* were introduced for their colorful shells. Exotic gastropods naturalize quickly and multiply. Degraded native habitats only make things worse by providing the conduit for dispersal and movement into new areas. These exotics can carry molluscan diseases and problematic parasites which can effect native land snails and other wildlife. Exotic slugs can be especially damaging pests in greenhouses and agricultural lands, costing millions of dollars worth of crop damage every year. Gardeners are most familiar with those primarily exotic species, the slugs, due to their insatiable appetites for tender vegetables. While foreign gastropods quickly become overpopulated without natural controls in place, native snails and slugs are rarely problematic, and most species actually become scarce or disappear entirely in areas where the natural vegetation has been eliminated. The best defense against exotic gastropod infestations is to keep natural forests intact.

A Human Connection

Mollusks are providing a number of life-supporting contributions to humans. Marine mollusks may help fight cancer; the drug Kahalalide F, a protein extracted from a species of mollusk that eats sea slugs in the Pacific Ocean, has shown great promise as chemotherapy for the treatment of liver cancer (Satheeshkumar *et al.* 2010). Lethal toxins produced by cone snails were used to develop a drug called Ziconitide for patients with cancer and AIDS who suffer from chronic pain that cannot be relieved by opiates, and Ziconitide is not addictive (Wallace *et al.* 2008). Slime from the land snail, *Cornu aspersum* (one of the commonly eaten snails referred to as escargot) is now used to treat many different types of skin disorders. This snail slime repairs skin damage from overexposure to the sun and reduces scarring caused by severe acne. Land snail mucus is also known to contain natural antibacterial properties and, some scientists speculate, may be the next generation of human antibiotics. In Central America, the Maya use gastropods to treat a number of ailments including skin disorders, glaucoma, and whooping cough (Dourson 2009).

The Diet of Land Snails

Most snails are dietary generalists (Burch and Pearce 1990) consuming a wide variety of herbaceous plant leaves or stems, decaying vegetation and leaf litter (detritus), wood or bark, and fungal fruiting bodies (such as mushrooms), animal scat, carrion, and shelf or bracket fungi (Burch and Pearce 1990; Dourson 2008). Land snails in the genus *Haplotrema*, *Ventridens, Vitrinizonites,* and *Mesomphix* are documented carnivores, feeding on a variety of other gastropods (Atkinson 1998). Gastropods sample and judge potential food by using the chemoreceptors located on the lower two tentacles (Shearer and Atkinson 2001). Once a food source is discovered, the snail begins the feeding process by first touching the food with its foot and mouth followed by rasping (rasping signs can be seen on the fungi illustrated to the right) and dislodging bits of food with the radula structure located in the mouth (Machensted and Markel 2001). The meal is then swallowed, and muscular contractions move the food along the esophageal tract mixing it with saliva. Feeding episodes can last anywhere from a few minutes to nearly an hour, depending on the durability of the food being consumed (Dourson 2008).

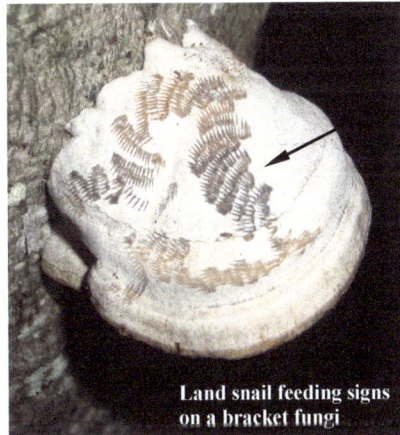

Land snail feeding signs on a bracket fungi

Fruiting bodies of fungi are a favorite food of land snails, especially native slugs. As slugs consume the tissues of mushrooms, toxins sequestered from the fungi are suspected to bolster defensive properties of their mucus. Slugs in the genus *Philomycus* have been observed feeding on the destroying angel, *Amanita virosa* (pers. obs.). This is one of the most poisonous mushrooms known in North America, and one that has caused the majority of mushroom fatalities in humans (Roody 2003), yet slugs seem impervious to the mushroom's lethal effects. Little is known of symbiotic relationships between mushrooms and slugs. What is known is that as slugs feed on the mushroom tissues, they ingest numerous spores, which are carried for a time and excreted elsewhere, spreading the fungi spores to new locations.

Above: a Brown-spotted mantleslug, *Philomycus venustus* feeding on Golden trumpets, Roan Mountain, North Carolina. Below: photo of three Glassy grape-skins, *Vitrinizonites latissimus* and a Black mantleslug, *Pallifera hemphilli* feeding on a fresh dead Balsam globe, *Mesodon andrewsae*, Roan Mountain, North Carolina.

Empty Snail Shells

When snails die, the shells do not immediately decompose. Some research suggests that shells can remain intact for years (Pearce 2008), and empty shells in some limestone locations can reach exceptional numbers. But these discarded shells are anything but vacant and actually provide a secure and protected refuge for a whole host of living micro-invertebrates including pseudoscorpions, ants, millipedes, tardigrades, and smaller land snails. Some invertebrates even deposit their eggs to be incubated in the security of shells. The translucent 3 mm wide shell of *Glyphyalinia indentata* (figure a) was the depository site for eggs of an unknown organism, not an uncommon event (pers. obs.).

Fluorescence in Land Snails

Not to be confused with bioluminescence (the natural light observed in fireflies), fluorescence is the term used to describe the absorption of light at one wavelength and its emission at another. Only one species of land snail is known to produce bioluminescence, *Quantula striata,* from Malaysia. Research has shown, however, that the mucus of several North American family groups of land snails including Discidae and Helicodiscidae have fluorescent slime under ultraviolet light. *Anguispira alternata* slime has a particularly brilliant, bluish fluorescence, and *A. kochi* slime has a similar fluorescence (Rawls and Yates 1971). Under laboratory conditions, the mucus on the foot and body and mucus trails of *Anguispira* and *Discus* species glowed brightly when exposed to UV light. *Discus patulus* and *Helicodiscus parallelus* mucus appears to have fluorescence that is unique to each species. Other species of polygyrids were tested and failed to exhibit any sign of fluorescence. The fluorescence observed in the specimens of the three genera noted above is extremely long-lived, being as bright and as distinctive in specimens preserved for twenty years or more as it is in living snails (Rawls and Yates 1971). I found that the crawling slime of *Anguispira jessica* (typically clear) did not fluoresce under UV exposure, but the defense slime, produced by the species under attack (which is orange-yellow), did fluoresce.

The function of fluorescence in land snails remains a mystery, and there is some speculation that fluorescent slime has no real function at all, being simply a random act of evolution. I propose a functional hypothesis for the fluorescence in land snails. It turns out that moonlight has a component of UV light and *Anguispira* species are most active at night. What if certain nocturnal land snail predators (i.e. snail-hunting beetles) are "avoidance conditioned" to the fluorescent-defense slime the same way that predators are trained to steer clear of the bright colors of coral snakes? Snail slime, especially that of *Anguispira* species, is pungent and distasteful to animals, including humans. *Anguispira* slime has a disagreeable numbing affect in one's mouth (author's personal experience). Like the coral snake which uses color to save its venom, snails may be using florescence to save precious mucus reserves. The more snails are harassed, the more mucus is produced, and losing copious slime (during an attack) could put the snail at risk of dehydration.

Above image of a flamed tigersnail, *Anguispira alternata,* in defensive posture with its aperture blocked by a noxious orange mucus. The bottom picture of same animal with UV light only. Through the camera's lens, the defense-slime appears as a bluish glow, but to the human eye, a fluorescent green. Images taken in the Red River Gorge, Powell County, Kentucky.

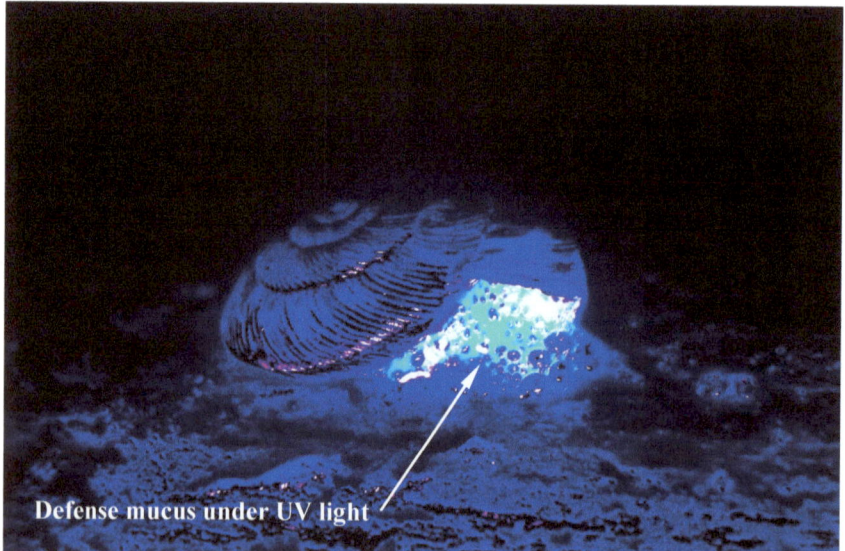

Shells, Ribs, Teeth, Hairs, Fringes, and Viscous Slime

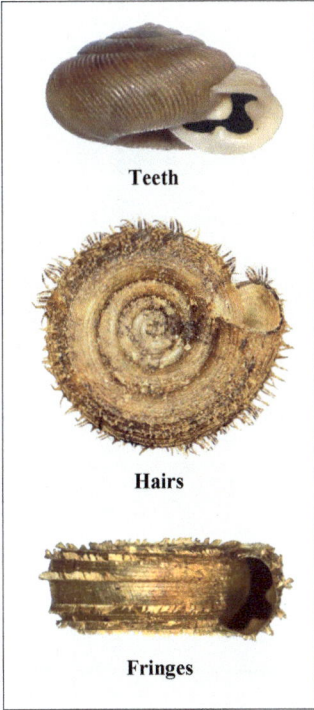

Teeth

Hairs

Fringes

Land snails use a variety of strategies to protect themselves from harm. The shell is the first line of defense and works well to ward off a number of predators. Opercula (snail doors) can prevent invertebrate attacks, but few snails in West Virginia possess these protective structures. Shells and opercula also aid in the conservation of critical body moisture. Teeth and lamellae are believed to act as barricades, preventing the entry of predators such as predacious beetle larva. Some research suggests that aperture teeth may also provide a calcium storehouse which can be used to repair damaged shells, act as pivotal points for balancing the shell during movement, or trap air if the animal becomes immersed in water, keeping it afloat (Emberton 1994). The periostracal hairs and fringes found on *Inflectarius, Stenotrema,* and *Helicodiscus* shells collect forest debris as the snails inch through the leaf litter. Several theories have emerged for shell ornamentation, including increasing shell-crypsis or protecting shells from fracture during falls. Hairs, fringes, and ribs do in fact function as water retention features, which hold and uniformly transfer (wick) water across the entire shell surface (pers. obs.). Although speculation, shells with these features may help facilitate snail movement across drier surfaces and/or help keep the snail cooler. Several species (like those in the genus Anguispira) contain colorful defense mucus which has an anesthetizing affect in the mouth (personal experience), no doubt an effective predator repellant. Although slugs are without protective shells, they are anything but defenseless. Slugs have copious, viscous, water-insoluble mucus (a), which can gum-up the mandibles of beetles or cause antagonistic garter snakes to vomit (pers. obs.).

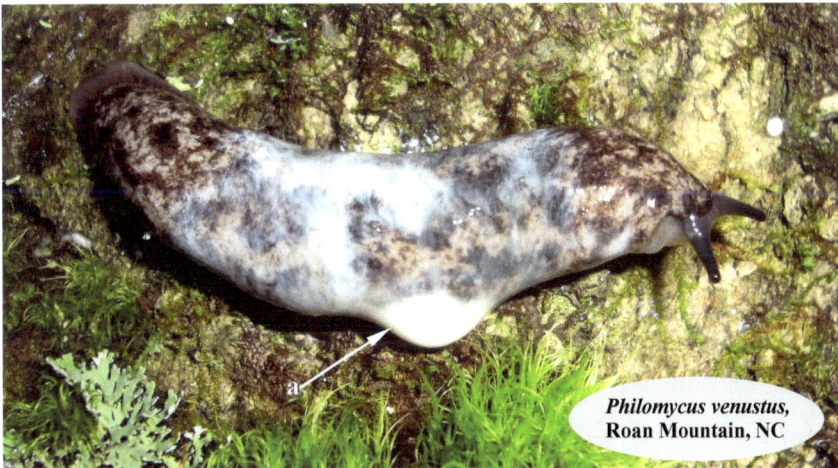

Philomycus venustus,
Roan Mountain, NC

23

Land Snails and Spider Webs

While crawling across a spider web, a Flat bladetooth, *Patera appressa,* appears to glean precious condensed water during a dry period in late July, Red River Gorge, Powell County, Kentucky. Land snails are often seen on spider webs, crawling effortlessly across the web's sticky surface without becoming entangled, an act that would imprison most other invertebrates of similar size. There is some evidence that snails are attracted to the silk nets that trap condensed moisture as a source of drinking water.

Land Snail Predators

With regard to land snail predators, small mammals such as shrews and moles are at the top of the list. Above image of a Hairy-tail mole preying on a Magnolia threetooth, *Triodopsis tridentata*, Roan Mountain, NC. Smaller than moles, shrews are tiny, venomous mammals that chew through the apex and sides of the shell, eating the snail flesh within. Piles of empty shells under rocks and logs are a sure sign of small mammal feeding. Salamanders are major predators on a variety of small to medium-sized land snails (Petranka 1998; Van Devender and Van Devender 2003), devouring the entire animal, shell and

A spotted salamander stalking a slitmouth, Red River Gorge, Kentucky

25

all. By and large, the shell passes through the salamander's digestive tract with little harm, the contents partially or entirely digested. Van Devender and Van Devender (2003) collected a number of land snail species from fecal matter of the Red eft, *Notophthalmus viridescens* and found the following genera: *Cochlicopa, Euconulus, Gastrodonta, Hawaiia, Glyphyalinia, Novisuccinea, Triodopsis, Ventridens, Vertigo, Vallonia,* and *Zonitoides. Vallonia excentrica* was a new record for North Carolina. Few studies have focused on the importance of gastropods in the diet of salamanders, especially juvenile salamanders which have easy and safe access to small land snail species (under 5 mm) living under the leaf litter. Spotted salamanders (opposite page) are also reported to include land snails as part of their diet (Van Devender pers. comm.).

In the above image, a snail-hunting *Cychrine* beetle dines on snail flesh of a golden dome, *Ventridens arcellus*; Spruce Knob, West Virginia. The head and thorax regions of beetles in this group are significantly smaller than other beetles of similar size, allowing the insect to enter through the aperture of the live snail to extract the slimy flesh with its large serrated jaws.

Reproduction

The majority of land snails found in the West Virginia are hermaphrodites, each individual having ovotestis in which both sperm and eggs are produced. When two individuals of the same species, in search of propagation, find one another (by following scent trails), they typically exchange sperm. Sperm can be stored for months to years by each individual snail until conditions are favorable for fertilization and egg laying, at which time, eggs are deposited under logs or in moist leaf litter. Interestingly, many land snails, including several native slugs in the genus *Philomycus,* are characterized by the presence of "love darts".

Dr. Ron Chase of McGill University, Canada, studied the complicated sex life of the European land snail commonly found on people's plates, *Cornu aspersum*. This species is one of the many employs calcareous love darts as part of its courtship. The love darts are not shot as one might imagine, but are forcefully expelled through the body wall of the partner. The dart itself appears in a variety of sizes and shapes, and most species have a single dart which is used only once (Chase *et al*. 2006). In his research on this fascinating sexual behavior, Dr. Joris Koene learned that some land snails employ multiple darts, while others use the same dart to stab partners repeatedly, as many as 3000 times (Koene *et al*. 2013). But what exactly is the function of the love dart?

Recent studies suggest that after copulation the dart functions to increase the reproductive fitness of the shooter. Snails are promiscuous and store sperm from multiple donors for several years before they use it to fertilize eggs. Thus, sperm donors (males) must compete to fertilize eggs. Dart receipt promotes the safe storage of the shooter's sperm, so there will be more sperm from successful shooters available for fertilization than from unsuccessful shooters. Since the female function chooses the sperm by a lottery-like mechanism, successful dart shooters sire more babies than unsuccessful dart shooters.

Chase and Blanchard (2006) tested whether the dart works by simply rupturing the skin or by injecting a bioactive agent. Just before the dart is thrust into a partner, it is covered with mucus (figure a, page 28) from a special gland located near the dart's launching site. Koene *et al.* (2013) conducted an interesting test in which needle stabbings were substituted for dart shooting. In one mating, saline was injected through the needle, in the other, mating mucus was injected. They found that the matings that were associated with mucus injections were responsible for more than twice the number of offspring as were the matings associated with saline injections. Thus, mucus is the agent of the dart's effect on reproductive success.

Mating Brown gardensnails, *Cornu aspersum*

Love dart

Chase and Blanchard (2006)

Copulating Plain buttons, *Mesomphix inornatus*, Cheat River Gorge, WV

Snail eggs

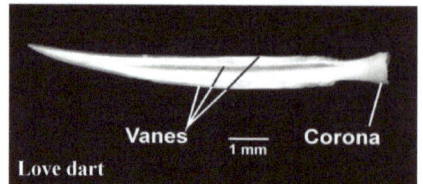

Vanes   1 mm   Corona

Love dart

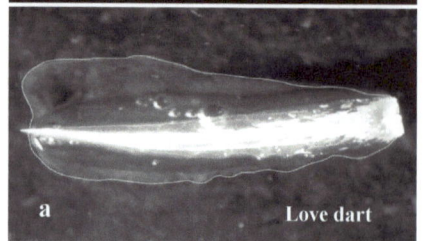

a   Love dart

Chase and Blanchard (2006)

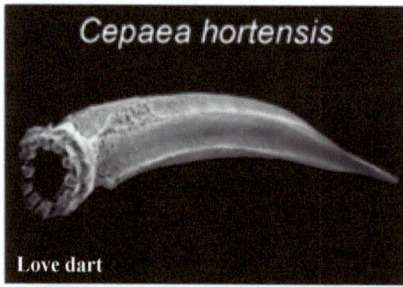

Cepaea hortensis

Love dart

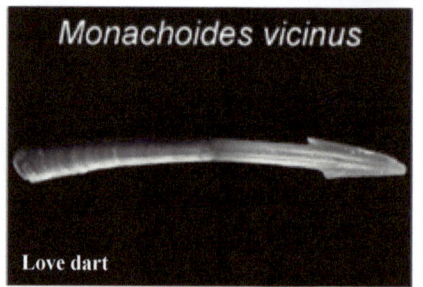

Monachoides vicinus

Love dart

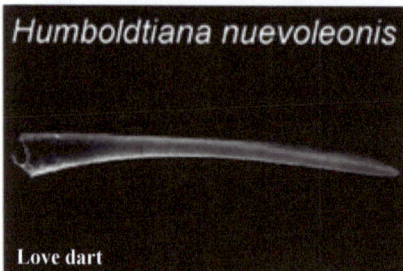

Humboldtiana nuevoleonis

Love dart

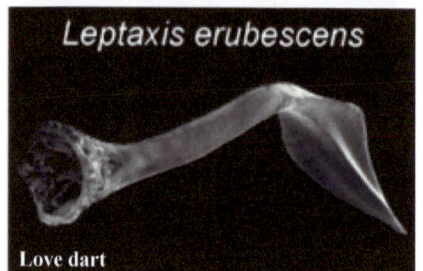

Leptaxis erubescens

Love dart

Koene and Schulenburg (2005)

28

Environment and Land Snail Distribution

In 1962, Burch affirmed that the most distinctive molluscan province east of the Rocky Mountains was the **Cumberland** which has several unique genera (*Clappiella, Gastrodonta, Pilsbryna* and *Vitrinizonites*) and many endemic species. This diverse geographic area includes eastern and southeastern regions of West Virginia, western Virginia, southeastern Kentucky, eastern Tennessee, western North Carolina, and small portions of northern Alabama, Georgia, and South Carolina (see next page). Today, 293 land snail species are reported within this territory, representing roughly 56 percent of the total land snail fauna of eastern North America. A surprising 114 species are restricted or nearly restricted to the Cumberland Province, and, consequently, endemic to the Southern Appalachian Mountains.

Two exceptionally high land snail endemism subregions exist within the Cumberland Province. The most northern of the two, the **Ridge and Valley Province,** occurs along the West Virginia and Virginia border, containing no less than 165 species, 18 of which are endemic. The **Blue Ridge Province** includes the higher mountain ranges of Tennessee and North Carolina, where 192 land snail species, 55 endemic, have been documented. The Blue Ridge Province is essentially the southern portion of a greater area extending as far north as southeastern Pennsylvania where it merges with the Ridge and Valley subregion. Positioned between the two provinces rests the **Great Appalachian Valley**, a lowland trough that begins in northern Alabama and ends in Canada; the whole conglomerate is sandwiched between the Appalachian Plateaus and the great costal plains of the Piedmont. While there are shared land snail species between the Ridge and Valley and Blue Ridge provinces, both contain numerous taxa exclusive to each. For its size, the Blue Ridge Province harbors the richest land snail fauna in North America, and new species continue to be discovered and described (Slapcinsky and Coles 2004; Dourson 2012).

These continental and global hot-spots have also been shown to harbor high salamander, invertebrate (crayfish and mussels), plant, lichen, bryophyte, fungi and slime-mold endemism as well. One reason the area holds a myriad of organisms is thought to be a result of its steep topography. Precipitous relief (contained within deep gorges to soaring peaks) produce two or three times the surface area of flat land of comparable size. As a result, this relief has more than doubled or tripled the biomass-holding capability of the region. Steep topography in concert with sundry geology and geophysical aspects of the land (caves, cliffs and talus), microclimates created by position (south and north facing slopes), all contribute to the area's remarkable biodiversity. Is it any wonder land snails have been so successful, finding refuge from the deepest caves to the highest mountain tops?

Without question, calcium carbonate is an essential mineral to land snails for regulation of bodily processes and reproduction, and, most importantly, shell-building (Burch 1962; Fournie *et al.* 1984; Hickman *et al.* 2003). Land snails obtain calcium in several ways including consuming soil particles from calcareous substrates, eating decaying leaf matter (Wareborn 1970; Burch and Pearce 1990; Nation 2005), almost certainly by ingesting Physarales slime

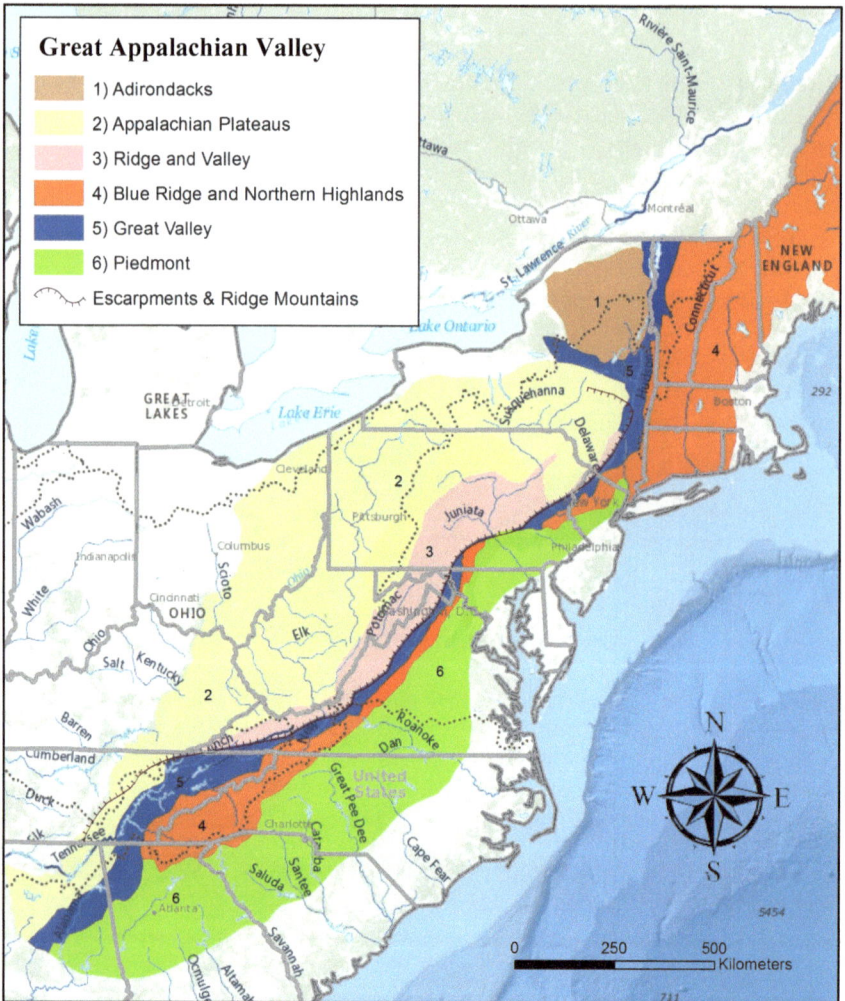

Great Appalachian Valley

1) Adirondacks
2) Appalachian Plateaus
3) Ridge and Valley
4) Blue Ridge and Northern Highlands
5) Great Valley
6) Piedmont

Escarpments & Ridge Mountains

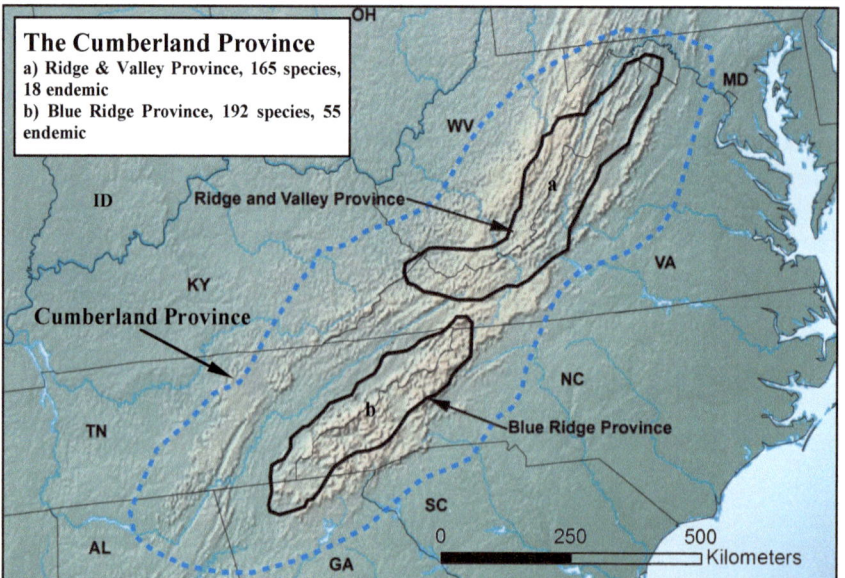

The Cumberland Province
a) Ridge & Valley Province, 165 species, 18 endemic
b) Blue Ridge Province, 192 species, 55 endemic

Ridge and Valley Province

Cumberland Province

Blue Ridge Province

molds (which precipitate amorphous calcium carbonate), and by gleaning calcium from the shells and bones of deceased animals. *Triodopsis platysayoides* (an endemic West Virginia land snail) has been documented feeding on the vacant shells of land snails, including its own kind (Dourson 2008).

Land snail abundance (number of shells) and land snail diversity (number of species) have long been associated with a variety of geological and ecological factors. Studies have shown, for example, that terrestrial gastropods living around carbonate cliffs can exhibit large and diverse populations (Nekola 1999), but show significant declines in abundance in as little as 50 m from a calcareous source (Kalisz and Powell 2003) or limestone cliffline (Dourson 2007). Other research has demonstrated that while limestone may impact abundance, it has little affect on diversity.

Land snail scarcity has reported associations with low soil pH (Burch 1955), declining soil cations, specifically calcium (Petranka 1982), increasing coniferous presence (Jacot 1935; Karlin 1961), and increasing elevation (Petranka 1982). The influence of pH on land snails is thought to be indirect, the main effect of low Ph being a lowering of the amounts of soil cations, principally calcium (Karlin 1961; Cameron 1970). However, low abundance in non calcareous (acidic) areas may only give the illusion of low diversity. Douglas (2011) found that land snail diversity on acid soils covered by old growth forests at Lilley Cornett Woods in Kentucky were analogous to limestone soils at Floracliff Nature Sanctuary along the Kentucky River Palisades.

In the Southern Appalachian Mountains, geological sources of calcium occur in limestone, dolomite, calcareous shale, sandstone, and the igneous-derived amphibolite, all which have calcium and/or Mg in varying amounts, supplying the necessary calcium for shell building. But what about land snails thriving on non-calcareous substrate like West Virginia's highlands? Where and how these land snails obtain sources of calcium are poorly understood. In regions lacking calcareous substrates, land snails may rely on abscissed leaves of deciduous trees or herbaceous plants (e.g. stinging nettle) as a primary source for calcium.

McHargue and Roy (1932), in a study of several species of deciduous forest trees, found that the amount of calcium in leaves expressed as a percentage of dry weight ranged from 1.64 to 7.8%, with the higher values occurring towards the end of the growing season. Other studies have shown that, unlike other macronutrients which are reabsorbed by trees prior to leaf abscission, foliar calcium concentrations in deciduous trees increase throughout the growing season and peak at senescence (Guha and Mitchell 1966; Potter *et al.* 1987). Calcium is a relatively immobile nutrient and the reabsorption of calcium may not be as high a priority to deciduous species as other nutrients. Further, Gosz *et al.* (1973) reported that dead birch leaves could form a significant pool of calcium on the forest floor because the concentrations of calcium remained high in dead leaves 12 months after abscission. Vesterdal and Raulund-Rasmussen (1998) found that the nutrient content of the forest floor under a single species of tree varied with the soil type, but they also found variation between different species on the same soil type. The idea that certain deciduous species contain higher levels of foliar calcium than others was supported by Arthur *et al.* (1993) who reported that yellow birch had relatively high levels of foliar calcium. Gosz *et al.* (1972) found foliar calcium concentrations were higher in yellow birch than in maple or beech. Ricklefs and Matthews (1982) looked at the leaf chemistry of 34 species of broad-leafed deciduous trees and found that yellow birch had higher than average calcium concentrations. Jenkins (2007) found that calcium levels in dogwood leaves growing in the Smoky Mountains were the most significant source of calcium in acidic forests, although many have since died due to dogwood anthracnose.

Additional factors such as gradient (slope), litter moisture, elevation, and microhabitat (leaf litter, moss and logs) can significantly affect the presence or absence of land snails. Coney *et al.* (1982) found more species of land snails on steep slopes than on more moderate ones. Petranka (1982) found that 15 of the 56 land snail species found on Black Mountain, Kentucky showed some preference for slope, with 9 species showing an affinity for increasing slope. The importance of leaf litter moisture (thought to be a factor of slope) to land snails was emphasized by Boycott (1934), Getz (1974), Pollard (1975), and others. Although aspect was reported to markedly affect microclimate (Geiger 1965; Braun 1940), Petranka's (1982) study found no environmental variable to be significantly correlated with aspect. With respect to elevation, Petranka (1982) reported that pH, potassium, calcium, and magnesium levels decreased with increasing elevation, and that the number of snail species and the number of individuals found per site decreased with elevation. In a study by Coney *et al.* (1982) conducted in the Hiwassee River Basin of Tennessee, the most important environmental factors influencing the presence or absence of land snail species was microhabitat (leaf litter, moss, and logs, $P<0.05$ for 27 species), followed, in decreasing order of importance, by slope ($P<0.05$ for 15 species), rock type ($P<0.05$ for 13 species), stages of forest succession ($P<0.05$ for 12 species), soil pH ($P<0.05$ for 8 species), elevation ($P<0.05$ for 7 species), and soil moisture ($P<0.05$ for 6 species).

Less well-known is how the convergence of large physiographic and geophysical landscape edges serve to bridge distinctive regions and their allied terrestrial gastropod communities (Dourson and Beverly 2009; Douglas 2013 *et al.*). Acting like travel corridors for dispersal, neighboring land masses may concentrate the distribution and mixing of snail faunas resulting in remarkably high land snail diversity in comparatively small places (Dourson 2007). Take for example, the two rather minor **Ridge and Valley** and **Blue Ridge Provinces** (page 30, bottom map, figures a & b), both positioned between the much larger Appalachian Plateaus, Great Valley and Piedmont Provinces. These two minor provinces, although relatively small, harbor around half (256 species) the reported land snail fauna east of the Rocky Mountains, and 75 endemic species.

Further evidence that converging physiographic and geophysical landscape edges are driving land snail diversity "hot spots" is supported by recent inventories at Bad Branch Nature Preserve in Letcher County, Kentucky and Breaks Interstate Park bordering Kentucky and Virginia. Bad Branch (approximately

2600 acres) harbors no less than 64 species, eleven representative of more southern mountain ranges, eight representative of West Virginia and Virginia, and nine having affinities with more northerly faunas (Dourson and Beverly 2009). Breaks Interstate Park, with its copious 81 species (above image of the 4600 acre park), represents the richest single site in North America and demonstrates still another great gastropod mixing pot located along a vast physiographic edge (Dourson and Beverly, unpublished data). Both Bad Branch State Nature Preserve and Breaks Interstate Park are positioned along the Pine Mountain massif of southeastern Kentucky, bordered to the south by the Ridge and Valley Province and to the north by the Cumberland Plateau. All three regions contain their own snail affiliations. Moreover, the amalgamation of these snail rich ecoregions provide a number of terrestrial gastropods an opportunity to coexist and represents a fascinating assemblage of species not occurring elsewhere in the states of Kentucky and Virginia.

The richest single "micro-site" reported in North America rests on a mesic hillside of Furnace Mountain in Powell County, Kentucky, located along the Central Knobstone Escarpment. Forming another large physiographic landscape edge, Furnace Mountain is positioned at the coalition of the Cumberland Plateau, the Knobs, and the Outer Bluegrass regions of Kentucky. The merging of these distinct regions brought together no fewer than 61 species, all packed into just 5 acres (Dourson 2007). A number of land snails documented at Furnace Mountain are established well beyond their eastern and western limits in the state (Branson 1973; Hubricht 1985; Branson and Batch 1988).

In West Virginia, land snails can be found in every type of habitat, even your own backyard. The most successful searches begin by turning over leaf litter; the more you uncover, the greater number of gastropods you'll find, especially

around outcrops of north-facing limestone (a). Using a three or four-prong garden hand rake works best and saves your finger tips from the sharp conveyers of pain (thorns and broken glass) and the occasional leaf-covered copperhead. Tree crotches that contain abundant litter-fill will often support large numbers and diverse populations of small species, where upwards of fourteen taxa have been found. The base of large diameter trees such as Black walnut, Butternut, Basswood, Buckeye, and Birch (the 5 "Bs") also appear particularly rich (pers. obs.). Prime real estate for native slugs is found in natural forests under exfoliating bark of rotting hardwood logs (b) in advanced stages of decay, particularly those logs that form log bridges over small ravines and streams.

How to Collect Land Snails

Collecting snails (shells or live animals) within the boundaries of West Virginia is prohibited without a valid permit. *The state offers opportunities for participation in Citizen Science activities including snail searches. For further information, contact West Virginia Division of Natural Resources.*

Selecting Sites to Survey

When surveying for land snails, there are specific habitats that should be targeted in order to comprehensively cover an area. These habitats include: under leaf litter, rocky outcrops, rock crevices, and logs; under exfoliating bark of standing and/or down dead trees; hollow trees like American beech and sycamore; damaged trees oozing sap which attracts snails (Dourson pers. obs.); under and on top of caps of fungi; under moss mats and the flaps of rock tripe; bases of black walnut trees; crotches of trees where leaf litter has accumulated; human-made features such as roadsides, steep banks, retaining walls, cement structures, spring houses, discarded bottles (neck facing upward), or other discarded refuse; cliffline features, caves, and rock talus.

Field and Lab Equipment Needed

Field equipment include:
- Ziploc bags and plastic vials
- Permanent marker
- GPS unit
- Field notebook
- Hand lens of 10X
- Quart-size litter bags

Collecting Methods

Samples of larger (macro) specimens from 5mm and greater should be collected and placed in plastic vials with date, site number, GPS coordinates, and collector name (preferably on the outside with a Sharpie). Do not put paper labels in bags with live snails; they will most likely eat the paper.

Samples of smaller (micro) specimens are best collected from leaf/soil collections; field sieves work well here. Sites that yield increased numbers of snails include the bases of black walnut and butternut trees, the bases of large mature hardwood trees, tree crotches and leaf litter along the edges of seeps. Optimal sites can be determined by collecting a handful of soil/leaf litter, then scanning the litter with a hand lens for evidence of micro specimens. If any snails are observed, a quart-sized cotton litter bag is filled with the material from the site, labeled with the date, site number, collector's name and GPS coordinates.

These leaf samples are taken back to the lab and dried for approximately two weeks. Dried samples should be sifted through a series of sieves ranging from 4.76 mm down to 500 micrometers. The subsequent debris that remains after this sifting process is then searched with the aid of an Optivisor or other magnification device. It will be necessary to use a zoom dissecting-scope to determine the species of these small snails. Many of them have microscopic orna-mentation that can be seen only under high magnification.

Field Notes and Labeling Samples

1). Date each daily entry and record the current weather conditions and last rain event.
2). Record for each area surveyed, the Forest Community Classification, main overstory tree species and dominant ground cover species, solar aspect, and elevation. Presence and abundance of rock or coarse woody debris (CWD) should be noted.
3). Record GPS position noting which datum used or mark precisely on a 7.5 minute quadrangle map.
4). Record the number of litter bags collected.

Preservation and Mailing Specimens

Generally, it is not necessary to collect live specimens due to the abundance of dead shells. If live snails are collected for the purpose of anatomical work, it is necessary to euthanize them in water for 24 hours so that they relax, then place them in a solution of 70-80% ethyl alcohol. It is best that the snail be preserved with the head and foot not retracted into the shell. Menthol or chlorotone crys-tals are sometimes used to relax land snails. Menthol cigarette tobacco can be soaked in water to provide a relaxant solution. Live snails can be sent through the mail by simply packing them in moist unbleached paper toweling. The pa-per will keep the snails hydrated and provide a food source. For experts in-volved in ongoing taxonomic studies, live snails provide valuable anatomical material.

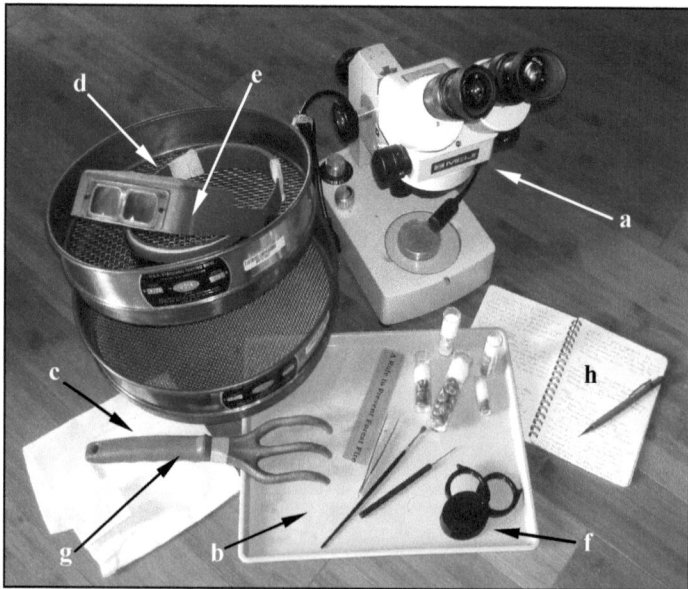

Necessary equipment for serious land snail work includes: a dissecting micro-scope (a), sorting tray (b), 1-liter litter bags (for the soil and leaf litter collec-tions) (c), a series of soil sieves sizes #4 (4.75mm), #10 (2.00mm), #18 (1.00mm), #14 (1.40mm) (d), magnifying head loop (e), forceps, vials, labels, paintbrush to pick up and transfer small snails (all pictured on tray), hand lens (f), hand rake (g), and notebook (h).

36

Shell Morphology

Pictured (right) are three standard shell views (a, b, & c) used by malacologists to identify land snails. Always observe snails in these views. The frontal view (a) shows the shell's general form and aperture shape. The bottom view (b) shows one of the most important diagnostic features of any land shell, the umbilicus region. The top view (c) shows the apex or embryonic whorl, the number of whorls, and the width of the whorls. In general, the frontal and bottom views are the most important diagnostic views of land snail shells. While the top view has limited value for separating snails within the same genus, it is a reliable view for separating snails of different families. Short black arrows or lines indicate key features of shells. All measurements used in this book are for greater diameter or height of adult shells.

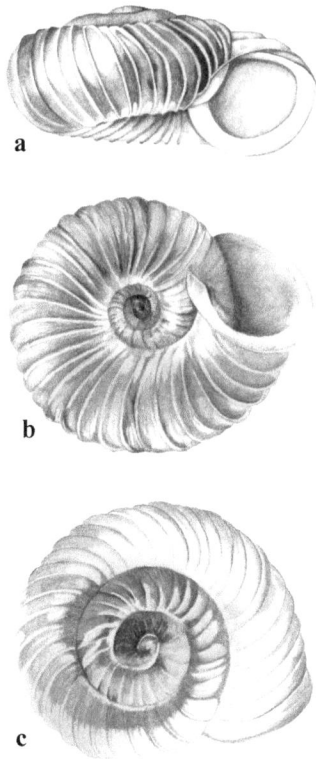

a

b

c

Terminology of the Shell

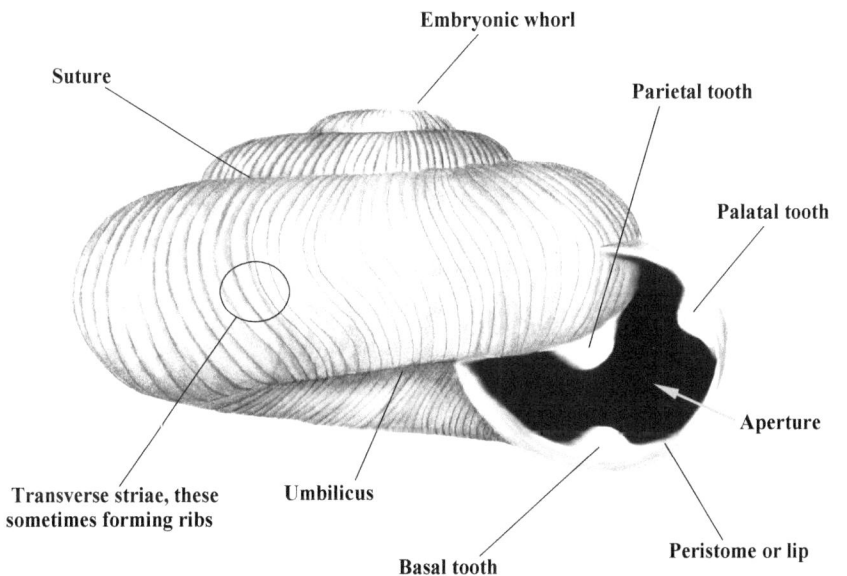

Embryonic whorl

Suture

Parietal tooth

Palatal tooth

Transverse striae, these sometimes forming ribs

Umbilicus

Aperture

Basal tooth

Peristome or lip

|||||||||||||||||||||||||||||||||||||||||||||||||||||||||||||||||||||||||||||||||||||||||||||||||||||||||||||||||||

0 cm   1   2   3   4   5   6   7   8   9   10   11

Umbilicus of the Shell (Burch, 1962)

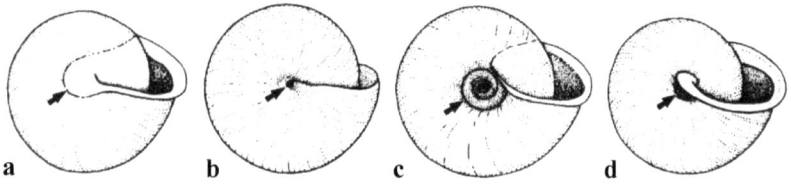

a   b   c   d

Figure a-Imperforate shell (closed umbilicus); Figure b-Perforate shell (small umbilical opening); Figure c-Umbilicate shell (wide umbilical opening); Figure d-Rimate shell (umbilical opening partially covered by aperture lip)

Shell Measurements

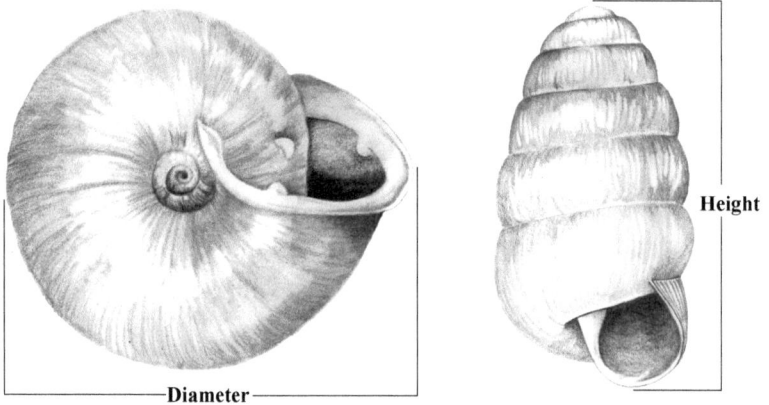

Height

Diameter

Shell Periphery

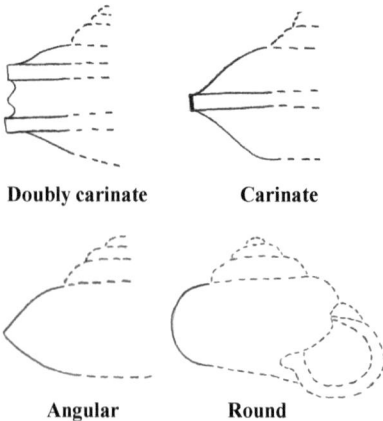

Doubly carinate   Carinate

Angular   Round

Peristome or Lip

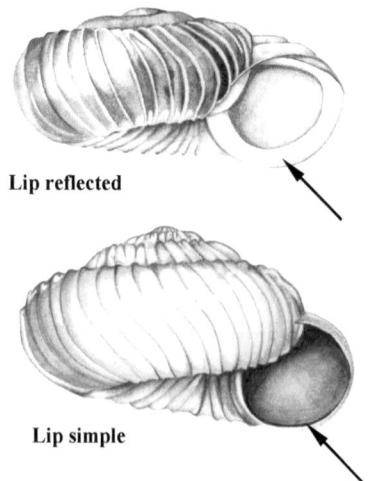

Lip reflected

Lip simple

38

Counting Whorls

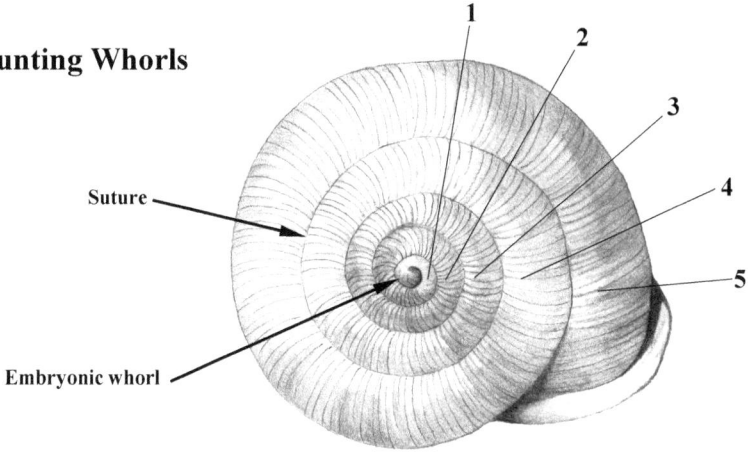

Suture

Embryonic whorl

1
2
3
4
5

Micro-features of the Shell Surface

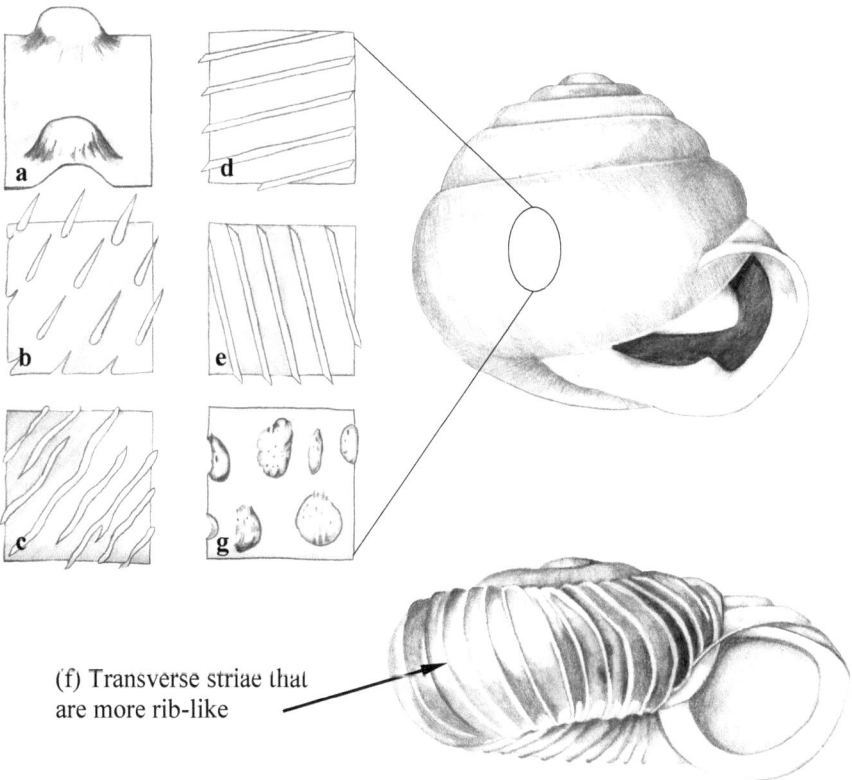

(f) Transverse striae that are more rib-like

(a) Papillae (raised bumps); (b) hairs; (c) wrinkles; (d) spiral striae (can be in-dented or raised) and run with the shell spire; (e) transverse striae (can be in-dented or raised) are perpendicular (vertical) to the shell spire; (f) in some shells the striae are more rib-like; (g) pits in the shell.

Shell Growth (author)

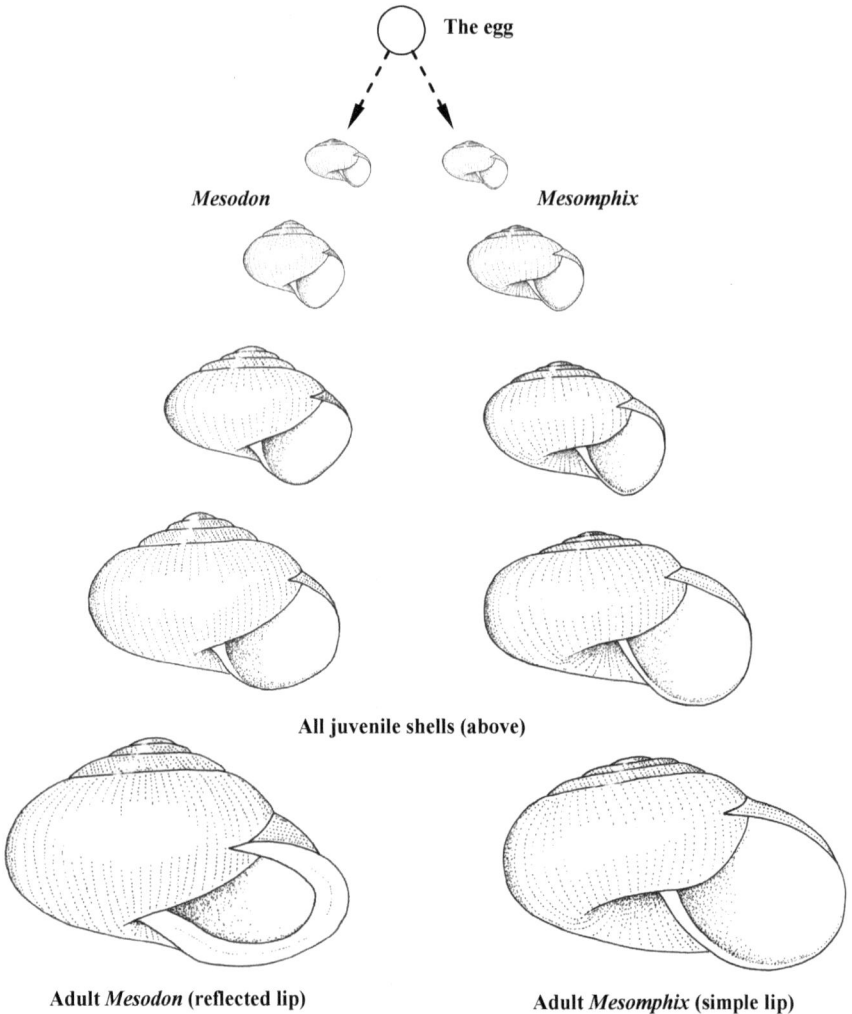

The egg

Mesodon

Mesomphix

All juvenile shells (above)

Adult *Mesodon* (reflected lip)

Adult *Mesomphix* (simple lip)

Immature shells of any species are difficult to identify. Determining the maturity of a shell can often be accomplished by examining the aperture and number of whorls. As shells mature, the shape of the aperture changes. Note the apertures of the juvenile shells of *Mesodon* and *Mesomphix*. The bottom of the aperture of each species appears to droop as if an invisible weight is attached. As shells of both species mature, the shape of the aperture changes to a more horizontally oval shape. For adult *Mesodon* species, a reflected lip forms and for adult *Mesomphix* species, the lip remains simple. Immature *Mesodon* species are easily confused with *Mesomphix* species. *Mesodon* and *Triodopsis* species do not form reflected lips until they reach maturity. Other species such as *Mesomphix, Anguispira* and *Ventridens* do not possess reflected lips at maturity. Finally, most mature shells should contain no less than 4 to 5 whorls at maturity.

Basic Land Snail Anatomy (redrawn from Burch, 1962)

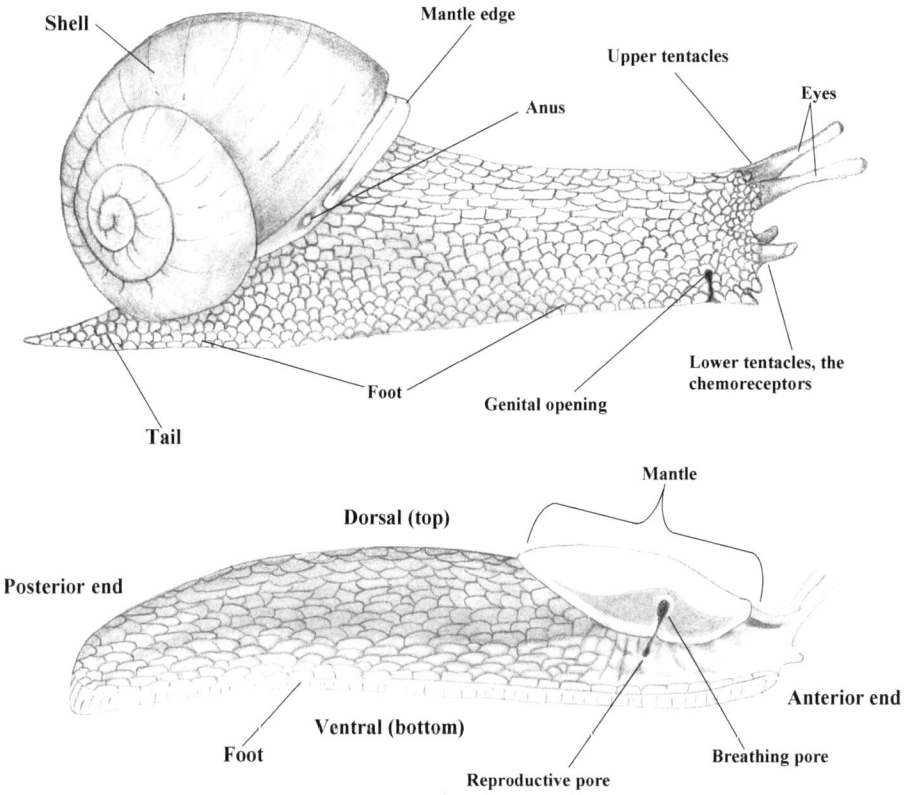

Shell

Mantle edge

Upper tentacles

Eyes

Anus

Lower tentacles, the chemoreceptors

Foot

Genital opening

Tail

Mantle

Dorsal (top)

Posterior end

Anterior end

Ventral (bottom)

Foot

Reproductive pore

Breathing pore

Internal Anatomy

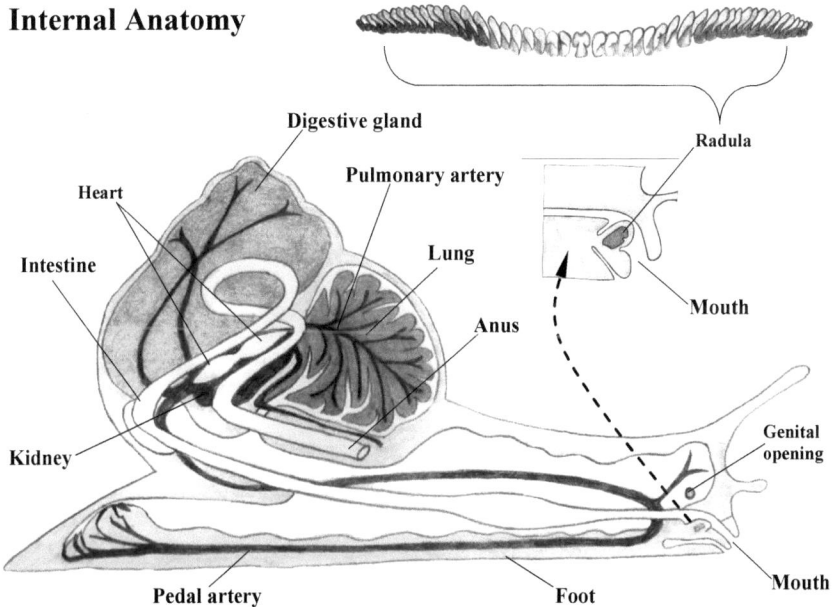

Digestive gland

Radula

Heart

Pulmonary artery

Intestine

Lung

Anus

Mouth

Kidney

Genital opening

Pedal artery

Foot

Mouth

Detailed Anatomy of a Land Snail (*Cornu aspersum*) Drawings
by Gordon Riley, 1979 (reprinted with permission)

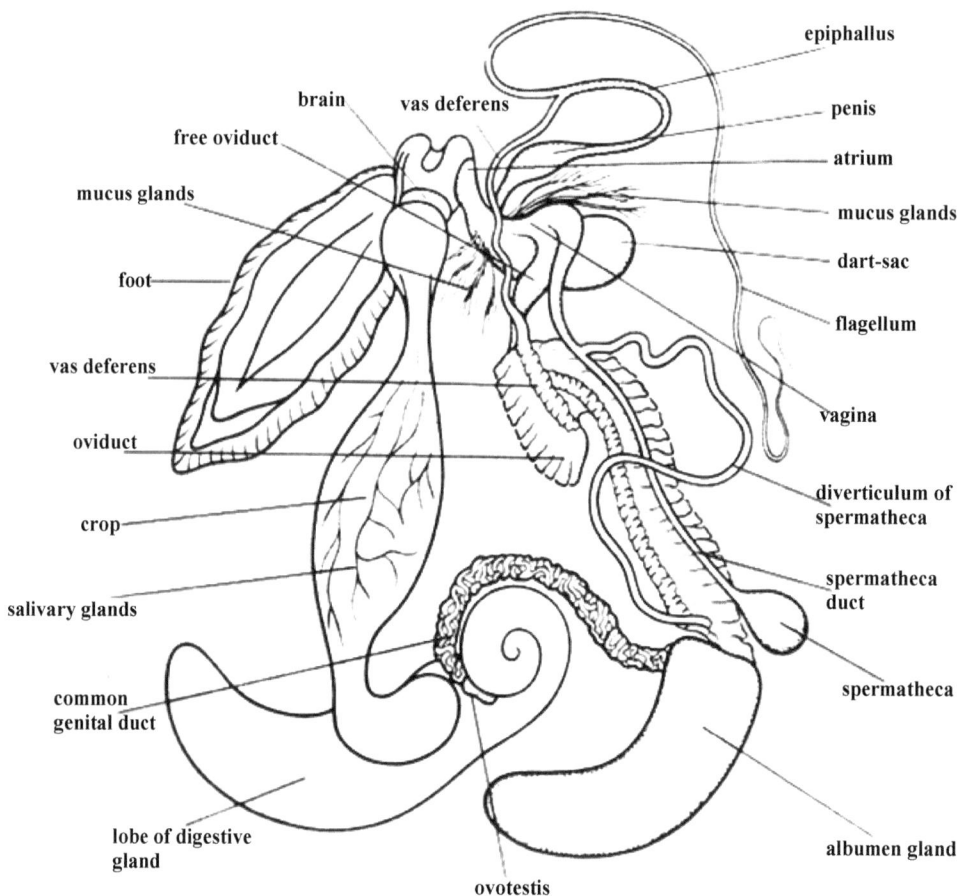

Basic Anatomy of a Slug

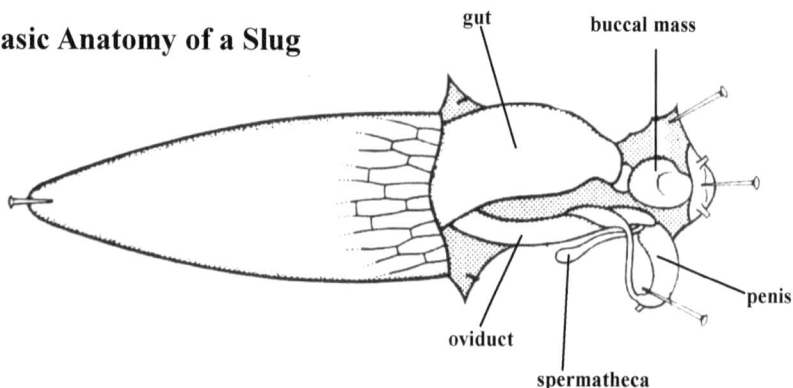

Identifying Land Snails

Land snails are a study in subtlety, and successful identification will depend on your aptitude for recognition of shell detail. The principal obstacle for beginners are the dichotomous keys. In an attempt to ascertain every species in one place, many keys overwhelm the reader with mind-numbing particulars. The reader can become frustrated quickly and lose interest. In this snail manual, the pictorial key takes you to genera only, keeping the finer details of each species in the text of the species accounts. Consequently, the keys are less cluttered with scientific muddle and, hopefully, easier to follow.

The keys are fundamentally based on two strategic characteristics: **shape** and **size** of the shell, with five additional shell discerning characters. These include: number of whorls, umbilicus, teeth or the absence of teeth, simple or reflected lip, and spiral striae or lack of this micro-feature. If you move through the keys but are uncomfortable with your final deliberation, go back and reconsider the shape (form) of the shell; this feature alone will most often lead to a misidentification. <u>While most shells fit easily into certain shapes or categories, some will prove more difficult to discern, so allow for these discrepancies</u>. It is not uncommon for the shell shape of some species to vary slightly from population to population and, on occasion, within the same population.

Although the colors of shells are stated throughout the text, keep in mind that color can vary greatly from site to site, and some of us (me included) are color-blind. Minute species were photographed through a microscope sometimes changing their natural color. Many shells used in the book have flaws such as cracks, blemishes or are bleached from weathering. To exemplify bottom and aperture views of some species, certain features were digitally enhanced in Photoshop. Remember, shell shape or form and micro-features are the most important diagnostic features. In old shells that are bleached, slightly wetting the surface will often bring out the micro-features.

Record observations 1-7, then proceed with the shell identification.

(1) **Shape**-Determine the general shape of the shell (see next page); same-species shells will sometimes vary slightly from site to site, and occasionally within the same population.

(2) **Greater Diameter or Height**-Measure the greater diameter for Heliciform shells and the height for Succiniform, Conical, and Pupa-shaped shells. Sizes may sometimes vary slightly from measurements given in species accounts.

(3) **Whorls** - Count the number of whorls (refer to pg. 39)

(4) **Umbilicus** -Is the shell imperforate, perforate, umbilicate, or rimate?

(5) **Teeth**-Make note of any presence or absence of teeth in the aperture and their size and spatial arrangement.

(6) Lip – Is the lip simple or reflected?

(7) Spiral Striae-To determine if spiral striae exist, examine shell surface under a dissecting microscope.

 This symbol indicates the use of a dissecting microscope is necessary or useful to view micro-features of the shell.

◆━━━▶ A scale bar is included near each shell to represent the actual size.

Basic Shell Shapes of Terrestrial Land Snails

Heliciform (globose)

Depressed heliciform

Shape of this shell lingers between depressed helici-form and heliciform; when this happens try both options.

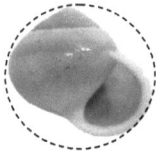

Roundish shape

Dome shape

Pill shape

Vitriniform

  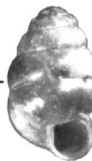

Pupa shape

Conical

Succiniform

General groupings of land snails in this book are based primarily on shape, lip and the presence or absence of teeth

	Page
*Shells taller than wide, simple lip (teeth absent)	54
*Shells taller than wide, reflected lip (teeth present)	66
*Shells wider than tall, simple lip	94
*Shells pill shape with narrow aperture	222
*Shells wider than tall, reflected lip	234
*Native slugs of West Virginia	310
*Exotic shelled snails and slugs (reported and potential)	340

Pictorial Key to Genera of West Virginia's Land Snails
Keys work for adult shells only
(Shells are generally arranged from smallest to the largest)

---

1) <u>Shell taller than wide, simple lip (teeth absent) 2-20 mm</u>

Pomatiopsis: shell conical, live snail with an operculum, 5 -7 mm tall, page 63

Operculum found in live individuals

Novisuccinea, Catinella, Oxyloma: shell succiniform and paper thin, without an operculum, 7-20 mm tall, page 55-59

Cochlicopa: shell pupa shape and translucent 4-7.5 mm tall, page 60-62

Columella: shell pupa shape, 2-3 mm tall, page 64

2) <u>Shell taller than wide, reflected lip (teeth present) under 4 mm</u>

Carychium: shell pupa shape, reflected lip, small parietal tooth, 1.5-2 mm tall, page 67-72

Gastrocopta: shell pupa shape, multiple teeth, lip notably reflected, 2-4.8 mm tall, page 73-80

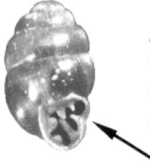

Vertigo: shell pupa shape, aperture containing multiple and sometimes crowded teeth, lip narrowly reflected with a slight indentation on the right side of aperture, 1.5-2.3 mm tall, page (73) 81-91

3) **Shell depressed heliciform, simple lip (usually without teeth) under 4 mm**

Punctum: shell minute with strong transverse and spiral <u>crisscrossing sculpture</u> (strong scope required), <u>umbilicate</u>, 1.5 mm, page (95, 100) 96-100

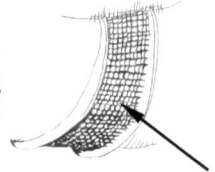

Striatura: shell minute with strong transverse and spiral sculpture (strong scope required), <u>umbilicate</u>, 1.5-3.5 mm, Page (95) 101-104

Hawaiia: shell minute with weak spiral sculpture (scope required), <u>umbilicate</u>, 2-3 mm; page (95) 105-107

Lucilla (Helicodiscus): shell minute with a nearly smooth surface (scope required), <u>umbilicate</u>, 2-2.5 mm, page (95) 109-110

Nesovitrea: shell minute with a nearly smooth surface, <u>umbilicate</u>, 3-5 mm, page (95) 111-112

Guppya: shell minute with a nearly smooth surface, but showing traces of spiral ornamentation, strongest around the apex of shell (scope required), <u>perforate</u>, 2 mm, page (95) 113

4) **Shell heliciform, paper-thin, simple lip (without teeth) 6 mm**

Vitrina: the shape of this small species is unlike any in its size class, 5-6 mm; page 114

5) Shell depressed heliciform, simple lip (without teeth) 4 to 12 mm

Glyphyalinia: shell thin with distinctive indented transverse striae (a), imperforate to umbilicate, 4-12 mm, page 121-138

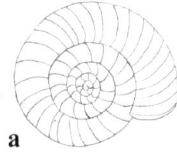

Zonitoides: shell surface smooth, transverse striae poorly developed, umbilicate, 4-8 mm, usually found under bark of dead trees, page 115-118

Discus: shell surface minutely ribbed (can be seen with hand lens of 10X), without teeth except for *Discus patulus*, widely umbilicate, 5-8 mm. page (150) 151-153

6) Shell domed-shaped, simple lip (teeth usually absent) 2-3.5 mm

Euconulus: shell dome or bee-hive shaped and thin, without teeth except for *E. dentatus* which has one or two inner teeth, perforate, 2-3.5 mm, page (214) 215-218

7) Shell heliciform to depressed heliciform, simple lip (teeth usually present, except for *Hendersonia*) 2-15 mm

Paravitrea: shell thin with either regular or irregular transverse striae on entire surface, strongest on the top, perforate to umbilicate, most species with internal teeth (b & c) at some stage of growth, 2-8 mm, page 139-149

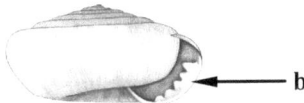

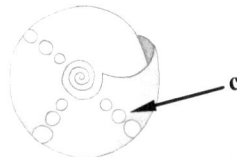

Gastrodonta: shell with two internal teeth, top surface boldly ribbed and visible with a hand lens of 10X, perforate, 6-7.5 mm, page (115) 119-120

Bold ribs on top of shell

47

Helicodiscus: shell flat with internal teeth, spiral striae (sometimes fringed or haired) are a strong feature, looking like the tread of a car tire, <u>widely umbilicate</u>, 2-5 mm, page (150) 154-161

Ventridens: shell can be flat or globose, glossy, with or without internal teeth, perforate to umbilicate, 5-15 mm, <u>important– note color of live animal</u>, page 163-178

Hendersonia: shell heliciform, thick for its small size with a callus on outer lip, imperforate, live snail with an operculum, 6-8 mm, page (214) 219-220

8) <u>Shell depressed heliciform, reflected lip (without teeth) under 3 mm</u>

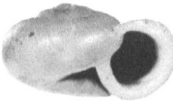

Vallonia: shell small with a widely reflected lip, widely umbilicate, 1.6-3 mm, page (236) 242-247

9) <u>Shell domed-shaped, reflected lip (lamellae present) under 3 mm</u>

Strobilops: shell small and beautifully sculptured with fine ribs, with two or more elongated lamellae, perforate, 2-2.8 mm, page (236) 237-241

10) <u>Shell pill-shaped, (aperture opening slit-like or more open) 5-12 mm</u>

Stenotrema: shell small and pill shape, lip opening restricted and narrow (d), <u>imperforate</u>, 5-15 mm, page 223-233

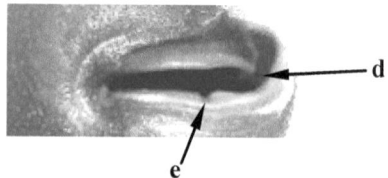

Euchemotrema: shell like *Stenotrema* but with a reflected lip, more open aperture and without a basal notch (e), <u>rimate</u>, 7-12 mm, page (223, 248) 249-250

11) <u>Shell depressed heliciform, reflected lip, imperforate, (large parietal tooth present) 7.5-14 mm</u>

Inflectarius: shells like *Stenotrema* but with a more open aperture, in WV parietal, basal and palatal tooth always present, <u>imperforate</u>, umbilicus completely sealed; 7-14 mm, page 254-256

12) <u>Shell heliciform to depressed heliciform, reflected lip, rimate (with or without teeth) 7.8-30 mm</u>

Euchemotrema: shells small like *Inflectarius* but <u>rimate</u>, umbilicus never completely sealed, 7.8-11 mm page (223, 248) 249-250

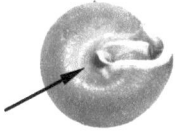

Mesodon (in part), shells globose, <u>rimate</u>, umbilicus never completely sealed, 15-30 mm, page (248) 251-252

13) <u>Shell heliciform to depressed heliciform, simple lip, perforate to um-bilicate, (without teeth) 15-36 mm</u>

Mesomphix: shell generally glossy, without teeth in all stages of growth, <u>perforate to umbilicate</u>, 15-36 mm, page 179-192

Anguispira: shell surface generally dull, without teeth in all stages of growth, <u>color features always present</u>, widely <u>umbilicate</u>, 15-31 mm, page 195-210

Haplotrema: shell glossy, without teeth, lip slightly re-flected in old adults, <u>umbilicate</u>, umbilicus the widest of any large snail found in West Virginia (a), 16-22 mm, page 211-212

a

14) Shell mostly depressed heliciform, reflected lip, imperforate, (long, low basal tooth usually present) 13-23 mm

Patera: shell depressed (except for *P. pennsylvanica*) with a large parietal tooth, imperforate, 13-23 mm, page 257-262

15) Shell depressed heliciform, reflected lip (three distinct teeth) 9-25 mm

Triodopsis: shell more or less depressed, usually containing three teeth in the aperture, the parietal tooth the largest of the three, umbilicate, 9-25 mm, page (263) 264-288

Xolotrema: shell more or less depressed, containing three teeth in the aperture, the parietal tooth the largest of the three, imperforate, 19-25 mm, page (263) 289

16) Shell large, depressed heliciform, reflected lip, umbilicate, 20-30 mm

Allogona: shell with or without multiple color bands, with a single basal tooth, around 30 mm, umbilicate, page (292, 293) 305

Appalachina: shell thin and somewhat fragile, with a small parietal and basal tooth, umbilicate, 20-30 mm, page (292, 293) 307

17) Shell large, mostly heliciform (globose), reflected lip, imperforate, 15-45 mm, but most shells here are greater than 20 mm

Mesodon: shell globose, with or without a small parietal tooth, imperforate, 15-38 mm, most greater than 20 mm, page (292, 293) 294-300

Neohelix shell more depressed, solid, with or without a large parietal tooth, 20-45 mm, imperforate, page (292, 293) 301-302

Webbhelix shell with multiple color bands, thin, without a parietal tooth, around 20 mm, imperforate, page (292, 293) 303

Blackwater Falls State Park, Tucker County, West Virginia

Part I
Land Snails of West Virginia

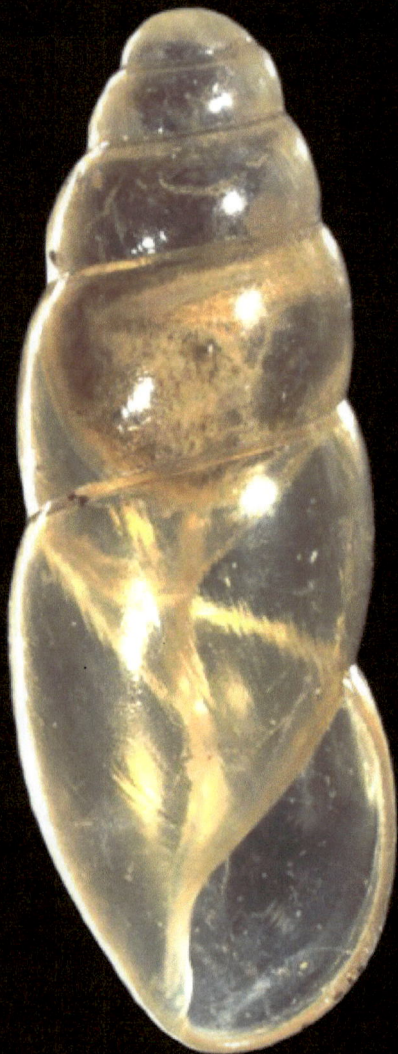

The study of snails is an inquiry into subtlety
and masterful sculpture.

The Junkyard Bug

The notorious "Junkyard Bug" is rarely observed because of its small size and cryptic behavior. Actually a larva of a lacewing in the family Chrysopidae, this interesting insect carries an unusual cargo of forest rubbish on its backside. The one pictured here was found in the Great Smoky Mountains National Park in 2006 and was hauling six species of land snails including: *Punctum vitreum* (new record for North Carolina), *Punctum minutissimum, Punctum blandianum, Carychium clappi, Gastrocopta contracta,* and *Gastrocopta pentodon* (figure a). Remarkably, one species, *G. pentodon,* was still alive but unable to detach itself from the back of the insect larva. The bug is reported to use forest litter, insect parts and now land snails to camouflage itself from potential predators. Figure (a) shows four of the six land snails (two remain hidden from view). Figure (b) is the naked lacewing larva showing the hairs that hold the "junk" in place. Other researchers have also reported finding lacewing larva with attached land snails.

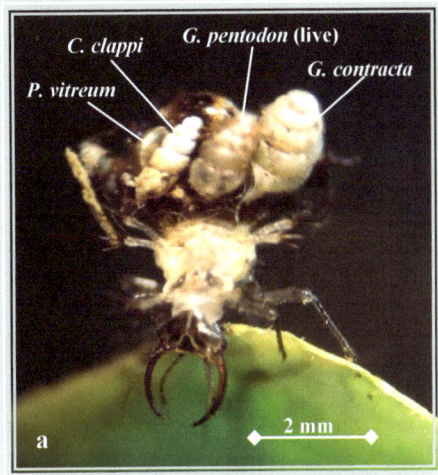

Land snails in this section are species that have shells taller than wide (Conical, Succiniform, or Pupa shape) and that have simple lips; although some may have a portion of the lip (usually on the left side) that is slightly reflected. They include small (under 5 mm), but mostly mid-sized (10-20 mm), snails and are found throughout West Virginia. At first glance, many of these snails look aquatic, and some are even referred to as amphibious, living close to, but not in water. Many have a glass-like luster, transparent and a near paper thin, but rigid shell. Consequently, after death these shells are quick to decompose and vanish from sight, therefore making their detection more challenging than other thicker-shelled species. All taxa in this section are without internal lamellae or teeth in the aperture.

Genera Included:
(in order of appearance in text)

Novisuccinea
Catinella
Oxyloma
Cochlicopa
Pomatiopsis
Columella

Columella

Oxyloma

Pomatiopsis

Oval ambersnail

Succineidae

Novisuccinea ovalis (Say, 1817)

Height: 10-20 mm, width 10.8-11 mm

Description: Succiniform; lip simple; shell with 2.5-4.5 whorls; imperforate; no teeth; shell paper thin, greenish-yellow to amber, often with darker streaks; shell surface weakly striate; sole of foot bluish-gray shading to orange at the edges.

Similar Species: *Catinella oklahomarum* is smaller, more compact and frequently covered in dried soil; *Oxyloma retusum* is more narrow in build with a notably more elongated aperture.

Habitat: This is the ambersnail of southern mountains, found from low wet valleys to high mountain tops (up to 2000 meters); it is often seen climbing on jewelweed, stinging nettle, and false hellebore.

Status: G5/S5; Fairly Common with well scattered records from across the state.

Specimen: North Carolina, Mitchell County, near Carvers Gap (author's collection).

Above illustration showing the live animal as seen through the translucent shell

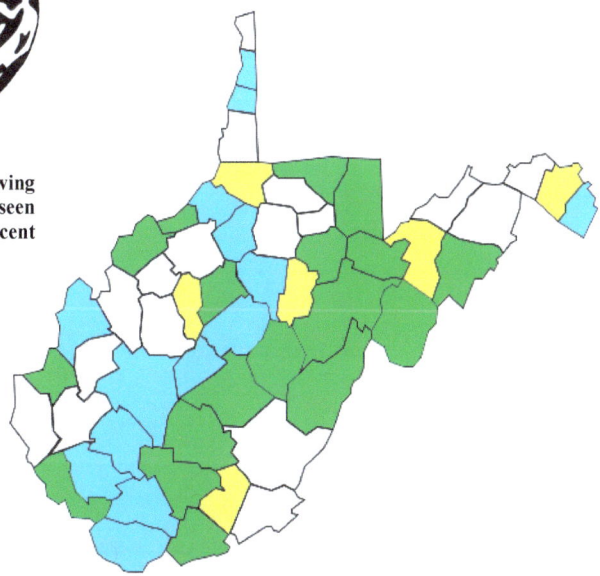

The live animal of oval ambersnail, *Novisuccinea ovalis* can be seen through its thin, translucent shell. Canaan Valley, West Virginia.

Detritus ambersnail

Succineidae

Catinella oklahomarum (Webb, 1953)

Height: 7-8.8 mm tall, width 4mm

Description: Succiniform; lip simple; shell compact and obese, with 2-2.5 whorls; no teeth or lamellae present in the aperture in any stage of growth; shell surface weakly striate and glossy, often coated with soil particles (as seen in specimens illustrated to the right); shell color is a greenish-yellow, but this feature is variable; sides of the foot smoky.

Similar Species: *Novisuccinea ovalis* is taller with a notably larger aperture; *C. vermeta* is larger and less compact in form; *Succinea forsheyi* from nearby states is larger with a capacious aperture.

Habitat: Usually found in the leaf litter of wooded hillsides or in pine woods, on acidic soils; the species is rarely very abundant (Hubricht 1985).

Status: G5/S3; Rare; reported from only a few counties in West Virginia, but is likely more widespread than current records indicate.

Specimen: Kentucky, Wayne County, Monticello (FM 236387).

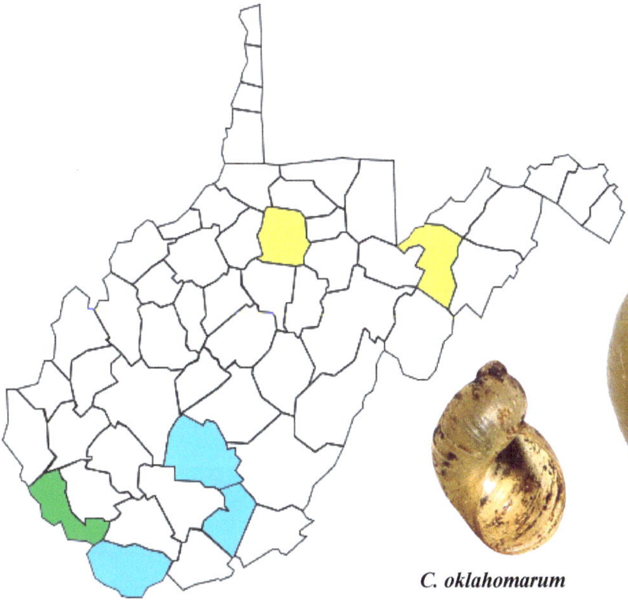

C. oklahomarum    N. ovalis

57

Suboval ambersnail

Catinella vermeta (Say, 1829)

Height: 7-13 mm tall, width 4-6.8 mm

Description: Succiniform; lip simple; shell with 2.5-3.5 whorls; perforate; shell surface weakly striate, pale yellow-olive, often coated with soil particles.

Similar Species: *Catinella oklahomarum* is more compact and generally found in drier situations; *Novisuccinea ovalis* is larger with a notably larger aperture.

Habitat: An amphibious land snail that appears restricted to wet ground found around ponds, marshes, muddy banks of open ditches, and swamps in both open and shaded conditions; in upland sites found under stones.

Status: G5/S3; Common; reported from many locations across the state; most authors (e.g. Burch 1962; Patterson and Burch 1966) recognized *C. vermeta* as the name applicable to specimens from the USA, which were formerly identified as *Catinella avara* (Turgeon *et al.* 1998).

Specimen: Pennsylvania, Allegheny County (CM 75744).

Succineidae

Kentucky, Fayette County, Raven Run Nature Reserve

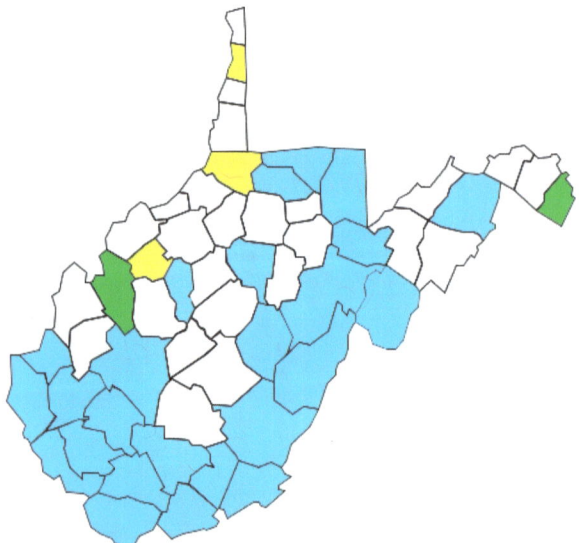

Blunt ambersnail

Succineidae

Oxyloma retusum (I. Lea, 1834)

Height: 7.5-20 mm, width 7.7-8.7 mm

Description: Succiniform; lip simple; shell with 2.5-4.5 whorls; perforate; no teeth or lamellae present in the aperture at any stage of growth; shell weakly striate, near paper thin and fragile; surface smooth; margin of genital aperture or opening not swollen as in *Novisuccinea ovalis*.

Similar Species: *Novisuccinea ovalis* has a wider shell and on the live animal, the margin of the genital opening is notably swollen.

Habitat: Found in low wet places around marshes and ponds crawling on mud or on the leaves of cattails and other aquatic plants.

Status: G5/S1; Rare, this species is known from only Wood and Pendleton counties; in Pendleton County, Hubricht collected 6 specimens crawling on watercress.

Specimen: West Virginia, Wood County, Muskingum Island (author's collection).

Bottom view   **Top view**

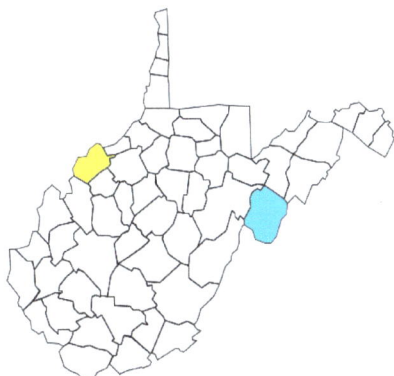

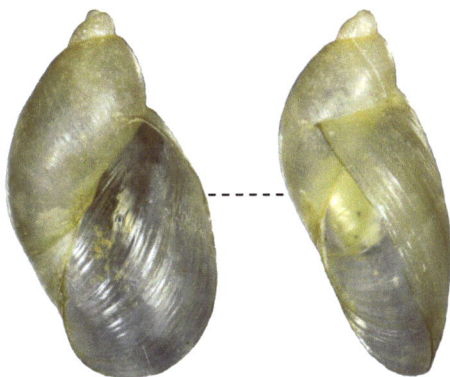

West Virginia, Pendleton County at a spring on watercress (FM 235482)

Appalachian pillar

Cochlicopidae

Cochlicopa morseana (Doherty, 1878)
Height: 6.6-7.2 mm width 2.3-2.5 mm
Description: Pupa shape; shell with 5.5 whorls; sutures slightly impressed; no teeth present at any stage of growth; shell smooth, glossy, translucent with a glass like surface; foot of live animal white with a grayish head.
Similar Species: *Cochlicopa lubrica* is similar in shape and luster, differing mainly in its broader stature, tighter less rounded whorls, and by its slightly wider and thicker callus rim of the outer lip; *Gastrocopta* and *Vertigo* species are much smaller and contain teeth in their aperture; *Columella simplex* is around 4mm smaller and does not have a translucent shell.
Habitat: A species that hides well in upland woods among deep moist leaf litter; it is rarely found on top of the litter even in wet weather, staying below or between the layers of leaves.
Status: G5/S5; Relatively Common and in general, under collected, it likely occurs in every West Virginia county.
Specimen: Kentucky, Powell County, Furnace Mountain (author's collection).

Edge of lip thin

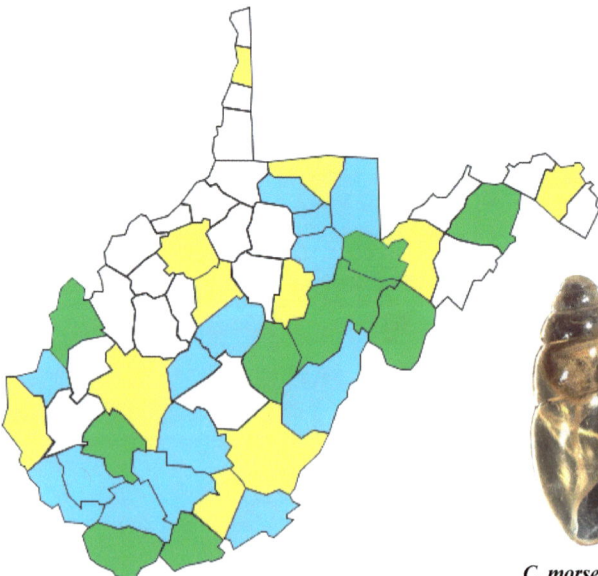

C. morseana   *Carychium exile*

60

Glossy pillar

Cochlicopidae

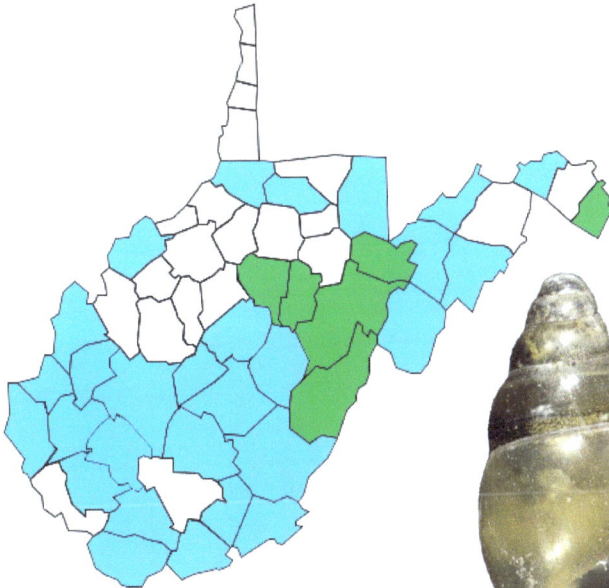

Cochlicopa lubrica (Müller, 1774)

Height: 5-7.5 mm, width 2.4-2.9 mm

Description: Pupa shape; shell with 5.5-6 whorls; sutures slightly impressed; no teeth present at any stage of growth; shell smooth, glossy and translucent with a glass-like surface.

Similar Species: *Cochlicopa morseana* is more narrow in form, its aperture more elongate with a thinner callus rim of the outer lip and is a lighter, more translucent honey color; *Gastrocopta* and *Vertigo* species are much smaller and contain teeth in apertures.

Habitat: Thought to be an exotic by some authors; an Holarctic species (Hubricht 1985); in North America, prefers disturbed sites (yards and road verges) and hard clay soils with thin litter (Nekola 2003).

Status: G5/SNA; Common; reported from across the state; the snail is most likely transported to new locations via potting soils of yard plants and by way of construction lumbers.

Specimen: Larry Watrous photo collection.

C. lubrica (two shell forms), both figures from West Virginia, Tucker County, Canaan Valley

61

Thin pillar

Cochlicopidae

Cochlicopa lubricella (Porro, 1838)

Height: 4.5-6.8 mm, width 2.1-2.5 mm

Description: Pupa shape; shell with 5.5-6 whorls; sutures slightly impressed; no teeth present at any stage of growth; shell smooth, glossy, translucent with a glass-like surface.

Similar Species: *Cochlicopa morseana* is larger, narrower in form, its aperture more elongate with a thinner callus rim of the outer lip and is a lighter, more translucent honey color; *C. lubrica* is larger and less compact in build.

Habitat: Thought to be an exotic by some authors; an Holarctic species (Hubricht 1985); in North America, favoring drier places than *C. lubrica*; such as limestone, grasslands, calcareous sand dunes and screes (Kerney and Cameron 1979); also a species of developed ground such as yards and road verges.

Status: G5/S4; Uncommon; reported from only six West Virginia counties.

Specimen: Kentucky, Wolfe County, yard in town of Zachariah (author's collection).

Edge of lip thickened

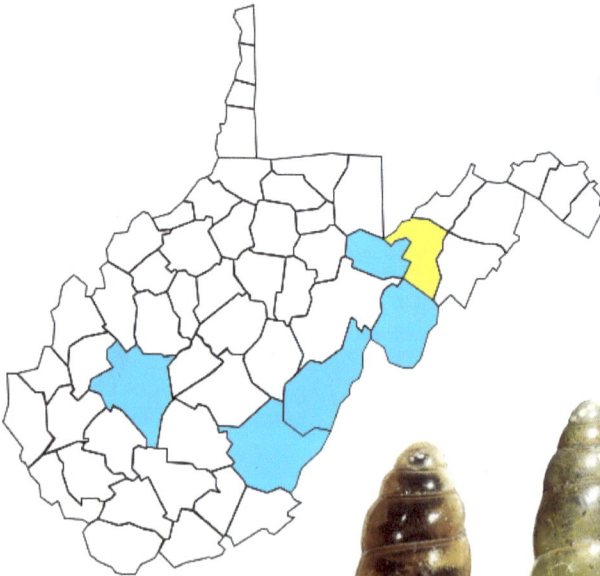

C. lubricella *C. lubrica* *C. morseana*

Slender walker

Pomatiopsis lapidaria (Say, 1817)

Pomatiopsidae

Height: 5-7 mm, width 3-3.5 mm

Description: Conical; lip simple; shell with 6-7 whorls; no teeth present at any stage of growth; fresh shells are a rich brown; live animal with an operculum (a), a calcareous door.

Similar Species: *P. cincinnatiensis* found in nearby states is smaller, has a more inflated profile and has one less whorl.

Habitat: A calciphile often referred to as amphibious or even aquatic; common on wet limestone rock faces, in dripping seeps, and in mats of algae along small streams but also found on hillsides in drier and more acidic situations.

Status: G5/S5; Common, found across the state in suitable habitat.

Specimen: *P. lapidaria* from Kentucky, Pike County, Breaks Interstate Park (author's collection); *P. cincinnatiensis* from Kentucky, Edmonson County, Mammoth Cave National Park (FM 229105).

a

Operculum

P. cincinnatiensis

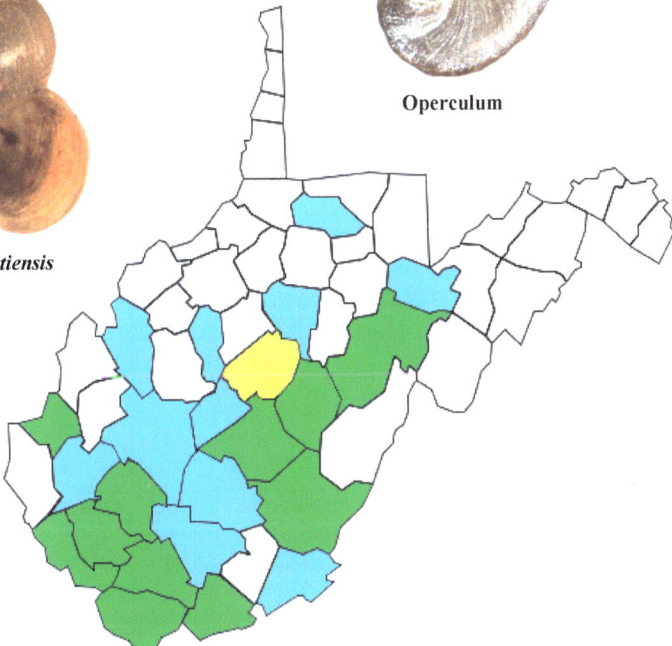

63

Toothless pupa

Pupillidae

Columella simplex (Gould, 1841)

Height: 1.75-2.5 mm, width 1.3 mm

Description: Pupa shape; shell with a slight taper; left side of lip slightly reflected, while the right side is simple; shell with 5-7 whorls that taper over the first 4-5 whorls; perforate; no teeth present in aperture; transverse striae weakly developed; two forms are illustrated, both found in West Virginia.

Similar Species: *Vertigo* and *Gastrocopta* species have teeth in their aperture and have notably reflected lips; *Carychium* are smaller, more narrow in form, and with small teeth.

Habitat: Found across a wide variety of open and forested habitats, ranging from subtropical to taiga, xeric to wet, and acidic to calcareous and on ferns and low shrubs, ubiquitous in most upland woods (Nekola and Coles 2010).

Status: G5Q/S5; Relatively Common; this species will likely be found in every West Virginia county.

Specimen: Figure (a) from Rock Creek, Cedar County, Iowa and figure (b) from Haywood Landing, Jones County, North Carolina (Nekola and Coles image 2010).

a

Two forms illustrated

b

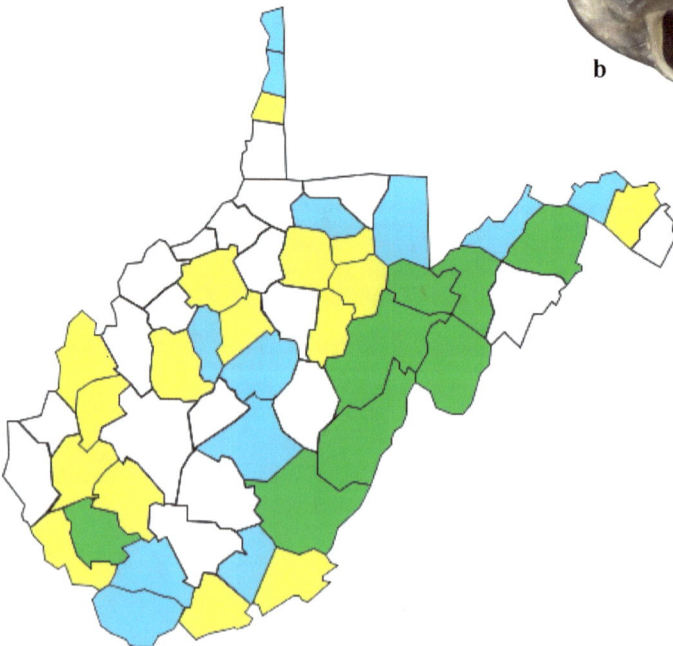

64

Seneca Rocks & North Fork of the South Branch of the Potomac River, Pendleton County, West Virginia

©2014 mdmpix.com

©2014 mdmpix.com

This section includes land snails that have shells that are taller than wide, pupa shape, and have reflected lips. These are the miniatures of terrestrial gastropods, most species under 3 mm as adults. In some habitats such as glades, dry outcrops of limestone or wet seeps, they may represent as much as 75% of the species found. <u>Consequently, gathering leaf litters at all sites is essential for a accurate register of species</u>. In general, fresh dead and live shells of this group are translucent and dead shells, with age, become bleached. *Vertigo* and *Gastrocopta* snails are cryptic, remaining well-hidden in the detritus and leaf litter. These two genera also have teeth in their apertures which are thought to prevent attacks from predators like predacious beetle larva. Although tiny, shells of *Carychium* species tend to be more easily spotted, a result of their lighter color.

Genera Included:
(in order of appearance in text)

Carychium
Gastrocopta
Vertigo

Gastrocopta

Carychium

Vertigo

Ice thorn

Ellobiidae

Carychium exile H. C. Lea, 1842

Height: 1.5-1.8 mm, width 0.6 mm

Description: Pupa shape; lip reflected and somewhat thickened within; shell with 4.5-5 whorls; there are two horizontal entering lamellae on the left side of the aperture (see below enlarged aperture view with lamellae); transverse striae are relatively well developed; no spiral striae; shells are translucent in live individuals and fresh shells.

Similar Species: *C. clappi* is taller, more narrow in build and the finer transverse striae are more closely spaced, although this distinction is variable and only seen under a strong lens; *C. nannodes* is smaller and lacks prominent transverse striae.

Habitat: The species is found living between moist, but not wet, matted leaves that accumulate in pockets of low wet depressions, next to logs and adjacent to headwater seeps.

Status: G5/S5; Common; the most common *Carychium* in West Virginia and should occur in every county.

Specimen: West Virginia, Greenbrier County, Greenbrier State Forest (author's collection).

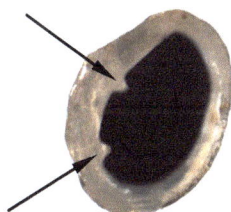

Aperture view of a *Carychium*

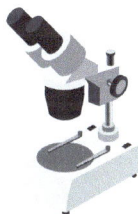

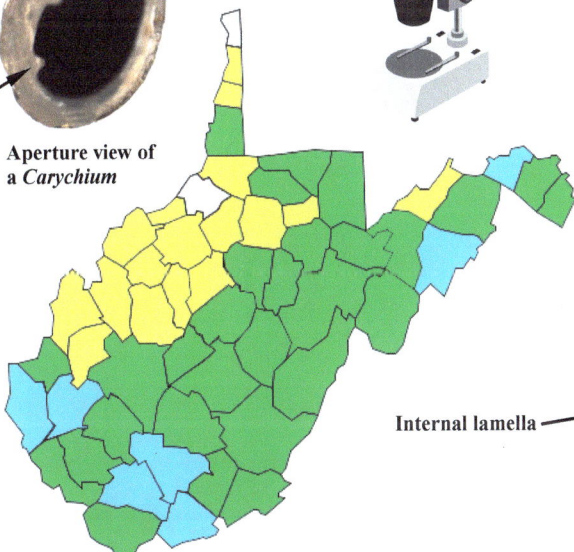

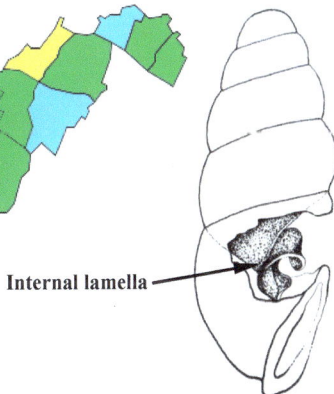

Internal lamella

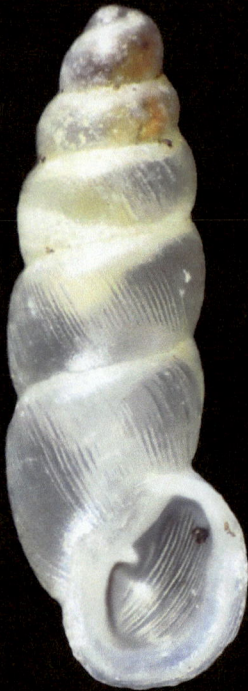

Carychium clappi, Pendleton County, WV

Carychium exile, Greenbrier County, WV

Appalachian thorn

Ellobiidae

Carychium clappi (Hubricht, 1959)

Height: 1.7-1.8 mm, width 0.8 mm

Description: Pupa shape; lip reflected; shell tiny with 4.5 whorls; two horizontally entering lamellae that can be seen on the left side of the aperture; transverse striae well-developed and closely spaced; shells translucent in live individuals, becoming bleached white with age.

Similar Species: *Carychium exile* is shorter in height but is wider in stature and has less distinct transverse striae on the first three whorls; *C. nannodes* is smaller and has a smoother surface.

Habitat: Found living between layers of moist leaves located in moist depressions on hillsides; it is often found with *C. exile* and *C. nannodes*.

Status: G5/S5; Relatively Common; found throughout most of the state.

Specimen: West Virginia, Pendleton County, UTM 625389-4283122 (author's collection).

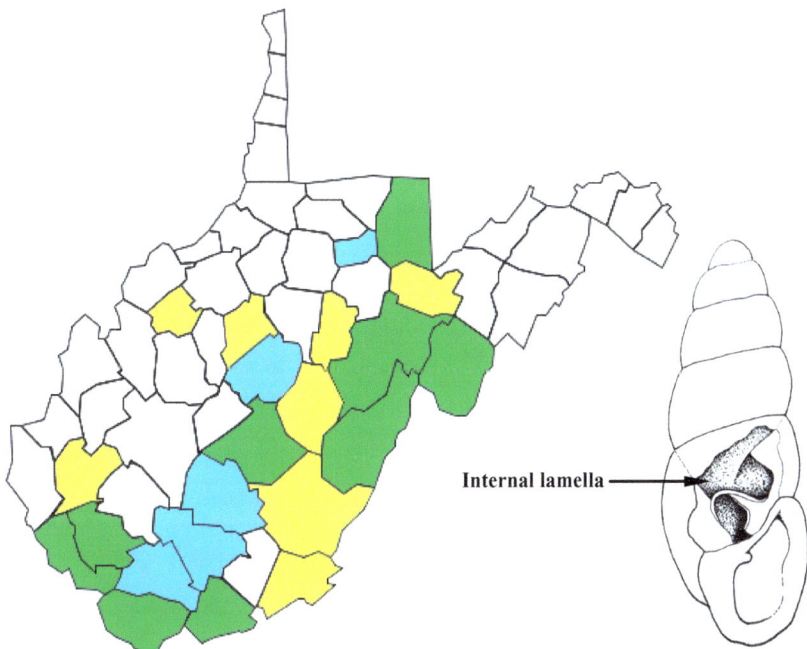

Internal lamella

Obese thorn

Ellobiidae

Carychium exiguum (Say, 1822)

Height: 1.5-1.6 mm, width 0.75 mm

Description: Pupa shape; lip widely reflected; shell with 4.5-5 whorls; there are two horizontal entering lamellae seen on the left side of the aperture; transverse striae are poorly developed and at times, hardly perceptible; no spiral striae; shells are translucent in live individuals.

Similar Species: *Carychium clappi* is slightly larger and more narrow in form with transverse striae which are more distinct; *C. exile* is around the same height, but is narrower (the most important feature separating the two species), and has more distinct transverse striae.

Habitat: Found in pockets of moist decaying leaves found in sinkholes or around the entrances of caves.

Status: G5/S3; Uncommon; found in scattered locations across the state.

Specimen: Tennessee, Blount County, Cades Cove, GSMNP (GSMNP collection).

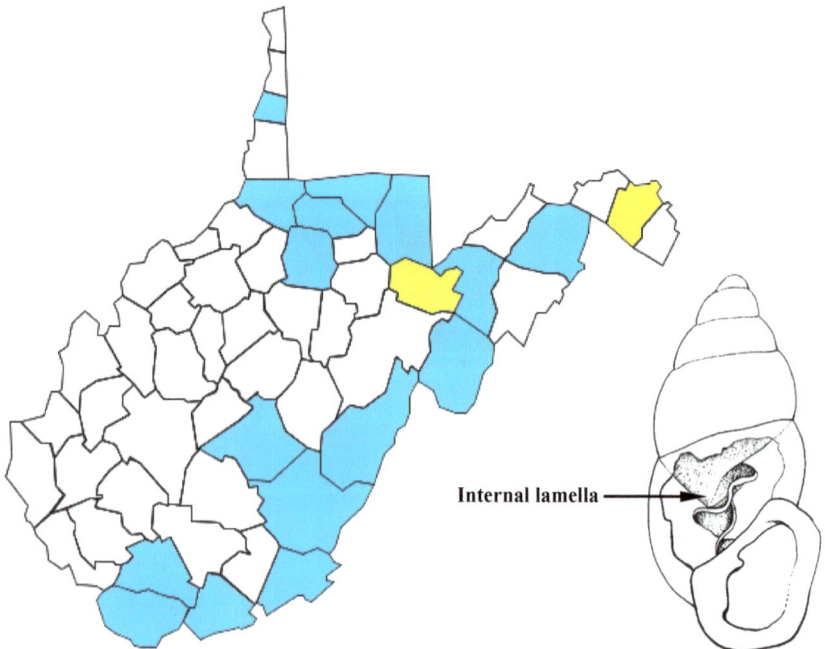

Internal lamella

File thorn

Carychium nannodes (Clapp, 1905)

Height: 1.3-1.5 mm, width 0.5 mm

Description: Pupa shape; lip reflected; shell with 4.5-5 whorls; sutures notably deep; there are two horizontal entering lamellae that can be seen on the left side of the aperture (strong lens required); transverse striae are poorly developed and at times, hardly discernible; shells are translucent when fresh but become bleached with age. This is the smallest *Carychium* species in West Virginia and the southern Appalachian Mountains.

Similar Species: *Carychium exile* and *C. clappi* are both larger with conspicuous transverse striae.

Habitat: Found in pockets of moist decaying leaves in low depressions and talus slopes; also common around large walnut, butternut, basswood, and buckeye trees.

Status: G5/S3; Relatively Common; throughout much of the state in appropriate habitat.

Specimen: West Virginia, Greenbrier County, near Davis Spring (author's collection).

Ellobiidae

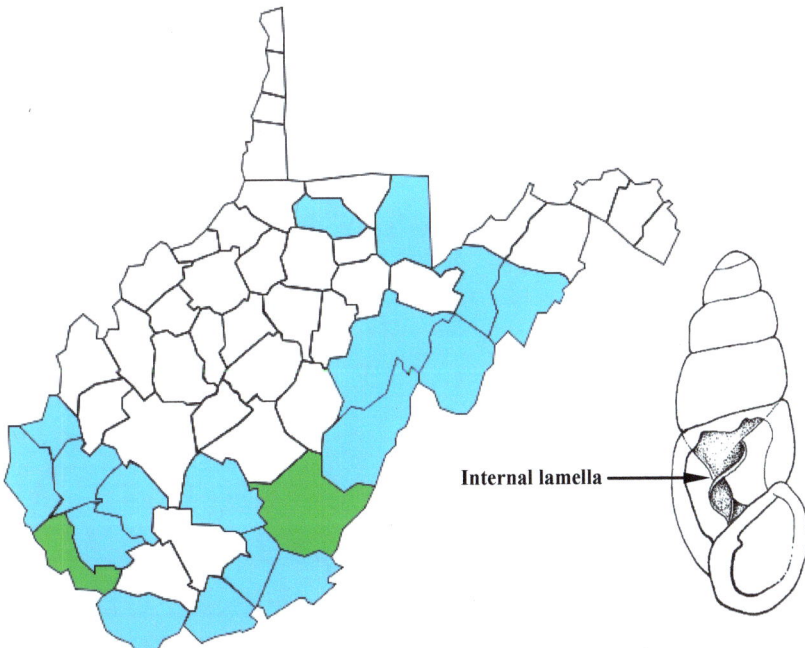

Internal lamella

West Virginia *Carychium* Compared (proportionate)

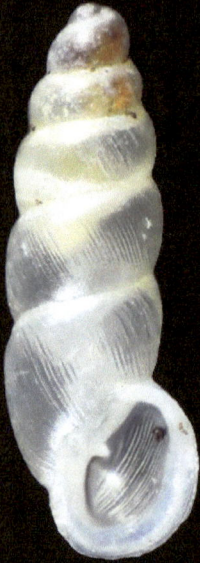

C. clappi

C. exile

C. exiguum

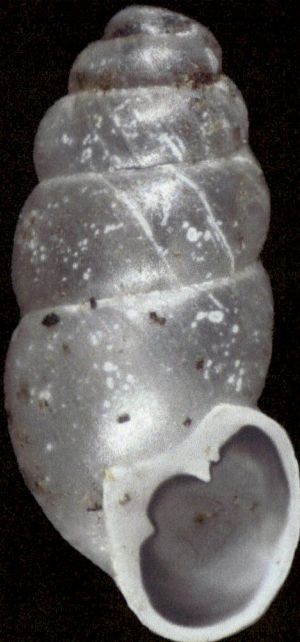

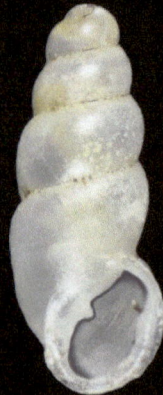

C. nannodes

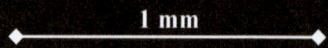

1 mm

Gastrocopta corticaria

Key to *Gastrocopta* and *Vertigo* Species

Gastrocopta and *Vertigo* are tiny snails with pupa shape shells with well-developed teeth in the aperture. In general, they differ from *Carychium* by their larger, more obese and toothier shells. *Gastrocopta* species are generally larger than *Vertigo*, usually have wider, more reflected lips (a) and are without a small indentation (b) in the right side of the aperture. The angular & parietal lamella are fused in *Gastrocopta*, simple in *Vertigo,* and most *Gastrocopta* have white shells when alive. Shell surface for either group can be smooth or with fine transverse striae, but are typically without spiral striae. The shell apertures in this section have been darkened behind the teeth to better illustrate the teeth and their spatial arrangement.

Shells proportionate

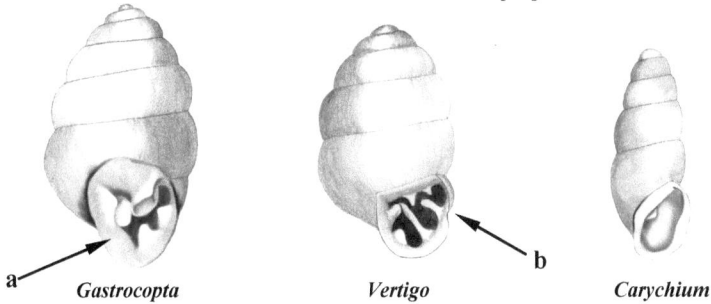

a *Gastrocopta*　　　*Vertigo*　　b　　*Carychium*

Terminology Used in Pupillidae Identification (Nekola and Coles 2010).

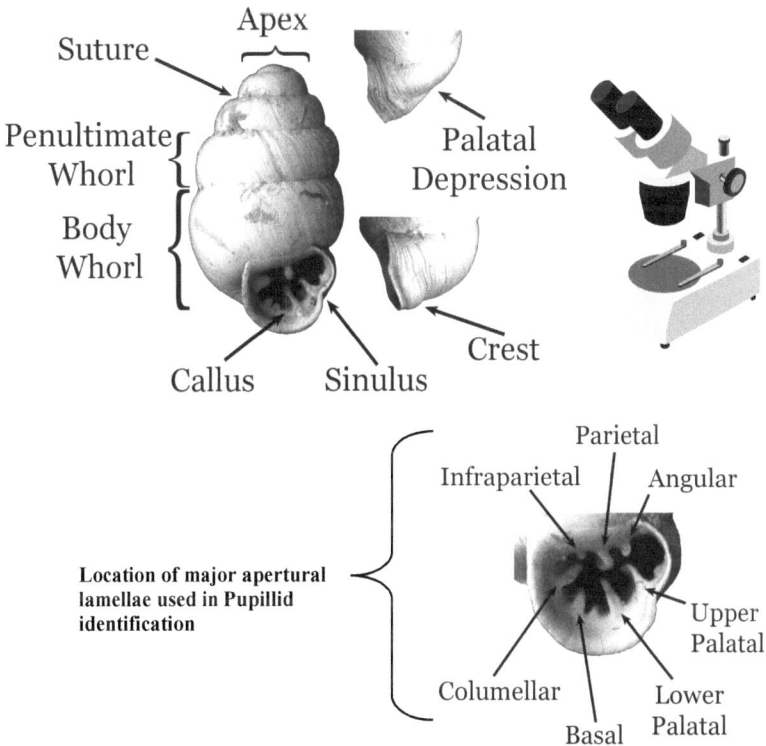

Apex

Suture

Penultimate Whorl

Body Whorl

Palatal Depression

Callus　Sinulus　Crest

Location of major apertural lamellae used in Pupillid identification

Parietal

Infraparietal　Angular

Columellar　Upper Palatal

Basal　Lower Palatal

73

Bark snaggletooth

Vertiginidae

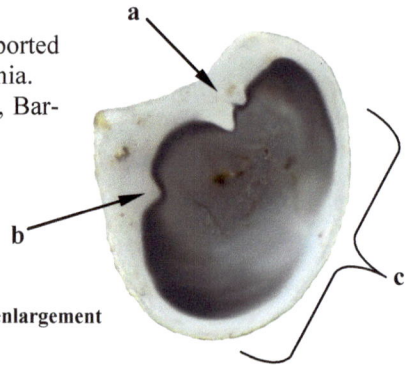

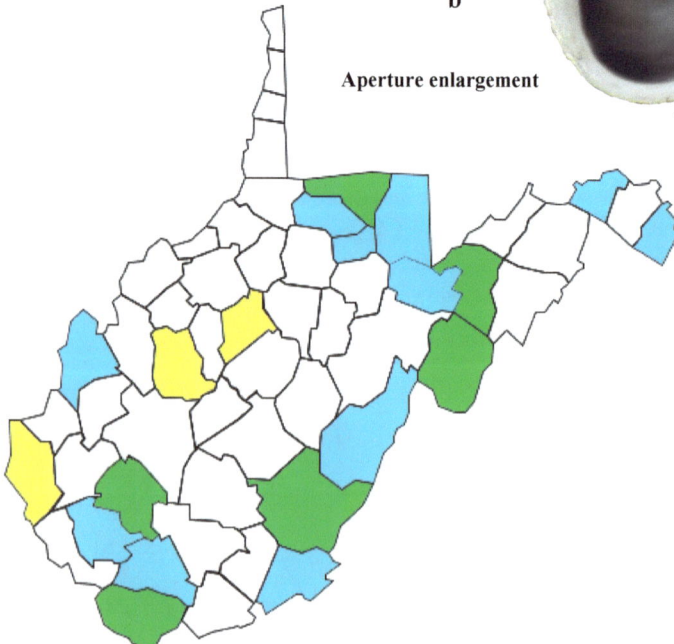

Gastrocopta corticaria (Say, 1816)

Height: 2.5 mm, width 1 mm

Description: Pupa shape; lip reflected; shell with 5.5 whorls; one small forked tooth present on the parietal wall (a) and one small columellar tooth (b); outer palatal wall without teeth (c); shells are somewhat translucent in live individuals and fresh shells.

Similar Species: This is the only *Gastrocopta* species in West Virginia to possess only two teeth, lacking any lamellae or teeth on the outer (palatal) wall of the aperture; *Columella simplex* is without any teeth in the aperture and has a simple non-reflected lip.

Habitat: A calciphile found in mixed hardwood forests with limestone outcrops but also found in acidic forests where it may be restricted to mossy tree trunks; in wet weather it can be found crawling on logs and the trunks of trees (Hubricht 1985).

Status: G5/S4; Relatively Common; reported from scattered counties across West Virginia.

Specimen: West Virginia, Roane County, Barnes Run (author's collection).

Aperture enlargement

74

Bottleneck snaggletooth

Gastrocopta contracta (Say, 1822)

Height: 2.2-2.5 mm, width 1.3-1.4 mm

Description: Pupa shape; lip widely reflected; shell with 5.5 whorls; with large teeth, the parietal tooth the largest (a); spiral of shell tapering; aperture more or less triangular in shape; shells are somewhat translucent in live individuals and fresh shells are frequently covered with dried soil; old shells quickly become bleached.

Similar Species: *Gastrocopta armifera* is larger, adding 2 additional whorls to its shell and the aperture teeth are less crowded; the lamellae configuration of *G. contracta* is also very different than *G. armifera* and has a tapered (not ovate) shell shape.

Habitat: This species is found in nearly all terrestrial habitats including, but not limited to, low wet places to dry mountainsides and acidic forests.

Status: G5/S5; Common; this common and widespread species in West Virginia is found in nearly every county; leaf litter collection will no doubt close in the remaining ten counties.

Specimen: West Virginia, Greenbrier County, Greenbrier State Forest (author's collection).

Vertiginidae

Side view

75

Armed snaggletooth

Vertiginidae

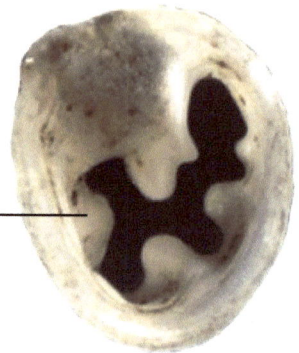

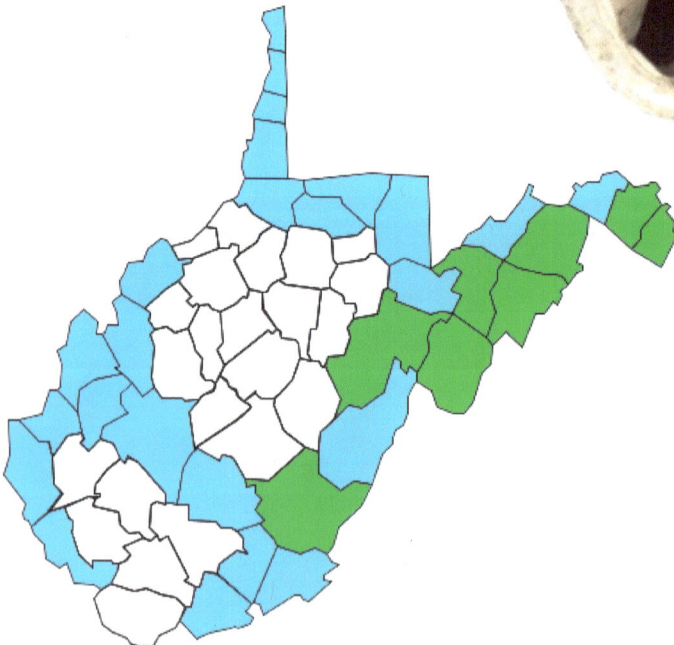

Gastrocopta armifera (Say, 1821)

Height: 3-4.8 mm, width 2.2 mm

Description: Pupa shape; lip reflected; shell with 6.5-7.5 whorls; aperture crowded with large teeth, the parietal tooth large and forked; columellar lamella with both forward and basally pointing components (a), appearing more or less pyramidal in apertural view (Nekola and Coles 2010).

Similar Species: This is a giant among *Gastrocopta,* usually double the size of most species in this genera; *Gastrocopta contracta* is smaller, having only 5.5 whorls and the aperture teeth are notably larger, choking the aperture; other *Gastrocopta* species are notably smaller.

Habitat: A calciphile found in more open, sunny areas such as cedar glades and limestone outcrops.

Status: G5/S4; Relatively Common; although not as common as *G. contracta,* it is a close contender in its abundance in the state.

Specimen: Vermont, Rutland County, Poultney River (Nekola and Coles image 2010).

a

White snaggletooth

Vertiginidae

Gastrocopta tappaniana (C.B. Adams, 1842)

Height: 1.6-2 mm, width 1.1-1.2 mm

Description: Pupa shape; lip reflected; shell with 4-5 whorls; outer lip with a distinct palatal callus or ridge (a) which support multiple teeth; shells are somewhat translucent in live individuals and fresh shells but become bleached with age.

Similar Species: *Gastrocopta pentodon* is only slightly smaller, less oval, thinner in form, and has a more tapered shell.

Habitat: A calciphile found around moist areas including marshes, floodplains of streams, and margins of ponds, but also in drift piles along small streams and in accumulations of decomposing leaf litter.

Status: G5/S3; Uncommon; only reported in scattered locations across the state, however, in the past, this species has been confused with *G. pentodon* and the two are suspected to intergrade where colonies overlap, making their easy separation nearly impossible for the beginner.

Specimen: Minnesota, Norman County, Faith Fen (Nekola and Coles image 2010).

a

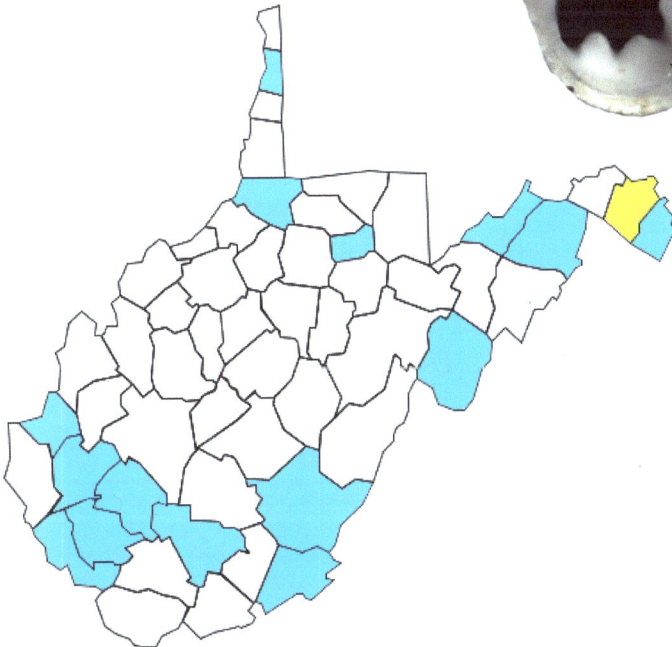

Comb snaggletooth

Gastrocopta pentodon (Say, 1821)

Height: 1.5-1.8 mm, width 0.8-1.1 mm

Description: Pupa shape; lip narrowly reflected; shell with 5 whorls; parietal tooth is the largest (a); outer lip with a distinct palatal callus (b) or ridge containing 2 to 6 teeth but sometimes up to 9 teeth; shells are somewhat translucent in live individuals and fresh shells are typically covered in soil.

Similar Species: *Gastrocopta contracta* and *G. armifera* are both larger, containing fewer but larger teeth in their aperture.

Habitat: A species that is found in a variety of habitats including dry upland hardwood forests, acidic forests such as pine savannas, around limestone outcrops, and occasionally, in low wet places.

Status: G5/S5; Common; by far the most common and widespread *Gastrocopta* in West Virginia and no doubt will be found in every county.

Specimen: New Jersey, Burlington County, Lebanon State Forest (Nekola and Coles image 2010).

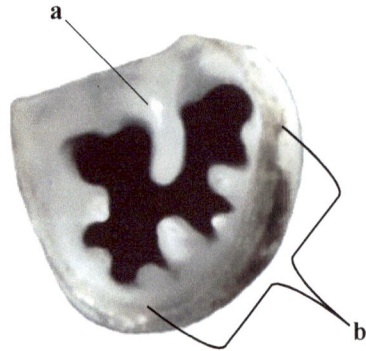

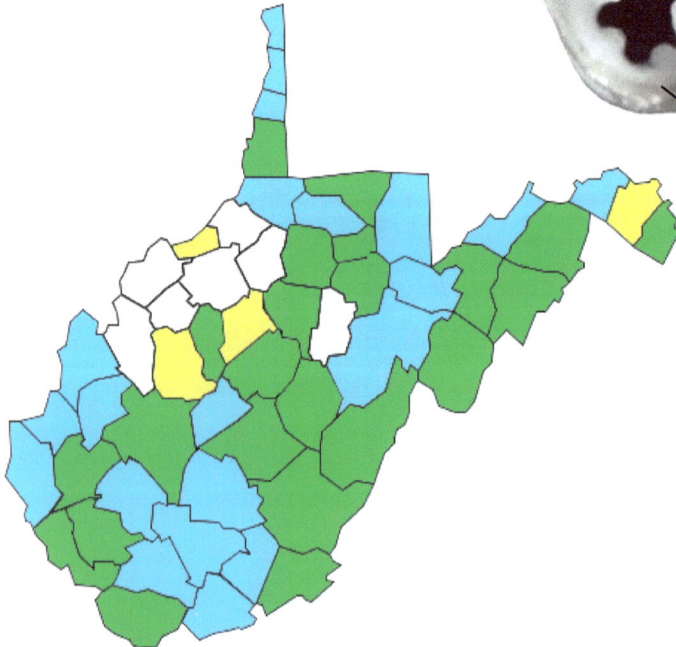

Wing snaggletooth

Gastrocopta procera (Gould, 1840)

Height: 2.2-3 mm, width 1-1.1 mm

Description: Pupa shape; lip reflected; shell with 5-6 whorls; angulo-parietal lamella strongly bifid at the lip; in frontal view a distinct spur (a) is evident on the side of the angulo-parietal lamella; five to six teeth present in aperture; shells are somewhat translucent and brownish in live individuals and fresh shells, but become bleached and white with age.

Similar Species: *Gastrocopta holzingeri* has a similar shell form but is smaller, whitish (not brown), and has a channeled not forked angulo-parietal; *G. rogersensis* (found in Ohio and westward) has the same shell form as *G. procera* but also has a channeled not forked anguloparietal;

Habitat: An obligate calciphile on dry ground where the vegetation is sparse such as cedar glades, also occurs in sandy river floodplains scrub and forest (Nekola and Coles 2010).

Status: G5/S2; Rare; in West Virginia reported from only four counties.

Specimen: Illinois, Monroe County, Fults Hill Prairie Nature Reserve (Nekola and Coles image 2010).

Vertiginidae

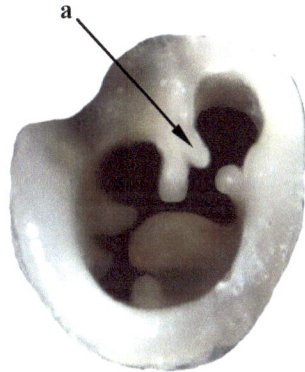

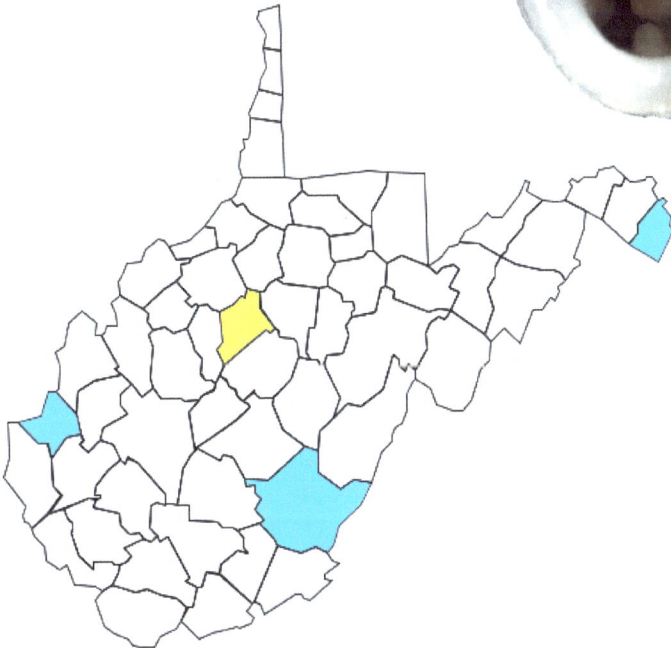

79

Lambda snaggletooth

Vertiginidae

Gastrocopta holzingeri (Sterki, 1889)

Height: 1.75 mm, width 0.8 mm

Description: Pupa shape; lip reflected; shell with five whorls; angulo-parietal lamella is grooved down the middle (a); aperture crowded with teeth, (typically six); shells are translucent in live individuals.

Similar Species: *Gastrocopta contracta* has a notably different shape and is larger; *G. procera* is larger having one less tooth on the outer palatal wall, is brown (not white) and the angulo-parietal is forked (not grooved as in *G. holzingeri*).

Habitat: A calciphile species frequently found around talus with sparse vegetation; in West Virginia the species is found on limestone river bluffs in the thin cedar detritus accumulations over bedrock.

Status: G5/S2; Rare; there are only a few records of this species by N.D. Richmond who assisted MacMillan (1949), but additional records were added by WVDNR surveys.

Specimen: West Virginia, Greenbrier County, cedar glade above Greenbrier River (author's collection).

a

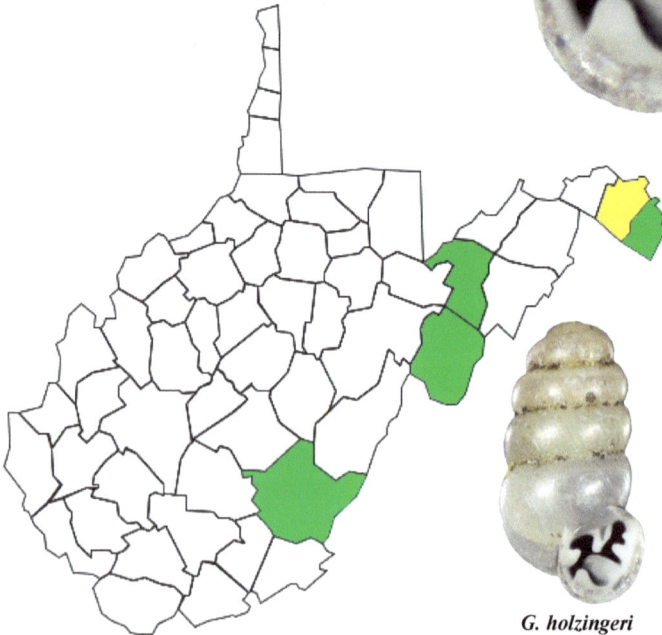

G. holzingeri   G. procera

Cupped vertigo

Vertiginidae

Vertigo clappi Brooks & Hunt, 1936

Height: 1.5 mm, width 0.8 mm

Description: Pupa shape; left side of lip reflected while the right side of the lip is simple containing a small indentation in the aperture wall (a); shell with 5.5 whorls; 6 well-developed teeth, the face of the upper palatal (b) is parallel to the front of the shell; in aperture (side) view the face of the upper palatal is seen, not the edge; shells are somewhat translucent in live individuals and fresh shells but become bleached and white with age.

Similar Species: *Vertigo milium* is around the same size but has a strongly curved lower palatal; *V. ovata* is larger and the lower and upper palatal lamella do not turn sideways as in *V. clappi.*

Habitat: Found in well decomposed leaf litter, fine soil on shaded boulders, talus, ledges and bases of forested bedrock outcrops and can be locally abundant (Nekola pers. comm. 2010).

Status: G1G2/SH; Rare; this species has a limited distribution in West Virginia.

Specimen: Tennessee, Monroe County, Tellico Gorge (Nekola and Coles image 2010).

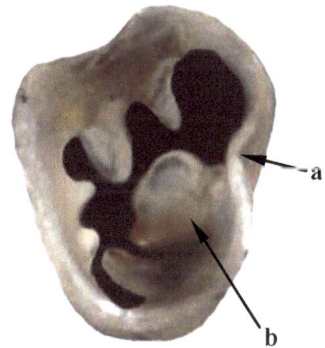

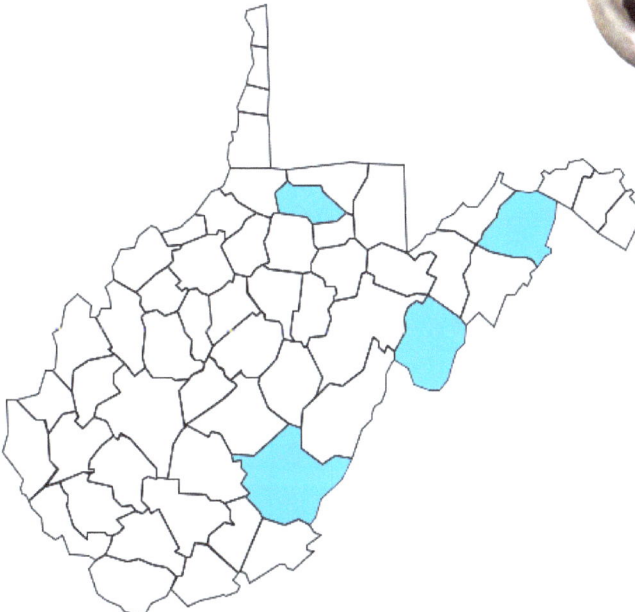

Tapered vertigo

Vertigo elatior Sterki 1894

Height: 2.1-2.2 mm, width 1.2 mm

Description: Pupa shape; left side of lip reflected while the right side of the lip is simple containing a small indentation in the aperture wall; shell with 5 whorls; 5 teeth; there is also a conspicuous callus deposit between the palatal lamellae (a); aperture small; shell surface has poorly developed transverse striae, so is nearly smooth; as with other *Vertigo* species live and fresh dead shells are translucent.

Similar Species: *Vertigo elatior* is most likely confused with a weakly toothed *V. ovata* which is larger and more ovate in shape (Nekola and Coles 2010).

Habitat: Found in well-developed damp leaf litter and graminoid thatch in a variety of open and wooded wetland habitats such as fens and wet meadows; in West Virginia limited to fens (Nekola and Coles 2010).

Status: G5/SH; Rare; in West Virginia reported from only two counties.

Specimen: Minnesota, Marshall County, Karlstad South (Nekola and Coles image collection 2010).

Vertiginidae

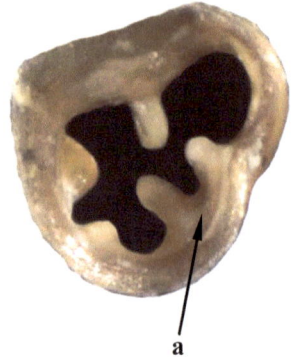

a

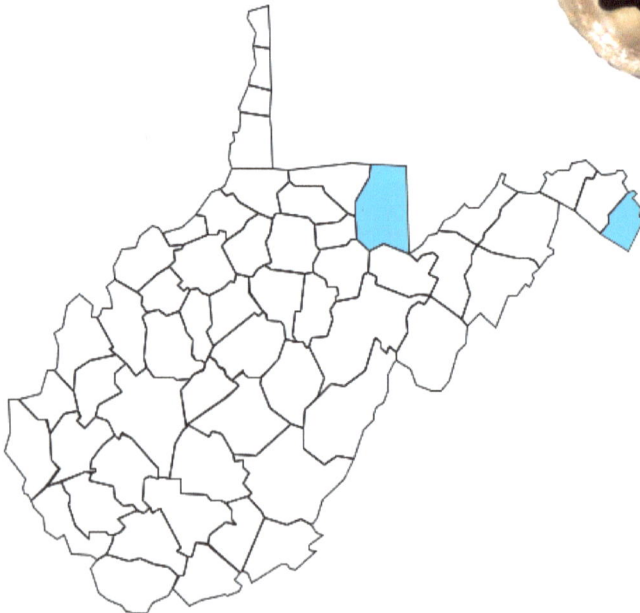

Five-tooth vertigo

Vertiginidae

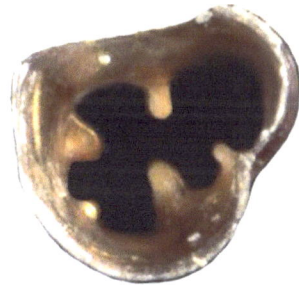

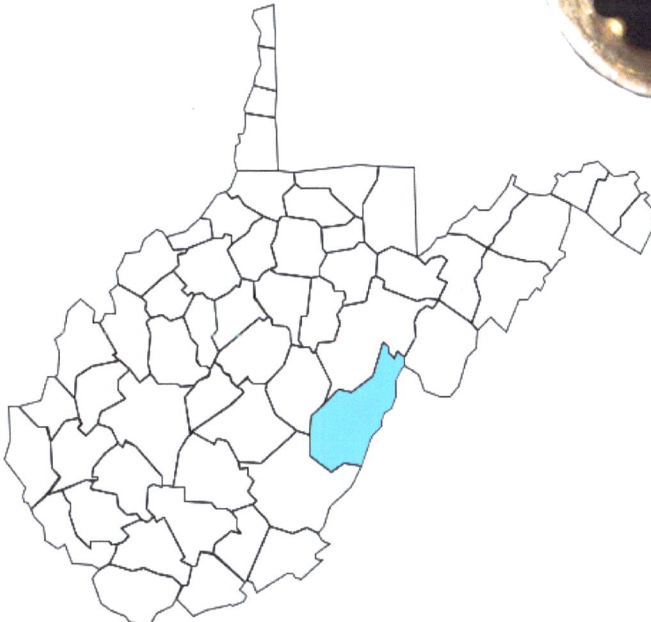

Vertigo ventricosa (E.S. Morse, 1865)

Height: 1.7-1.95 mm, width 1-1.2 mm

Description: Pupa shape but compact; left side of lip reflected while the right side of the lip is simple containing a small indentation in the outer aperture wall; shell with 4.5-5 whorls; shell surface has poorly developed transverse striae so is nearly smooth; the 5 teeth are smaller than seen in other *Vertigo* species and nearly equal in size.

Similar Species: *Vertigo ventricosa* differs from *V. ovata* in its constantly smaller size, absence of an angular lamella, and smaller number of teeth; *V gouldii* is striate, has a very different shape, and is not nearly as transparent.

Habitat: Occurs in accumulations of humid, well decomposed graminoid and broadleaf plant litter in moderately to highly acidic wooded and open wetlands, in particular lowland northern white cedar and red maple forests, sedge meadows, Sphagnum peatlands, and poor fens. (Nekola and Coles 2010).

Status: G5/SH; Rare; reported from only one county in West Virginia (Nekola, pers. comm. 2014).

Specimen: Maine, Aroostook County, Portage Lake (Nekola and Coles image 2010).

Crested vertigo

Vertiginidae

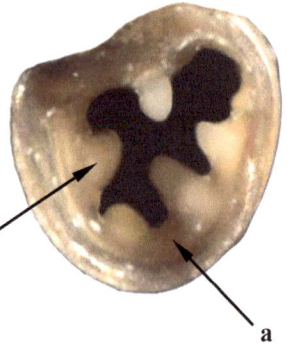

Vertigo pygmaea (Draparnaud, 1801)

Height: 1.8-2 mm, width 1 mm

Description: Pupa shape; left side of lip reflected while the right side of the lip is simple containing a small indentation in the aperture wall; shell with 5 whorls; 5 peg-like teeth that are thickened and stand on a strong callus rim (a); aperture opening reduced.

Similar Species: Unlike most *Vertigo* species, *V. pygmaea* has a large crest with a white-calcareous callus on the palatal wall of the aperture; it lacks the glassy luster of *V. elatior* (Nekola and Coles 2010).

Habitat: Found in well-developed humid leaf litter and graminoid thatch in a variety of anthropogenically disturbed grassland including roadsides, old fields, yards, and abandoned quarries (Nekola and Coles 2010).

Status: G5/SH; Rare; in West Virginia reported from only four counties; Nekola and Coles (2010) suspect populations of this species represent Eurasian waifs brought to North America over a century ago; DNA identical to western European populations. Possibly exotic.

Specimen: Wisconsin, Manitowoc County, Kingfisher Farm (Nekola and Coles image 2010).

Teeth are thickened and peg-like

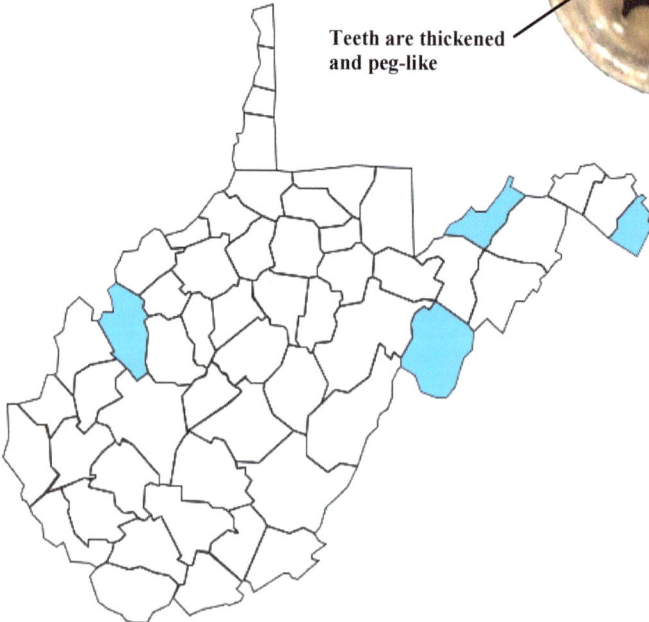

a

84

Crested vertigo

Vertiginidae

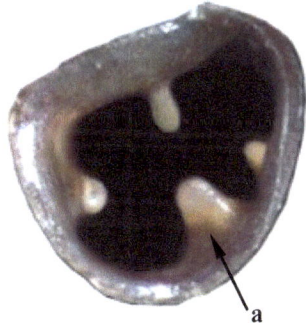

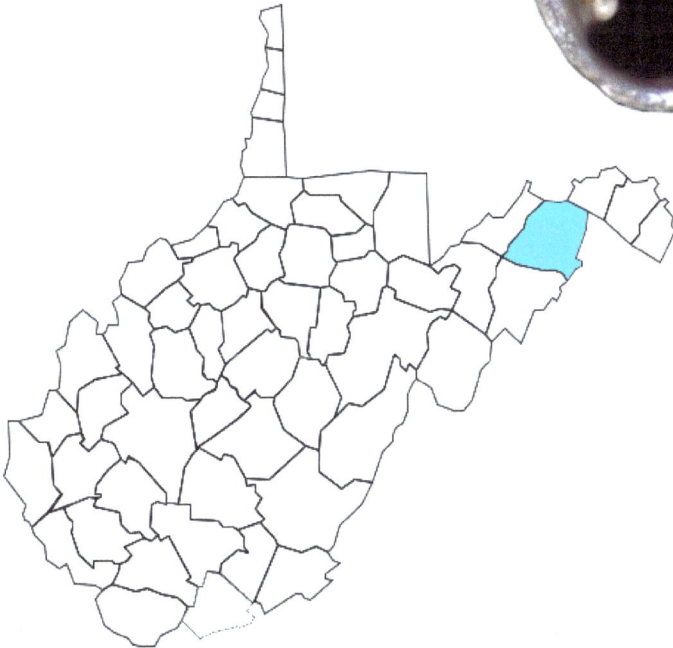

Vertigo cristata (Sterki, 1919)

Height: 2.1 mm, width 1.2 mm

Description: Pupa shape; left side of lip reflected while the right side of the lip is simple; with or without a small indentation in the outer aperture wall; shell with 4.5-5 whorls; notably transverse striae on all but the first two whorls; the 4 teeth are fairly welldeveloped, the lower palatal being the largest (a), the remaining 3 teeth of equal or nearly equal size.

Similar Species: Has the sculpture of *V. gouldii,* but without the small basal columellar.

Habitat: Found in well decomposed leaf litter in a wide variety of northern forest habitats, ranging from wetlands to dry upland rock outcrops; it is particularly common in base-poor sites such as pine and spruce forest heaths, and *Sphagnum* dominated peatlands (Nekola and Coles 2010); limited in WV to cold acid pine litter at Ice Mountain.

Status: G5/S1; Rare; this northern species was first reported in West Virginia from Ice Mountain in Hampshire County by Hotopp and Pearce (2008).

Specimen: Wisconsin, Oneida County, Sugar Camp Bog (Nekola and Coles image 2010).

Ovate vertigo

Vertiginidae

Vertigo ovata (Say, 1822)

Height: 2.2-2.3 mm, width 1.4 mm

Description: Pupa shape; left side of lip reflected while the right side of the lip is simple containing a small indentation in the outer aperture wall; shell with 4.5-5 whorls; there are 6-9 teeth, 4 which are well-developed, the remaining 5 teeth are petite in build, together crowd the small aperture.

Similar Species: *Vertigo ovata* does not have a long, curved lower palatal like *V. milium*; it is strongly ovate, shell volume easily 5X that of than *V. milium* (Nekola and Coles 2010).

Habitat: Found on cattail leaves in swamps, sedge meadows, wet and mesic prairie, low calcareous meadows, river banks, lakeshores, roadside ditches, bedrock outcrops and upland forests (Nekola and Coles 2010); it can also ascend vegetation to around one meter off the ground.

Status: G5/SH; Uncommon; reported from scattered locations across the state.

Specimen: Iowa, Dubuque County, Epworth Fen (Nekola and Coles image 2010).

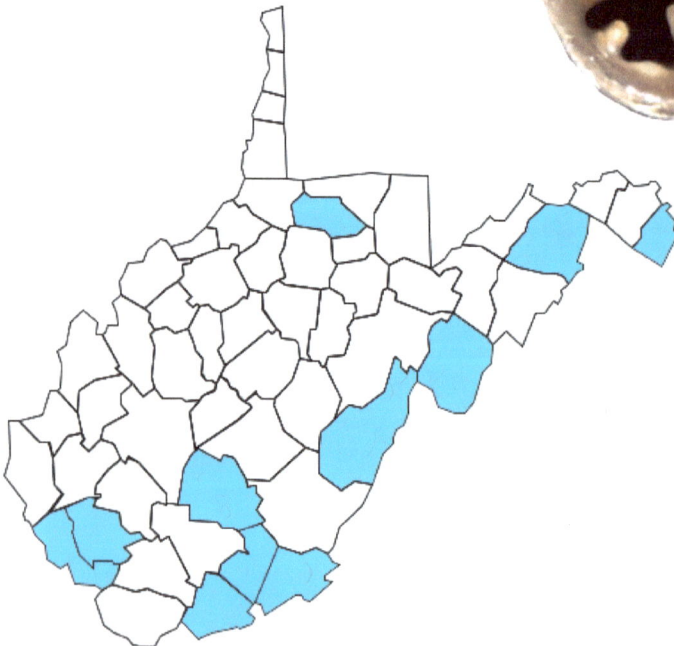

Blade vertigo

Vertiginidae

Vertigo milium (Gould, 1840)

Height: 1.4-1.8 mm, width 0.81 mm

Description: Pupa shape; small indentation in the outer aperture wall; shell with 4.5-5 whorls; usually with 6 teeth that crowd the aperture, one of the palatal teeth (a) especially long with a wide entering base; shell surface glossy, weakly striate; shells are somewhat translucent in live individuals and fresh shells, but become bleached white with age.

Similar Species: The inner teeth are relatively long and fill the aperture; the long, curved lower palatal and the broad inward-sloping columellar plate of *V. milium* are very different from the peg shaped columellar of *V. ovata*.

Habitat: Found in damp, well-developed thatch and leaf litter across a wide variety of mesic to wet sites including rocky woodland, riparian woodland, cliffs, wet prairie, sedge meadows, roadside verges, fens, and swamps (Nekola and Coles 2010).

Status: G5/S2; Uncommon; reported from ten counties in West Virginia.

Specimen: Wisconsin, Green Lake County, Berlin Fen (Nekola and Coles image 2010).

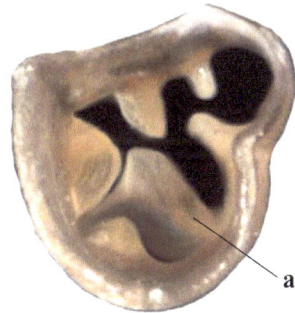

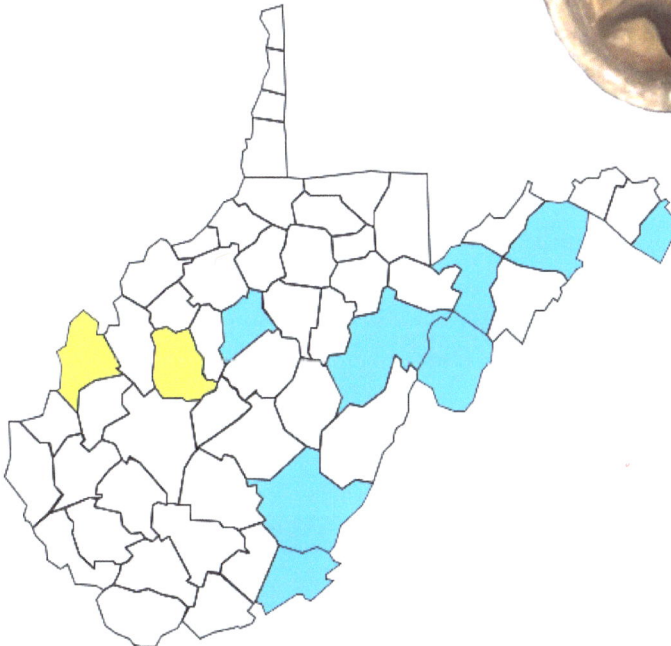

Honey vertigo

Vertiginidae

Vertigo tridentata (Wolf, 1870)

Height: 1.8-2.3 mm, width 1.1 mm

Description: Pupa shape, but with a taper; outer lip containing a small indentation in the outer aperture wall (a); shell with 5 whorls; 3 or 4 short teeth, aperture semicircular in shape; shells are smooth, lacking sharp striations, somewhat translucent in live individuals with a honey-yellow color shading to somewhat browner below, like most *Vertigo* species the shells becoming bleached with age.

Similar Species: The shell color of *V. tridentata* is honey-brown whereas *V. gouldii* is a dark-brown; the parietal lamella (b) is pointing at the lower palatal in *V. tridentata*, not the upper palatal as seen in *V. gouldii* (page 91), figure a (Nekola and Coles 2010).

Habitat: A species found climbing on herbs in the Lamiaceae family (mints) of low sunny places (Hubricht 1985), bedrock glades, in well decomposed leaf litter accumulations on shaded cliff ledges and talus (Nekola and Coles 2010).

Status: G5/S3; Uncommon; found throughout the state in somewhat scattered localities.

Specimen: Iowa, Dubuque County, Little Maquoketa River (Nekola and Coles image 2010).

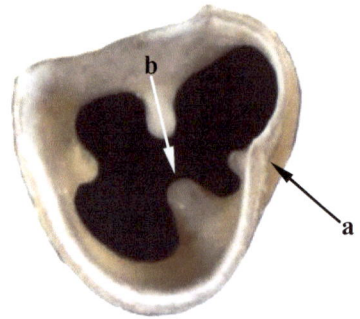

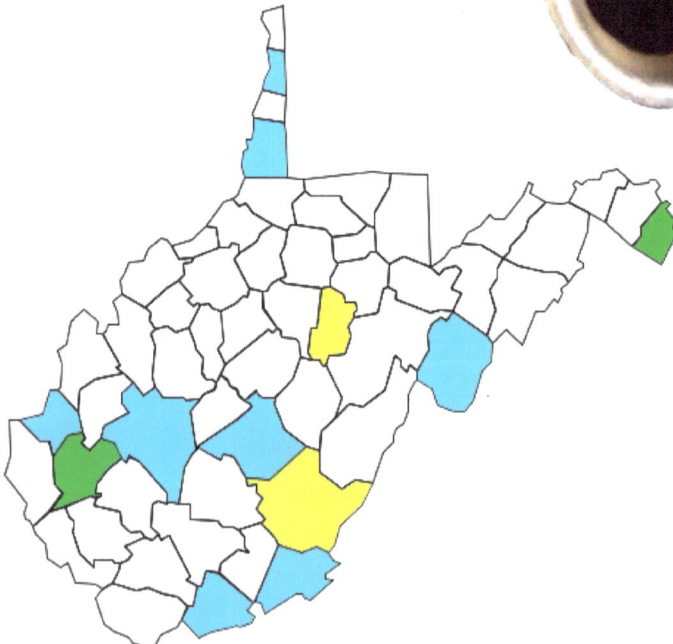

Capital vertigo

Vertiginidae

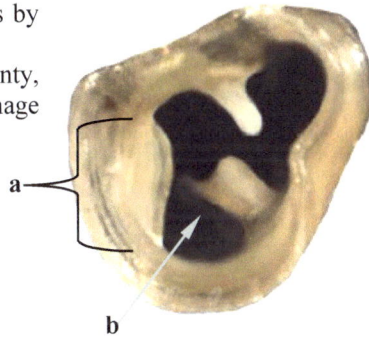

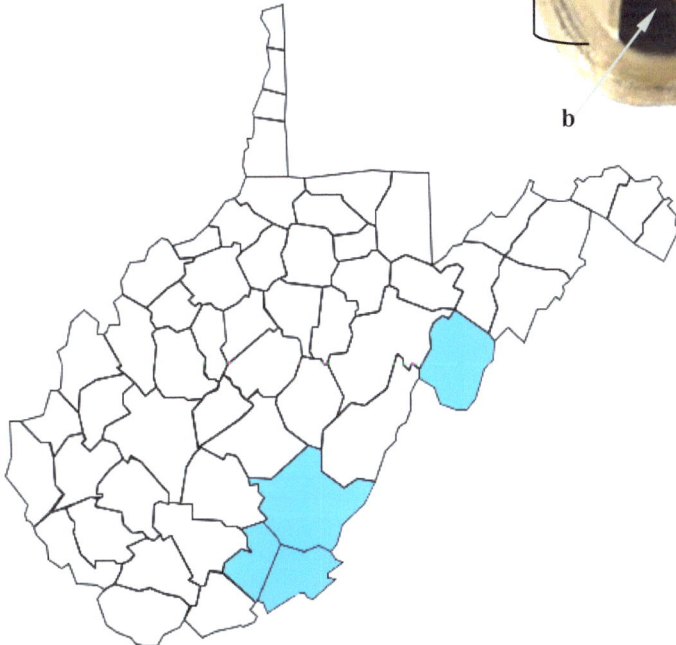

Vertigo oscariana (Sterki, 1890)

Height:1.4-1.6 mm, width 0.85 mm

Description: Pupa shape; small indentation in the outer aperture wall; shell with 4.5-5 whorls; 3 medium sized teeth, columellar lamella blunt and wide (a); palatal fold (tooth) short, thick and set deeply within the aperture (b); shells are translucent in live individuals while old shells become bleached white.

Similar Species: *Vertigo parvula* is similar but has a peg-shaped, not sheet-like columellar; *V. oscariana* is the only *Vertigo* which has a body whorl (c) much less wide than the penultimate (d) whorls (Nekola and Coles 2010).

Habitat: Found in well decomposed accumulations of broadleaf and pine litter in mesic, wet woodlands and shaded rock outcrops; in western VA this species is found most often in dry upland pine litter (Nekola and Coles 2010).

Status: G4/SH; Rare; in West Virginia this tiny *Vertigo* is reported from only four counties by Hubricht (1985).

Specimen: South Carolina, Berkley County, Wadboo Creek (Nekola and Coles image 2010).

89

Smallmouth vertigo

Vertigo parvula Sterki, 1890

Height: 1.4-1.6 mm, width 0.85 mm

Description: Pupa shape; the lip indentation weakly developed; shell with 5 whorls; 3 rather large teeth; columellar lamella small; the shells are somewhat translucent in live individuals.

Similar Species: *V. parvula* looks like a small *V. tridentata*, but is consistently lacking the upper palatal, has a short, elongated lower palatal (a), and is also much smaller than *V. tridentata* (Nekola and Coles 2010); *V. oscariana* has a columellar lamella that is notably wider and the palatal tooth is shorter and sits a little deeper in the aperture.

Habitat: Found in accumulations of well decomposed leaf litter in base rich cove forests, rock outcrops, and talus slopes at mid to low elevations (Nekola and Coles 2010).

Status: G3/S2; Rare; reported from eight counties in the state by MacMillan (1949) and during the recent (2012-2014) WVDNR statewide survey, documented in McDowell and Grant Counties.

Specimen: Tennessee, Washington County, Buffalo Mountain (Nekola and Coles image 2010).

Vertiginidae

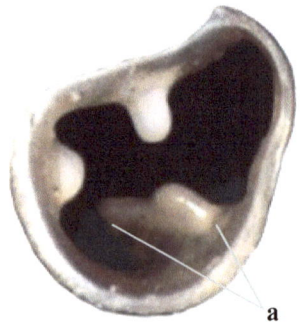

a

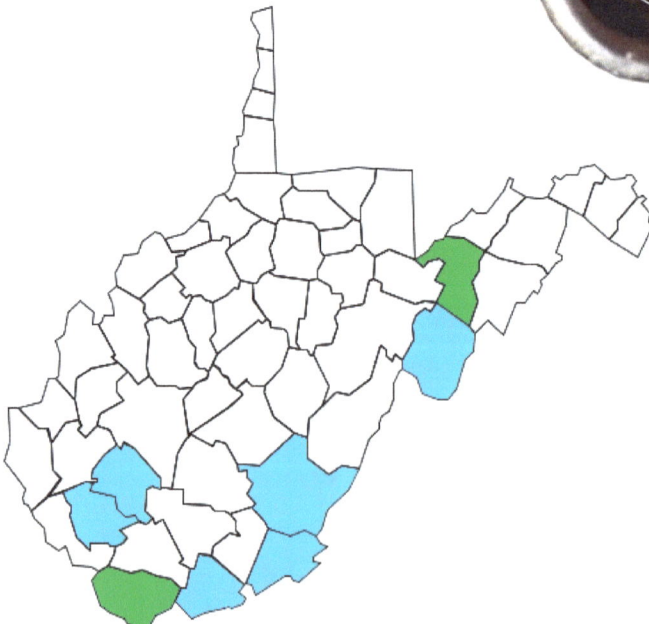

Variable vertigo

Vertiginidae

Vertigo gouldii (A. Binney, 1843)

Height: 1.5-2.1 mm, width 1 mm

Description: Pupa shape; the indentation in the outer aperture wall is weakly developed or imperceptible; shell with 4.5-5.5 whorls; 4 to 5 teeth, all around the same size; parietal lamella and palatal folds well-developed; shell surface has well-developed transverse striae; shells are somewhat translucent in live individuals and fresh shells but become bleached with age;

Similar Species: *Vertigo tridentata* lacks sharp striae, has a more tapered shell, and the parietal lamella points at the lower palatal, whereas in *V. gouldii,* the parietal lamella points at the upper palatal (a); *Vertigo bollesiana,* a species reported from West Virginia but not validated by actual specimens (pers. comm. Nekola 2014) are the same size as small *V. gouldii*, but differs in having a more tapered outline and stronger palatal depression.

Habitat: Limited to forested sites and is most common on wooded bedrock outcrops.

Status: G5/S5; Common; one of the most common *Vertigo* in West Virginia.

Specimen: Tennessee, Monroe County, Tellico Gorge (Nekola and Coles image 2010).

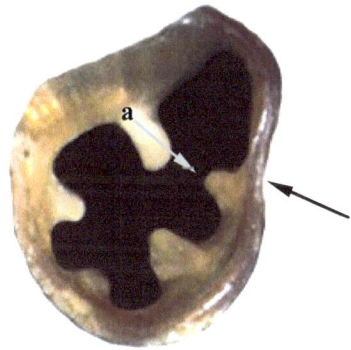

V. gouldii   *V. bollesiana*

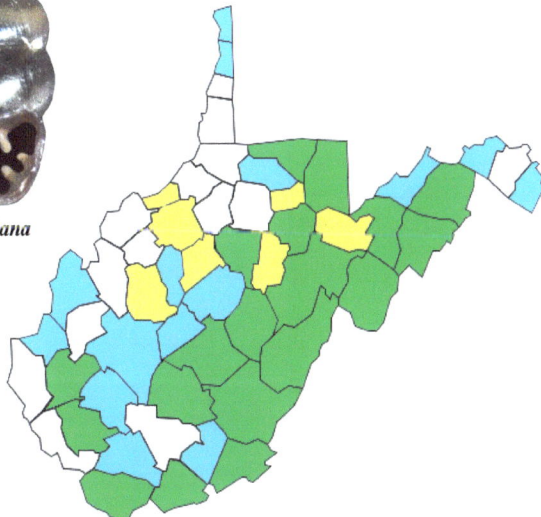

White-lip dagger

Pupoides albilabris (C. B. Adams, 1821)

Height: 4.2-5 mm, width 2 mm

Description: Pupa shape; lip reflected; shell with 6-6.5 whorls; with or without a small angular lamella in the form of a callus (a) located in the upper right hand corner of the aperture; surface of shell mostly smooth or with weak transverse striae; without spiral striae.

Similar Species: In West Virginia, other pupa shaped shells such as *Carychium, Vertigo, Gastrocopta* are notably smaller; the shell of *Cochlicopa* species are larger and have a glass-like, translucent surface.

Habitat: A species of bare ground around natural glades but also found in waste places such as old quarries and parking lots; in wet weather, the snail can be observed crawling up the stems of plants; may be especially abundant in limestone areas (Hubricht 1985).

Status: G5/S3; Uncommon to Relatively Common; in scattered locations across the state.

Specimen: Figure (b) from Ron Caldwell collection (CMRC 1594) and figure (c) from Missouri, Wright County, Gettle Farm (Nekola and Coles, 2010).

b

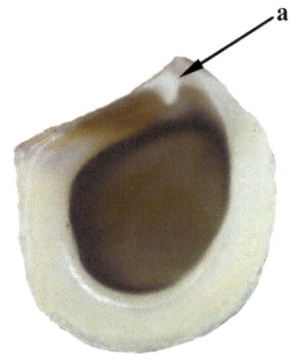

a

c

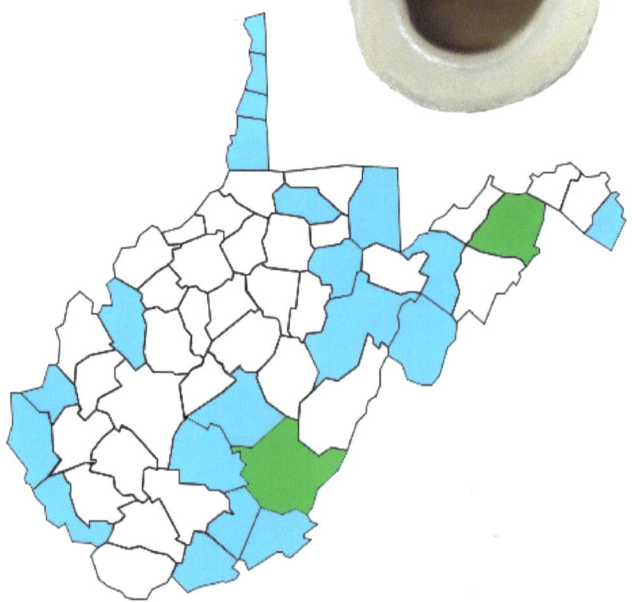

Widespread column

Pupilla muscorum (Linne, 1758)

Height: 2.8-4 mm, width 1.7 mm

Description: Pupa shape; lip reflected; shell with 5.5-7.5 whorls; with or without a small angular lamella in the form of a callus (a) located in the upper right hand corner of the aperture; transverse striae fairly well developed; without spiral striae.

Similar Species: In West Virginia, other pupa shaped shells such as *Carychium, Vertigo, Gastrocopta* are notably smaller and are not cylindrical; *Cochlicopa* species are larger and with a glass-like surface.

Habitat: Generally occurs in disturbed anthropogenic habitats such as road verges, vacant lots, abandoned quarries, old fields and concrete culverts (Hubricht 1985).

Status: G5/SNA; Rare; this species is a European exotic, with at least the Iowa haplotypes identical to Moravian populations; also a number of undescribed native species, more closely related to the Eurasian *P. 'muscorum'* = exotic *P. muscorum* & native *P. hudsonianum* (in north) and *P. hebes* or *P. blandii* (in west), which have been misidentified as *P. muscorum* in North America, but none occur in eastern North America. (Nekola and Coles 2010)

Specimen: Minnesota, Beltrami County, Lake Bemidji State Park (Nekola and Coles image 2010).

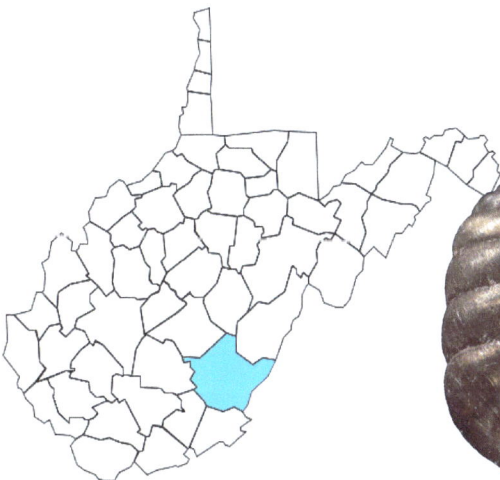

P. muscorum   *P. albilabris*

Land snails in this section have shells that are wider than tall, either heliciform or depressed heliciform and have simple lips (although a few may have slightly reflected lips). They include minute snails from 1.5 mm to snails well over 35 mm including many of the mid-size species occurring in West Virginia. Several of the smaller species in this section such as *Glyphyalinia, Helicodiscus,* and *Striatura* are in company with some of the more beautifully sculptured gastropods in North America. Many have glasslike, transparent shells while others display extraordinary micro-ornamentation. This outstanding craftsmanship, however, can only be appreciated under the strong lens.

Genera Included:
(in order of appearance in text)

Punctum
Striatura
Hawaiia
Lucilla
Guppya
Vitrina
Zonitoides
Nesovitrea
Gastrodonta
Glyphyalinia
Paravitrea
Discus
Helicodiscus
Ventridens
Mesomphix
Anguispira
Haplotrema
Euconulus
Hendersonia

Land snails less than 3.5 mm, umbilicate (*G. sterkii* is perforate), simple lip, typically without teeth or lamellae

Land snails in this category are less than 3.5 mm in diameter, without reflected lips, and generally found in moist leaf litter and detritus. To separate these small snails, it is essential that a strong dissecting scope be employed, otherwise a correct ID will be nearly impossible. Each genus has its own unique ornamentation. These microscopic features, although sometimes hard to see, are key for a successful identification. If you have small specimens collected from leaf litter that fit the below Groupings 1 through 6 then proceed with the following separations.

Grouping 1: shells under 1.5 mm, umbilicate, whorls tightly coiled (a), with crisscrossing striae (b), pg. 96-100
Punctum minutissimum
Punctum vitreum
Punctum blandianum (rare)
Punctum smithi (small tooth)

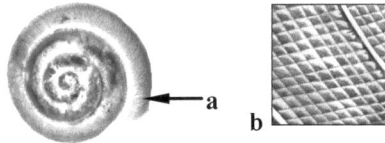

Grouping 2: shells between 1.5 mm to 3.5 mm, umbilicate, whorls loosely coiled (c), transverse striae strongly developed (a), pg. 101-104

Striatura meridionalis
Striatura milium (rare)
Striatura exigua
Striatura ferrea

Grouping 3: shells between 1.5 mm to 2.8 mm, umbilicate, surface without crisscrossing striae, with a dull sheen, pg. 105-107

Hawaiia minuscula
Hawaiia alachuana

Grouping 4: shells around 2.2 mm, umbilicate, surface without crisscrossing striae, more or less smooth and polished, pg. 109-110

Lucilla (Helicodiscus) scintilla (rare)
Lucilla (Helicodiscus) singleyana (rare)

Grouping 5: shells around 3.5-5 mm, umbilicate, shell surface more or less smooth, polished and translucent, pg. 111-112

Nesovitrea electrina
Nesovitrea binneyana

Grouping 6: shells between 1.2 to 1.3 mm, perforate, surface with fine spiral striae, translucent and very glossy, pg. 113

Guppya sterkii (very common)

Small spot

Punctidae

Punctum minutissimum (I. Lea, 1841)

Diameter: 1.1-1.3 mm, height 0.75 mm

Description: Depressed heliciform; lip simple; aperture roundish, not reflected; shell with 3.5-4.5 whorls; umbilicate; pale brown to corneous; no teeth present; transverse and spiral striae that crisscross each other (a), this amazing micro-feature is so small it can only be seen under the lens of a strong dissecting scope; fresh and live shells are thin and translucent.

Similar Species: *Punctum vitreum* has a smaller umbilicus and has more widely spaced ribs; *P. blandianum* has a larger umbilicus and flatter shell, micro-surface features are the same as other *Punctum* species.

Habitat: Live animals live between the layers of moist, matted leaves of hardwood forests where leaves accumulate next to logs or near headwater seeps.

Status: G5/S5; Common, the most common *Punctum* species in West Virginia, occurring at all elevations and likely in the two remaining undocumented counties.

Specimen: Kentucky, Letcher County, Bad Branch NP (author's collection).

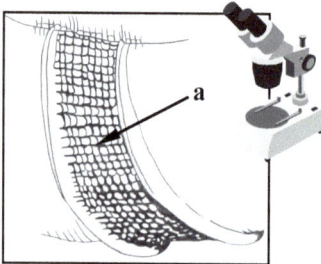

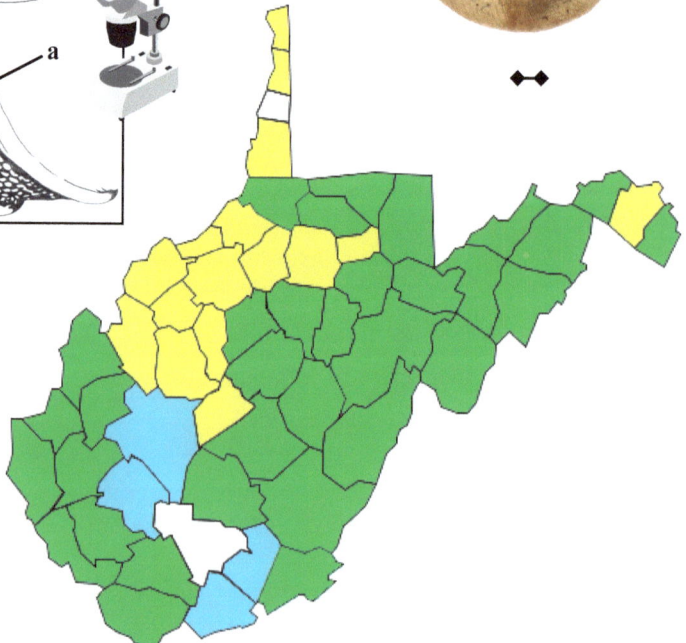

Glass spot

Punctum vitreum (H. B. Baker, 1930)

Diameter: 1.2-1.4 mm, height 0.8 mm

Description: Depressed heliciform; lip simple; aperture roundish; shell with 4-4.5 whorls; umbilicate; corneous to colorless; no teeth present; transverse and spiral striae that are more rib-like but paper thin and crisscrossing each other (a); fresh shells are thin, fragile and more or less translucent.

Similar Species: *Punctum minutissimum* has a slightly wider umbilicus and has less widely-spaced transverse striae; *P. blandianum* has a notably larger umbilicus but its micro-surface features are more or less the same as seen in *P. minutissimum*.

Habitat: Found between the moist, matted leaves around small seeps in mixed hardwoods.

Status: G5/S2; Uncommon; this species has frequently been confused with *P. minutissimum* by past collectors.

Specimen: North Carolina, Haywood County, Purchase Knob (GSMNP collection).

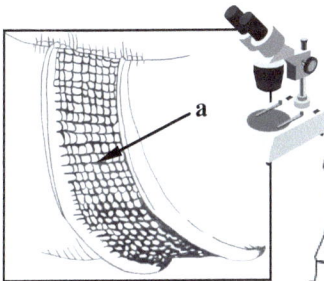

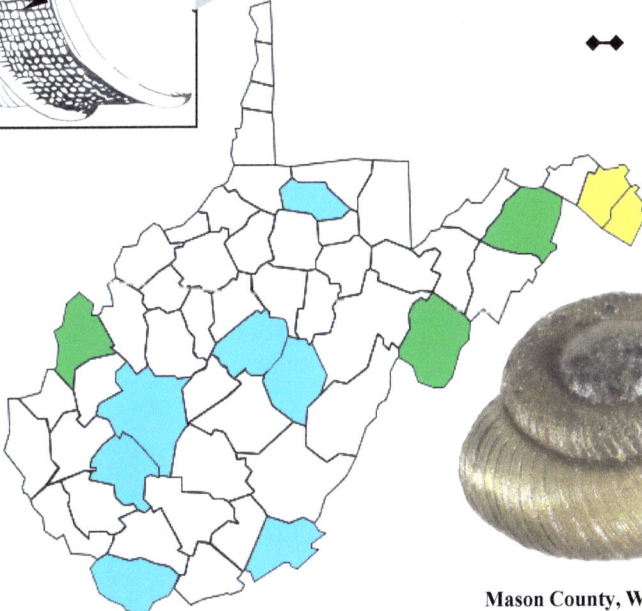

Mason County, West Virginia

Brown spot

Punctidae

Punctum blandianum (Pilsbry, 1900)

Diameter: 1.1-1.3 mm, height 0.6 mm

Description: Depressed heliciform; lip simple; aperture roundish to slightly oval shaped; shell with 4 whorls; widely umbilicate; pale brown; no teeth present; transverse and spiral striae are well-developed and crisscrossing (a); fresh and live shells thin and translucent, but older shells will bleach with time.

Similar Species: *Punctum minutissimum* is the same size but has a notably smaller umbilicus; *P. vitreum* has more widely spaced transverse striae than is seen in *P. blandianum,* a smaller umbilicus and a more elevated, compact shell.

Habitat: Found in deep moist pockets of leaf litter in depressions or around logs, but also between the layers of moist, matted leaves adjacent to small headwater seeps.

Status: G4/SH; Rare; this mostly southern Appalachian species is reported from only one West Virginia county.

Specimen: Tennessee, Rutherford County, Murfreesboro (FM 235261).

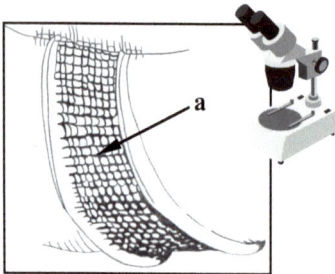

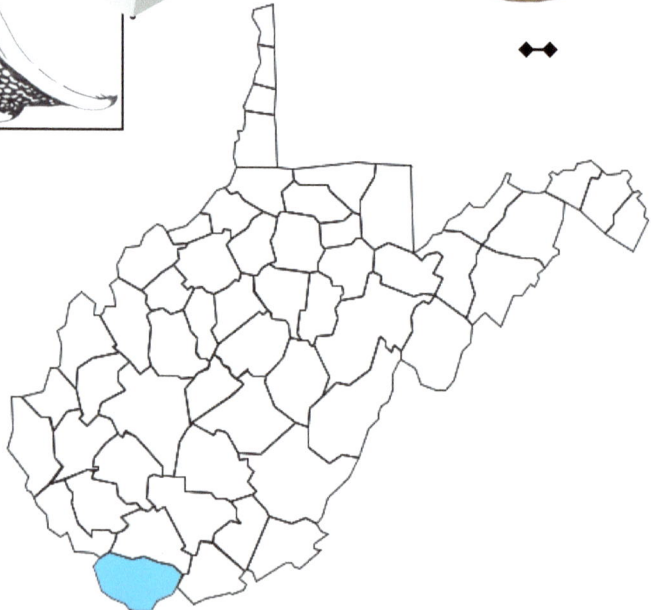

Lamellate spot

Punctum smithi Morrison, 1935

Diameter: 1.1-1.2 mm, height .62 mm

Description: Depressed heliciform; lip simple; aperture roundish; shell with 4-4.5 whorls; umbilicate; pale tan or brown; one basal tooth (a), sometimes forming a lamella (b) positioned just inside the lower aperture; in some populations, lamella can be exceedingly long; transverse striae are fine and crossed with spiral striae (c).

Similar Species: No other small species having the shape and size of *P. smithi* has a single tooth in the aperture; other *Punctum* species have more notable micro-sculpture and are without internal armature; *Paravitrea* species typically have multiple teeth or lamellae.

Habitat: The species resides in deep moist pockets of leaf litter found in depressions or around rotting logs in ravines of mixed hardwood forests.

Status: G4/S2; Uncommon; although historic records show only one West Virginia county, recent surveys by WVDNR have added twelve new counties.

Specimen: Kentucky, Jefferson County, near McNeely Lake Dam, Okolona, (FM 234970)

a

◆◆

c

b

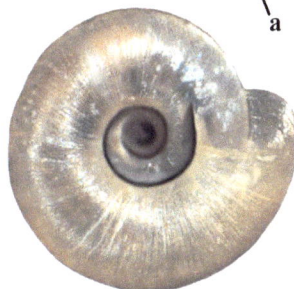

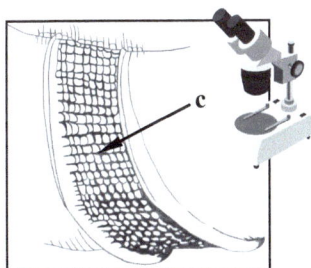

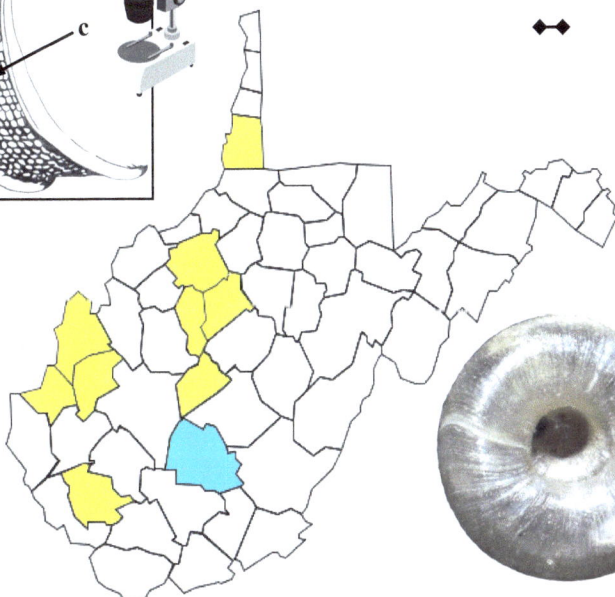

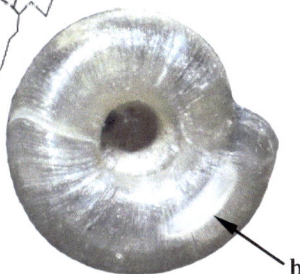

Black Oak Road, Mason
County, West Virginia

Punctum species Compared

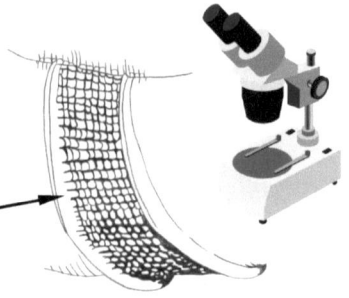

The unique crossing micro-sculpture
found in *Punctum* species

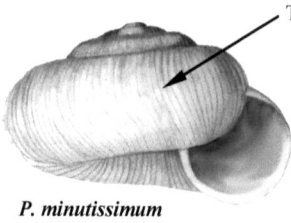

Tight striae

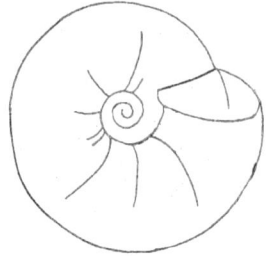

P. minutissimum

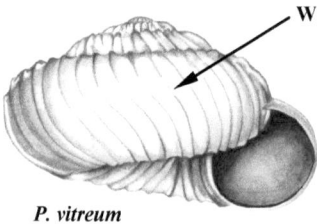

Wide striae

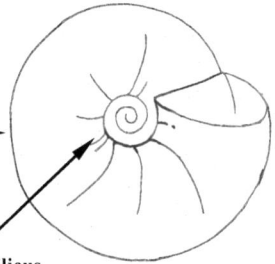

Narrow umbilicus

P. vitreum

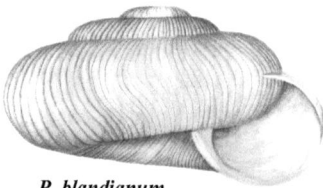

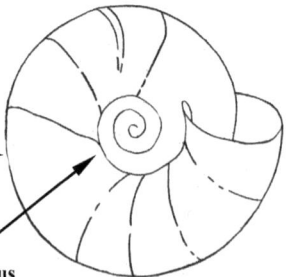

Wide umbilicus

P. blandianum

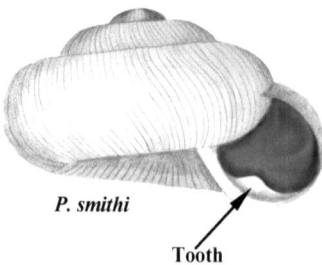

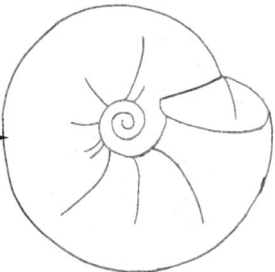

P. smithi

Tooth

Median striate

Gastrodontidae

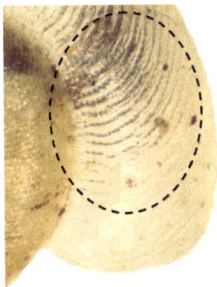

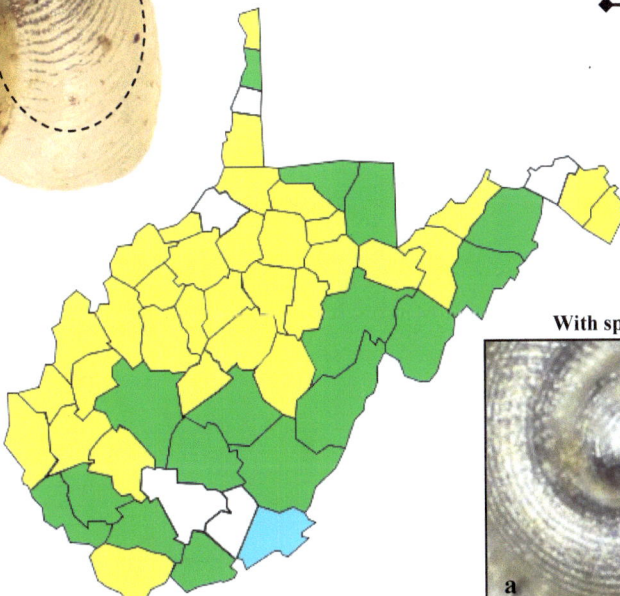

Striatura meridionalis Pilsbry and Ferriss, 1906

Diameter: 1.7-1.8 mm, height 1 mm

Description: Depressed heliciform; aperture roundish; lip simple; shell with 3-3.5 whorls; widely umbilicate; corneous with a greenish cast; shell translucent in live snails and fresh dead; nuclear whorl with spiral striae, figure (a), but without transverse striae, the transverse striae developing in later whorls (b) forming minute low riblets and continuing onto the base where they become somewhat weaker.

Similar Species: *Striatura ferrea* is larger in diameter, has a larger aperture and a notably smaller umbilicus.

Habitat: Found in mixed hardwood forests on hillsides and ravines living between the layers of moist leaves; also found at the base of butternut and black walnut trees.

Status: G5/S5; Common; a species that will likely be found in every West Virginia county, especially if leaf litter samples are secured during surveys.

Specimen: North Carolina, Macon County, Nantahala National Forest (author's collection).

b

With spiral striae

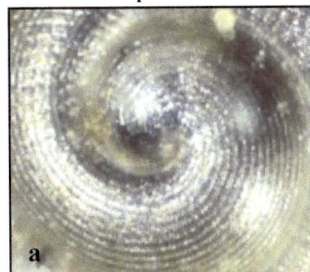

a

101

Fine-ribbed striate

Gastrodontidae

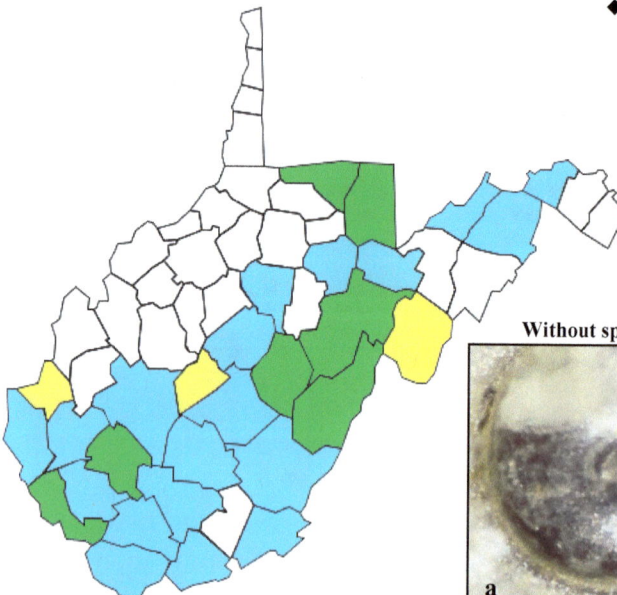

Striatura milium (Morse, 1859)

Diameter: 1.5 mm, height 0.8 mm

Description: Depressed heliciform; aperture roundish; lip simple; shell with 3-3.5 whorls; widely umbilicate; yellowish-corneous or gray; the live animal is white with dark spots on the head and tentacles; shell translucent in live snails and fresh dead; nuclear whorl smooth (a) and without spiral striae; in the remaining whorls the spiral striae not prominent; fine transverse striae well-developed forming minute, low riblets; base nearly smooth.

Similar Species: *Striatura meridionalis* is larger in diameter, has spiral striae on the nuclear whorl, and has coarser riblets; *S. ferrea* is larger with a proportionally smaller umbilicus; *S. exigua* is larger with wider and taller riblets.

Habitat: Found in mixed hardwood forests on hillsides and in ravines living between the layers of moist leaves.

Status: G5/S3; Fairly Common; reported from multiple locations across West Virginia.

Specimen: Iowa, Fayette County (FM 250966).

Without spiral striae

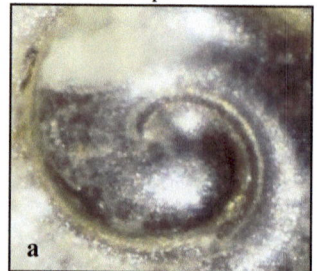

a

Ribbed striate

Striatura exigua (Stimpson, 1847)

Diameter: 2.2-2.4 mm, height 1.25 mm

Description: Depressed heliciform; lip simple; aperture roundish; shell with 3.5 whorls; widely umbilicate; a coppery color; shell translucent in live snails and fresh dead; without teeth; transverse striae are well-developed forming minute paper-thin riblets which are the most widely spaced of any *Striatura*; spiral striae also well defined but an extremely faint micro-feature of the shell surface.

Similar Species: *Striatura ferrea* is larger in diameter, has a much smaller umbilicus, and has finer shell surface features.

Habitat: Living on wet ground around head water seeps in upper elevation sphagnum bogs (over 1000 meters) that are usually surrounded by northern hardwood forests.

Status: G5/S2; Uncommon; in West Virginia this mostly northern species appears restricted to the higher elevations.

Specimen: North Carolina, Swain County, Nantahala National Forest (all specimens in author's collection).

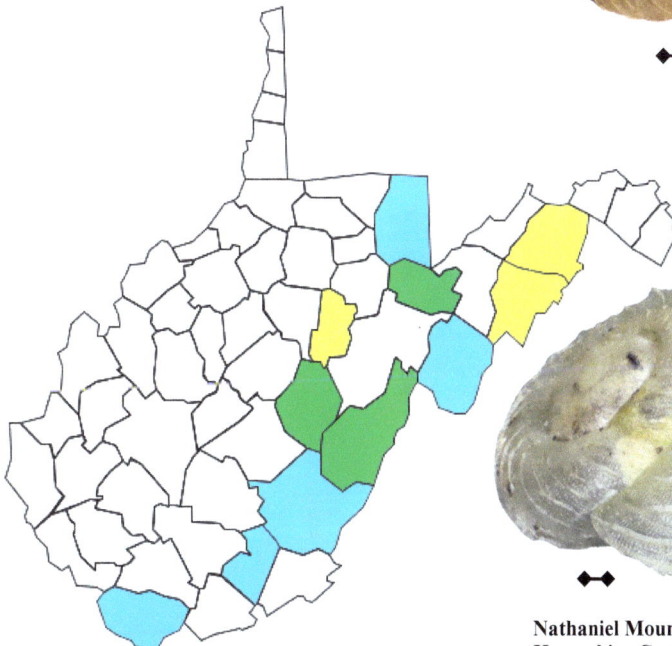

**Nathaniel Mountain WMA
Hampshire Co, WV**

Black striate

Gastrodontidae

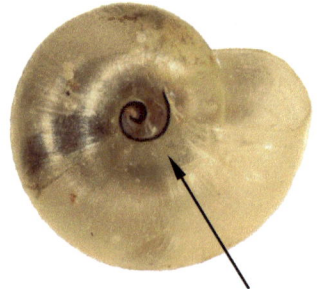

Striatura ferrea Morse, 1864

Diameter: 2.5-3.4 mm, height 1.5 mm

Description: Depressed heliciform; aperture oval; lip simple; shell with 3.5-4 whorls; narrowly umbilicate; shell grayish and rather dull; transverse and spiral striae not as well developed (a) as in *S. meridionalis* spiral but are a constant and reliable diagnostic feature; without teeth; deceased and dried animal can be seen through the translucent shell (bottom view) giving it a darker appearance in some spots.

Similar Species: *Striatura meridionalis* is smaller in diameter, has a smaller aperture, and a notably larger umbilicus; *Punctum* are smaller and have a proportionally wider umbilicus.

Habitat: Found in mixed hardwood forests on upper elevation mountainsides and in ravines living under layers of moist leaf litter.

Status: G5/S3; Uncommon to Common; leaf litter sampling will no doubt turn up additional sites in West Virginia.

Specimen: North Carolina, Cherokee County, Nantahala National Forest (author's collection).

Umbilicus small

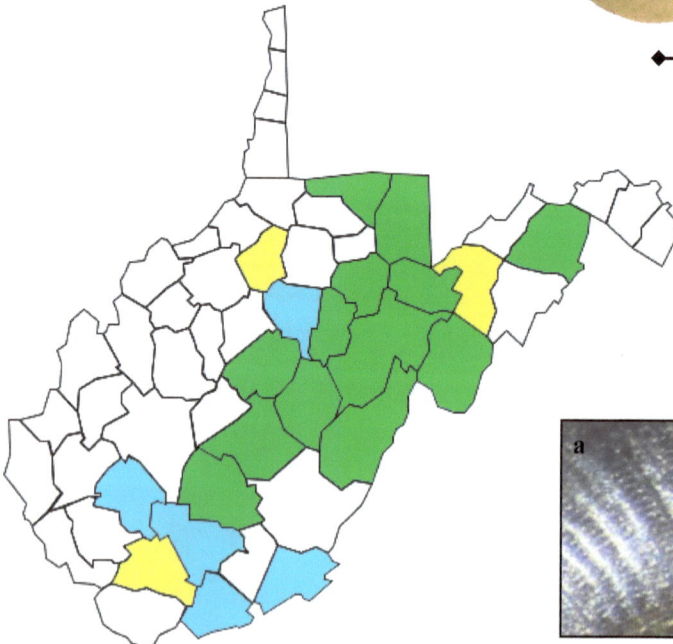

104

Minute gem

Pristilomatidae

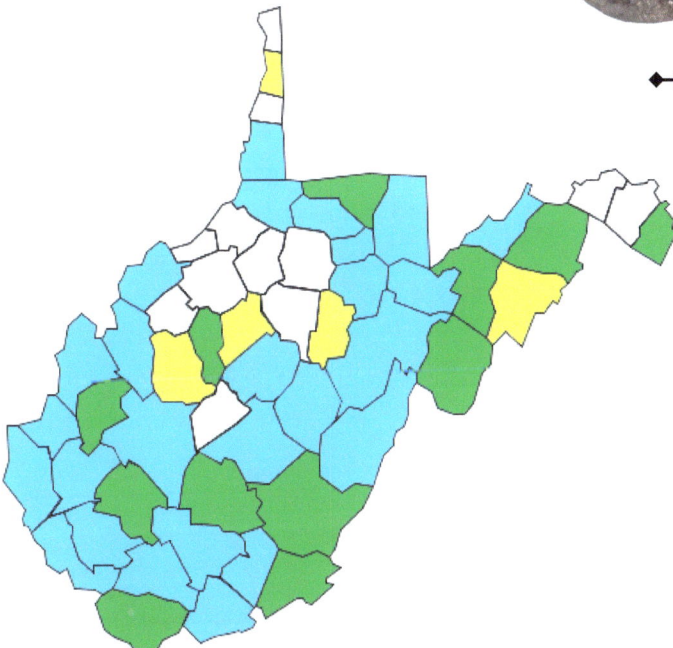

Hawaiia minuscula (A. Binney, 1840)

Diameter: 2-2.8 mm, height 1.2 mm

Description: Depressed heliciform; lip simple; shell with 3.5-4.5 whorls; widely umbilicate, 0.8 mm wide; live shells thin, pale gray and translucent, dead shells quickly turning white with age; weakly developed transverse striae are irregularly spaced, and there are always faint spiral striae present under a strong lens.

Similar Species: Both *Lucilla scintilla* and *L. singleyana* have a much less distinctive sculpture and, more importantly, a smaller umbilicus; *Punctum* species have a crisscrossing sculpture; the spiral striae of *Striatura* species are more well developed.

Habitat: A snail of bare ground occurring in floodplains, meadows, roadsides, and in urban areas; the snail also occurs in the mountains in mixed hardwood forests under leaf litter and detritus.

Status: G5/S5; Common; a species that likely occurs in every West Virginia county.

Specimen: Kentucky, Letcher County, Bad Branch NP (author's collection).

Southeastern gem

Pristilomatidae

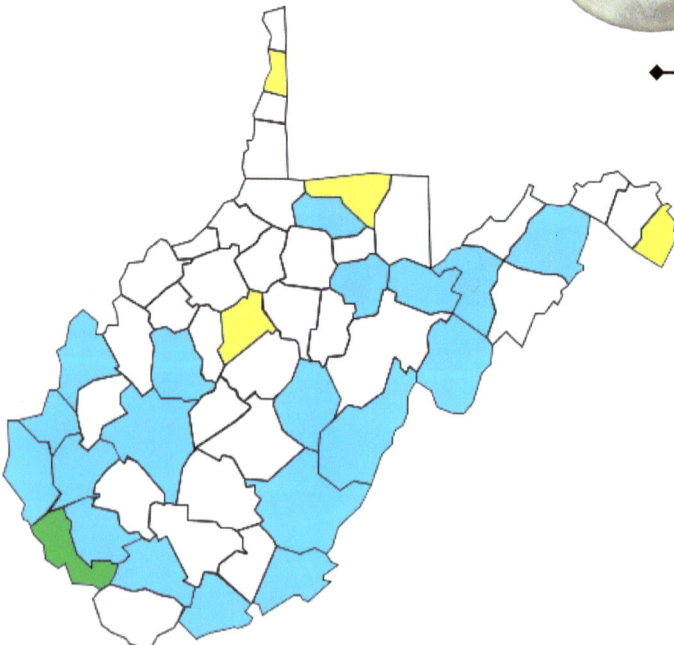

Hawaiia alachuana (Dall, 1885)

Diameter: 2-3 mm, height 1 mm

Description: Depressed heliciform; lip simple; shell with 4-5.5 whorls; widely umbilicate, 1.3 mm wide; live shells thin, pale gray and translucent, dead shells quickly turning white with age; transverse striae are irregularly spaced and there are always faint spiral striae under a strong lens.

Similar Species: *Hawaiia minuscula* is smaller, has a higher shell profile, and has a notably smaller umbilicus (see page 107) , but this feature is highly variable in both *Hawaiia* species; *Lucilla inermis* and *L. singleyana* lack distinctive sculpture.

Habitat: A calciphile species of leaf litter and detritus and waste places.

Status: G4/S3; Uncommon; a species that has been confused with *H. minuscula,* a result of the two species having a wide range of umbilicus widths.

Specimen: West Virginia, Gilmer County, (author's collection).

Hawaiia Shells Compared (Larry Watrous photo collection)

1mm

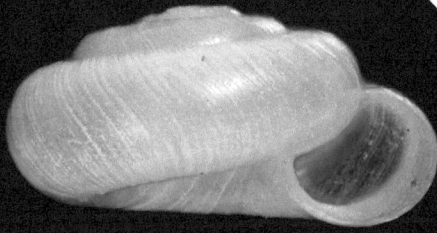

Minute gem, *Hawaiia minuscula* (A. Binney, 1840)

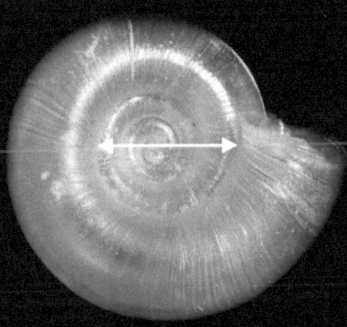

Southeastern gem, *Hawaiia alachuana* (Dall, 1885)

Some Rather Remarkable Predators of Land Snails

Above image of a Speckled snail-sucker, *Sibon nebulata,* eating a Banded cone, *Drymaeus sulfureus,* Bladen Nature Reserve, Belize, Central America. In North America, brown snakes are known to hunt and consume slugs, especially the exotic *Arion* species (pers. comm. John MacGregor). It is unknown if other small West Virginia snakes, such as ringneck and red-bellied snakes include terrestrial gastropods as part of their overall diet. Another fascinating predator of land snails is the ants. As with our native snakes, ants may also be hunting minuscule snail species in the leaf litter, but this remains entirely unstudied. Below picture of African weaver ants carrying a land snail to their nest for consumption.

© Piotr Naskrecki

Smooth coil

Helicodiscidae

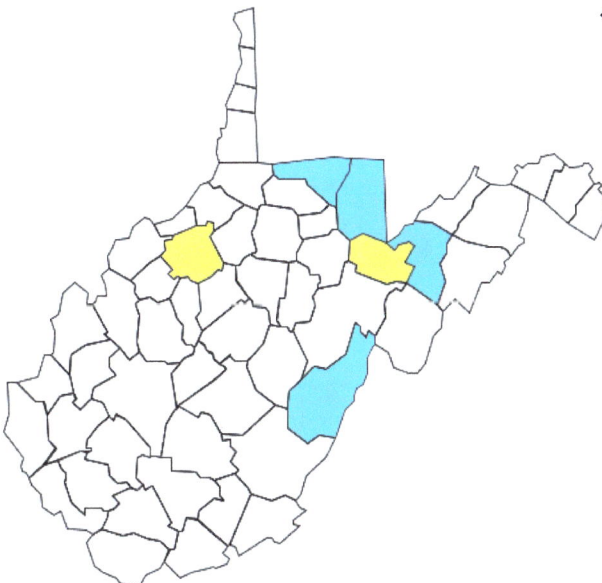

Lucilla singleyana (Pilsbry, 1889)

Diameter: 2.4-3 mm, height 0.9 mm

Description: Depressed heliciform; lip simple, aperture egg shaped; shell with 3-4 whorls; fresh shells are glossy, translucent and nearly smooth; umbilicate; no teeth present; scant traces of spiral ornamentation under a strong lens, but this faint feature is usually present in fresh shells only; deceased and dried animal seen in frontal and bottoms views, making the shell slightly darker in places.

Similar Species: Bottom and top views of *Lucilla* species are nearly indistinguishable, except that *L. inermis* is smaller and displays no microscopic spiral striae under a strong lens; *Hawaiia* species are very similar in size and build, but have transverse striae that are notably more developed, when compared to *Lucilla* species.

Habitat: Usually a species of open grassy areas, roadsides, along railroads, and occasionally washed into caves during rain events.

Status: G5/S2; Rare; likely more common than current records indicate, formally in the genus *Helicodiscus*.

Specimen: Tennessee, Franklin County, Dry Cave (author's collection)

Oldfield coil

Helicodiscidae

Lucilla scintilla (Lowe, 1852)

Diameter: 2.2 mm, height 1.24 mm

Description: Depressed heliciform; lip simple, rounded; shell with 3-4 whorls; fresh shells are yellowish corneous with a dull sheen and are translucent, but turn white with age (a bleached specimen shown here); umbilicate; no traces of spiral ornamentation.

Similar Species: *L. singleyana* displays fine microscopic spiral lines, is more compressed, and has an oval-shaped aperture (see below); be forewarned, separating *Lucilla* species is no joyride and can lead to "SB", a debilitating mental condition known as "Shell Bewilderment."

Habitat: Usually a species of open grassy areas, meadows, old fields and roadsides.

Status: G4/SH; Rare; likely more common than current records indicate; formally in the genus *Helicodiscus*.

Specimen: Texas, Val Verde County, Pecos River and US Route 90 (FM 240318)

L. singleyana

Pilsbry 1946

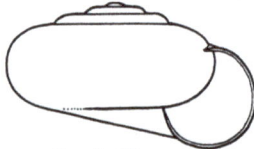

L. scintilla

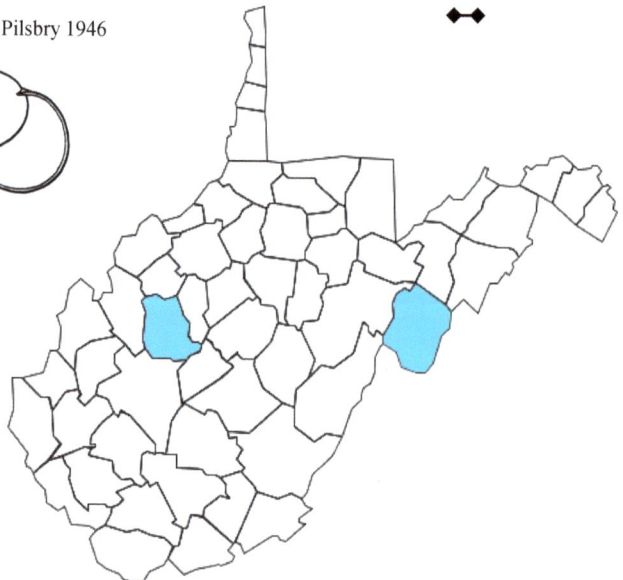

110

Amber glass

Oxychilidae

Nesovitrea electrina (Gould, 1841)

Diameter: 4.6-5.2 mm, height 2.5-2.8 mm

Description: Heliciform; lip simple; shell with 3.5-4.5 whorls; umbilicate; no teeth present; transverse striae are crowded but poorly developed, not reaching the base, which is generally smooth; without spiral striae or very rarely very weak and seen only as mere traces; shell faintly yellow or pale greenish and glossy, translucent; periphery is rounded, specimen shown here with an epiphragm (figures a and b); this is a snail's way of conserving moisture during dry periods; deceased and dried animal seen through the translucent shell in all three views.

Similar Species: Separation between *Glyphyalinia* and *Nesovitrea* can be an arduous task; the primary difference between the two genera is that *Nesovitrea* species are usually more compact in form; *N. electrina* most resembles *G. wheatleyi,* but is without the fine spiral striae.

Habitat: A species of low wet ground, margins of ponds and marshes and floodplains of rivers.

Status: G5/S3; Common; found throughout most of West Virginia.

Specimen: New York, Schoharie County, 0.7 miles E of Sharon, roadside (FM 240400).

a

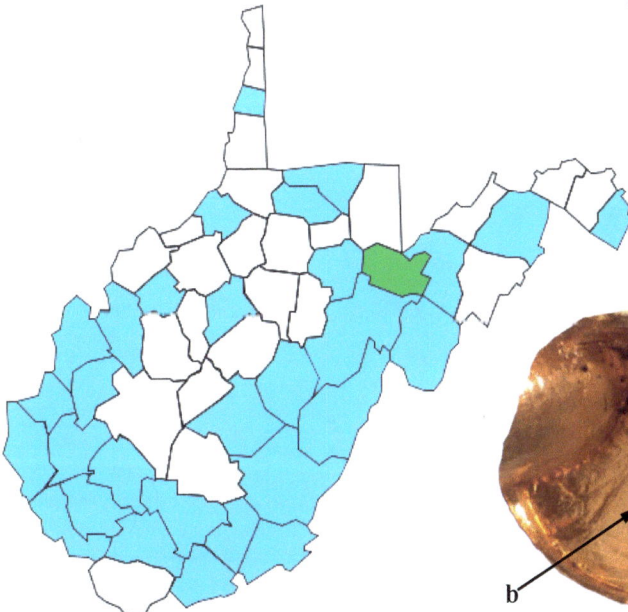

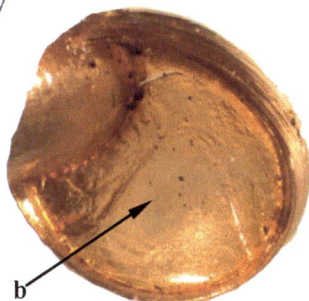

b

111

Blue glass

Nesovitrea binneyana (E. S. Morse, 1864)

Diameter: 3.5-4.3 mm, height 1.75 mm

Description: Heliciform; lip simple; shell with 3.5-4 whorls; umbilicate; no teeth present; transverse striae are closely spaced but poorly developed, missing on the base; no spiral striae present; shell nearly colorless or with a greenish tinge, translucent; live animal almost white with darker tentacles; periphery is rounded; deceased and dried animal clearly seen through bottom shell view.

Similar Species: *Nesovitrea binneyana* resembles *N. electrina,* but is one third smaller and has a different color being nearly white with a greenish tinge; *G. wheatleyi* is less compact, larger, and has fine spiral striae.

Habitat: A species of upland mixed hardwood forests living among the leaf litter.

Status: G5/S?; although this northern species was widely reported in West Virginia, Hubricht (1985) did not report it south of Pennsylvania, consequently <u>records in West Virginia are highly questionable and are in need of further investigation.</u>

Specimen: Michigan, Keweenaw County, Isle Royale NP (FM 58444).

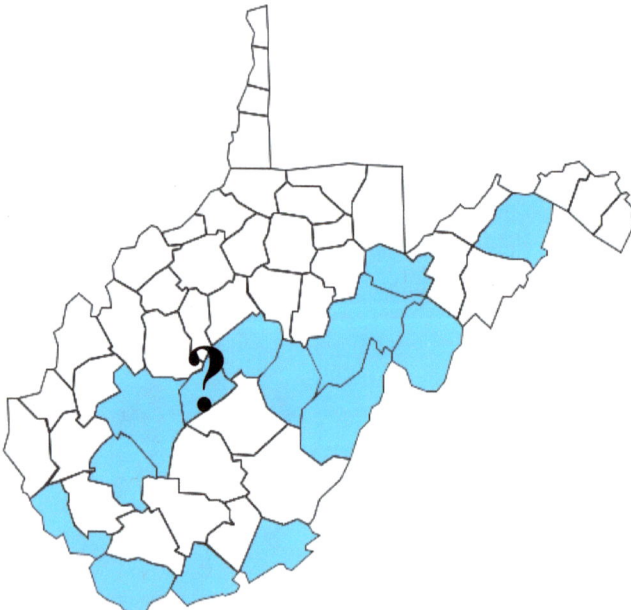

Brilliant granule

Euconulidae

Guppya sterkii (Dall, 1888)

Diameter: 1.2-1.3 mm, height 0.75 mm

Description: Depressed heliciform; lip simple; shell with 3.5-4 whorls; minutely perforate, the umbilicus not completely closed; shell yellowish and highly translucent in live snails and fresh dead shells; transverse striae weak (strong light and scope required); spiral striae always present, but sometimes a hard feature to detect, best seen around the first whorls but also on the base; tilting the shell slightly should bring out this minute feature.

Similar Species: This species differs from other minute snails by having a nearly closed umbilicus (a).

Habitat: Found in mixed hardwood forests at all elevations; this is likely one of the most common minuscule land snail residents of leaf litter.

Status: G5/S5; Common; one of the most frequent tiny land snail species found in leaf litter collections.

Specimen: North Carolina, Swain County, Nantahala National Forest (author's collection).

a

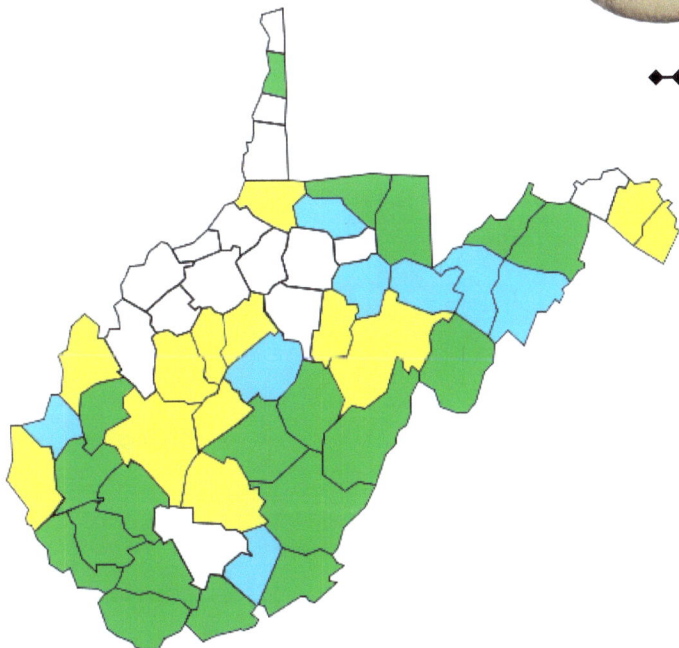

Eastern glass-snail

Vitrinidae

Vitrina angelicae Beck, 1837

Diameter: 5-6 mm, height 4.5 mm

Description: Heliciform; lip simple with a slight reflection near the columellar margin; shell with 2-3 whorls; imperforate; shell surface clear, glasslike, transparent in live and fresh dead; paper thin and very fragile; faint sculpture of low growth wrinkles.

Similar Species: No other West Virginia land snail has the delicate build and transparent shell of *V. angelicae*.

Habitat: A species of open, grassy places, meadows, roadsides; in West Virginia known only from Canaan Valley from low moist margins of wetlands.

Status: G5/S1; Rare; this apparently scarce species in West Virginia was only recently documented in Canaan Valley NWR in 2008 by the author and Marquette Crockett.

Specimen: West Virginia, Tucker County, Canaan Valley (author's collection).

Pilsbry, 1946

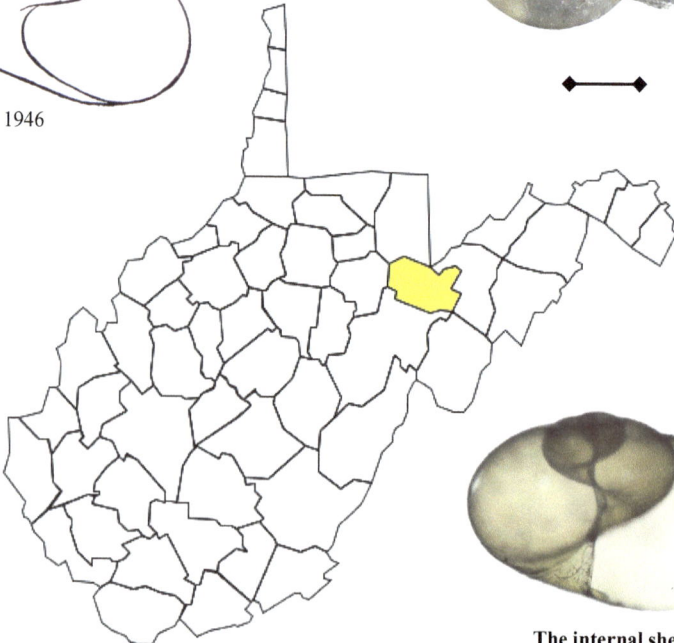

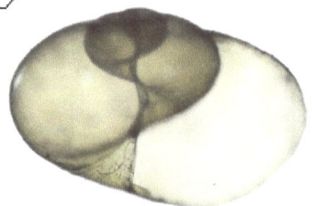

The internal shell

114

Shells 5-8.4 mm, umbilicate or perforate, simple lip

Zonitoides species are small colonial snails often associated with rotting hardwood in advance stages of decay, either standing or downed trees. They are sometimes found in large numbers (I have found upwards of 50 on one log). Key features to this group of gastropods are the open umbilicus and absence of any teeth in all stages of growth. They are occasionally confused with *Glyphalinia* and *Paravitrea* species but, in general, *Zonitoides* species is larger, have much thicker shells, and are less finely sculptured. Pay close attention to any spiral ornamentation or the absence of it under the lens. At times, this micro-feature is hard to see, depending on the lighting source and angle of the shell. It is best viewed on the top surface and will take some experience to key in on this finer feature. Slightly tilting the shell (while under the scope) will sometimes help reveal this micro-feature; wetting the surface of the shell will also aid in sculpture enhancement, especially if the shells are bleached or badly weathered. There are three species reported from the state.

Grouping 1: shells 5-8.4 mm, umbilicate (a), whorls moderately to loosely coiled (b), spiral striae usually a weak feature or absent, pg 116-118

Zonitoides elliotti (rare)
Zonitoides arboreus
Zonitoides nitidus

Gastrodonta species differ from *Zonitoides* in being more tightly coiled (c), perforate (d), having coarse ribbing (f), and are always with well-developed teeth (e). Two species are reported from West Virginia.

Grouping 2: shells 6-7 mm, perforate (d), with tightly coiled whorls (c), internal teeth (e) and strongly developed ribbing on top of the shells (f), pg 119-120

Gastrodonta interna
Gastrodonta fonticula (rare)

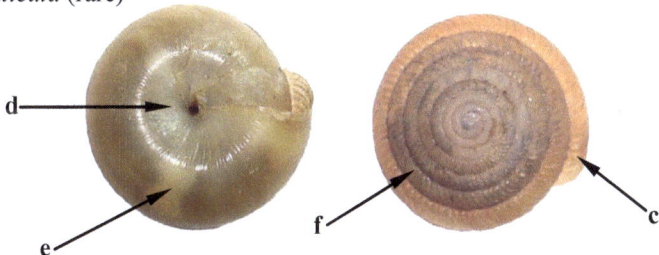

Green gloss

Zonitoides elliotti (Redfield, 1856)

a

Zonitidae

Diameter: 7.5-8.4 mm, height 4– 4.9 mm

Description: Depressed heliciform; lip simple and slightly reflected, behind which is a small callus or thickening of the peristome (a); aperture square shaped; shell with 5-6 whorls; umbilicate; greenish-horn color and glossy; transverse striae are poorly developed; spiral striae present, but are so faint that detection is often eye straining; without teeth.

Similar Species: The slightly reflected lower lip and thicker shell of *Z. elliotti* are key characteristic features in separating it from all other snail shells found in this size class.

Habitat: Found in mixed hardwood and pine woods on hillsides and ravines; in well rotting wood that is partly embedded in the soil, or under rotting hardwood logs; the species is often found in colonies, upwards of twenty-five on one single log (pers. obs.); the species appears to indicate a rich snail fauna (pers. comm. Ron Caldwell).

Status: G4/S2; Rare; in West Virginia records for this species are scattered across the state.

Specimen: North Carolina, Swain County, Nantahala National Forest (author's collection).

Quick gloss

Zonitidae

Zonitoides arboreus (Say, 1816)

Diameter: 5-6 mm, height 2.5-2.8 mm

Description: Depressed heliciform; lip simple not reflected; shell with 4.5–5 whorls; umbilicate; olive buff and glossy; transverse striae are poorly developed; spiral striae present, but a weak, eye-straining feature to see; without teeth; periphery roundish; when crawling its shell is held rather high, figure (a).

Similar Species: *Z. elliotti* is larger, has a thicker shell and a slight reflection of the lower lip; *Z. nitidus* is slightly larger and taller with a darker shell, no spiral striae, and a more roundish aperture.

Habitat: A common species found on or under exfoliating bark of standing or down rotting trees in advanced stages of decay; very common on old timbers inside abandoned coal mines (MacGregor, pers. comm. 2010); usually found in small colonies.

Status: G5/S5; Common; this is the most common and wide ranging land snail in North America occurring in all lower 48 states and south to Panama in Central America; in the Rockies it has been collected at 10,000 feet.

Specimen: Kentucky, Powell County, Furnace Mountain (author's collection).

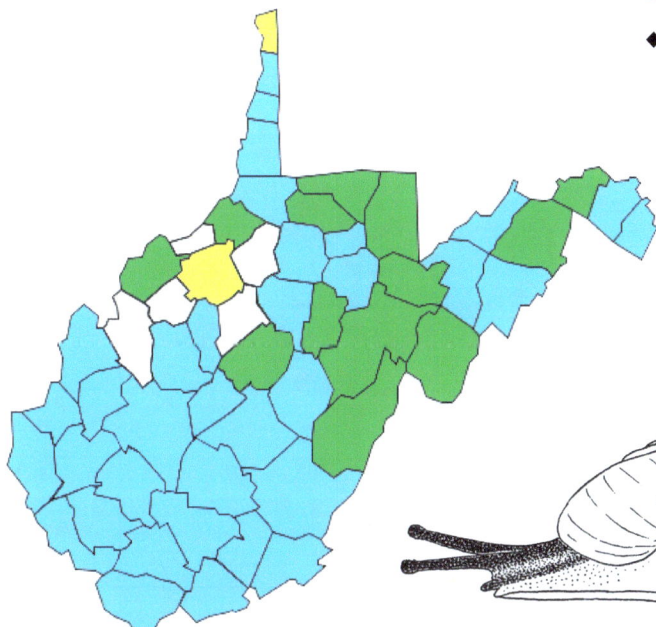

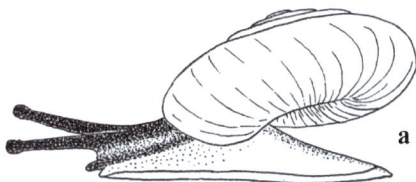

a

Pilsbry, 1946

Black gloss

Zonitoides nitidus (Müller, 1774)

Diameter: 6-8 mm, height 3.6-4 mm

Description: Heliciform; lip simple not reflected; shell with 4.5–5 whorls; umbilicate; olivaceous-yellow, semi-transparent and very glossy; transverse striae are poorly developed; without spiral striae; without teeth in any stages of growth; shell periphery roundish.

Similar Species: *Z. arboreus* is smaller, typically more depressed (although this is somewhat variable among populations), glossier with spiral striae, and the aperture is more oval in shape, but this feature also varies considerably between populations.

Habitat: A species of low ground; found around logs and leaf litter near floodplains, marshes, and wet roadsides; a Holarctic species (Hubricht 1985); may be transported on treated lumber to new locations.

Status: G5/S5;Common; reported from most counties in West Virginia.

Specimen: Massachusetts, Suffolk County, Great Brewster Island (author's collection).

Zonitidae

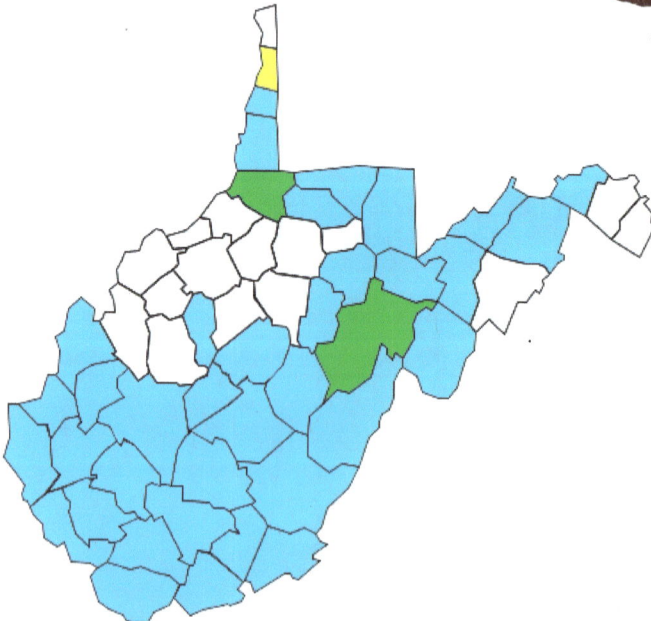

118

Brown bellytooth

Gastrodonta interna (Say, 1882)

Diameter: 6.5-7.4 mm, height 4.5-5.4 mm

Description: Heliciform; lip simple; shell with 8-9 tightly coiled whorls; perforate (a); cinnamon-brown; shell solid (thick) for its small size; transverse striae are distinct, rib-like and regularly spaced on top (a) but much weakened on the base; two teeth are present at all stages of growth and can be seen through the bottom of live and fresh shells (b).

Similar Species: *Gastrodonta fonticula* has a larger umbilicus, is smaller, and in general, has a flatter, darker shell.

Habitat: A colonial species found under leaf litter, very common under loose bark of decaying wood on hillsides and sandstone ridgetops of mixed hardwood forests.

Status: G5/S3; Relatively Common; in West Virginia reported from 14 counties.

Specimen: Kentucky, Powell County, Tunnel Ridge Road, Red River Gorge, Daniel Boone National Forest (author's collection).

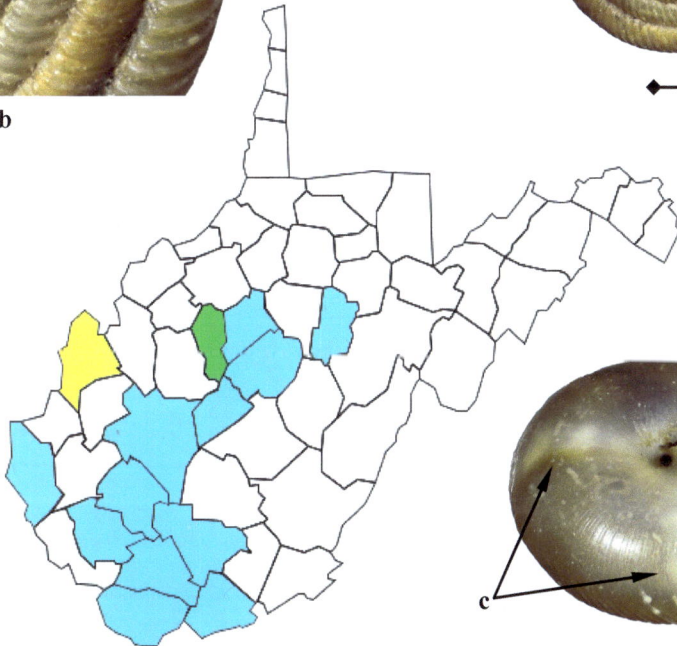

Gastrodontidae

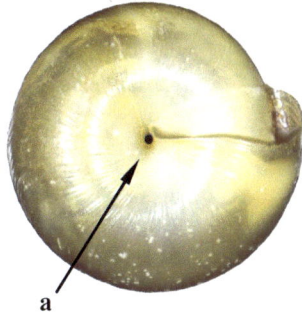

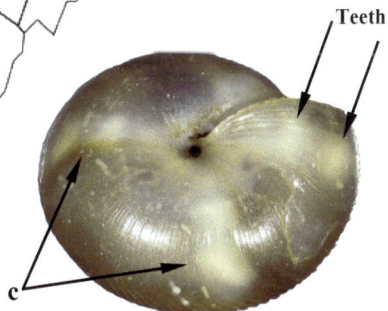

a

b

Teeth

c

119

Appalachia bellytooth

Gastrodontidae

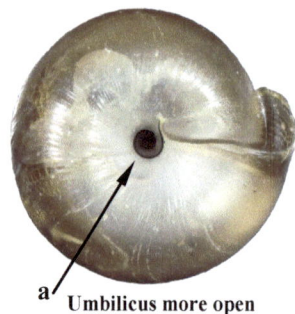

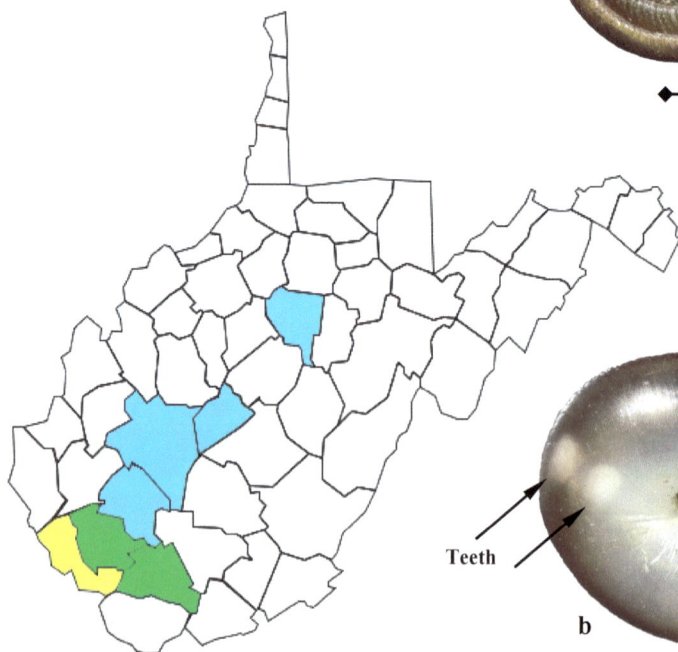

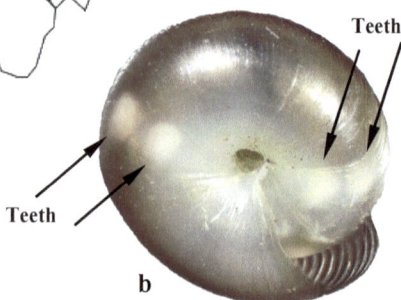

Gastrodonta fonticula (Wurtz, 1948)

Diameter: 5.7-6.9 mm, height 4.5 mm

Description: Heliciform to depressed heliciform; lip simple; shell with 7-8 tightly coiled whorls; cinnamon-brown; shell solid (thick) for its small size; umbilicate (wider than seen in *G. interna*) (a); transverse striae are distinct, rib-like and regularly spaced on top but indistinct on the base; one or two pairs of teeth situated on a callous ridge which are present at all stages of growth, these teeth usually visible through the bottom of live and fresh dead shells, figure (b).

Similar Species: *Gastrodonta interna* has the same shell form or with a slightly higher profile, this feature varying considerably, but most importantly has a smaller (pin-size) umbilicus.

Habitat: A species found under leaf litter and detritus close to rotting hardwood logs in advanced stages of decay in upland mixed hardwood locations.

Status: G3G4/S2; Uncommon; this primarily southern Appalachian species is reported from 7 counties in West Virginia.

Specimen: West Virginia, Mingo County, (author's collection).

a Umbilicus more open

Teeth

Teeth

b

Shells 4-12 mm, without teeth, simple lip, loosely coiled whorls, unique indented transverse striae (*Glyphyalinia*)

Glyphyalinia species can be a troublesome group of land snails for the novice. Even for the experienced malacologist, they are challenging. Key features for this group of gastropods are the small size, the loosely-coiled whorls, the last whorl usually widely expanded (a), the glossy and translucent shells and most importantly, the indented transverse striae (c), which can vary from closely-spaced (d) to widely spaced (e). The somewhat unique indented or etched sculpture on the shell surface is the genera's signature and is easily viewed under a hand lens of 10X. Although *Paravitrea* species have indented sculpture as well, they are tightly coiled, the last whorl is not expanded (b) and they usually have internal armature at some stage of growth. *Glyphyalinia* species never contain internal teeth. Give special attention to any spiral sculpture under the scope. At times, this micro-feature is hard to see, depending on the light source and angle of the shell. It is best viewed on the top surface and will take some experience to key in on this fine feature. Slightly tilting the shell (while under the scope) will sometimes help reveal this micro-feature. Wetting the surface of the shell will also bring out these features, especially if the shells are bleached or badly weathered. *Glyphyalinia* snails here are arranged in three groups; **(1)** shells that are imperforate **(2)** shells that are perforate or rimate and **(3)** shells that are umbilicate.

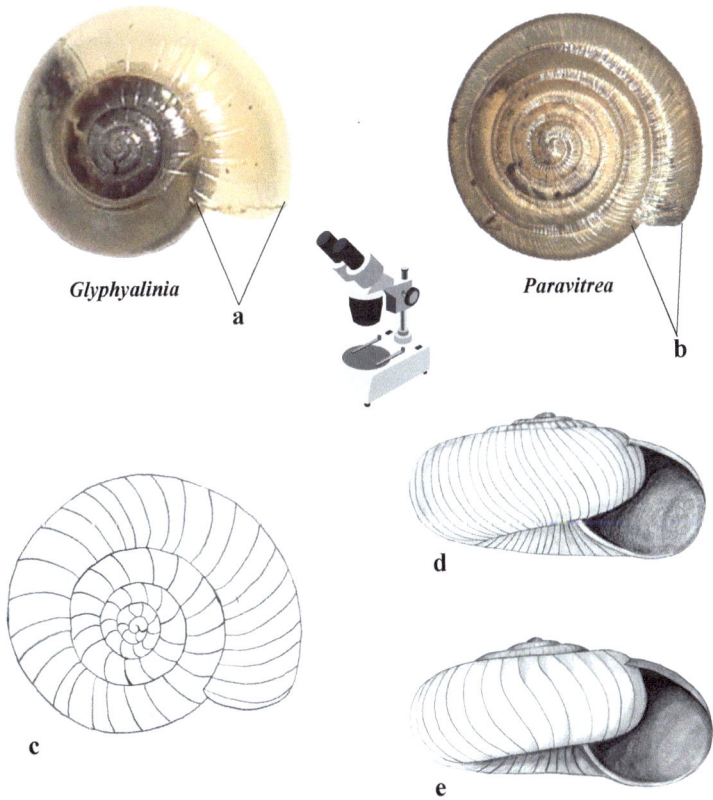

Glyphyalinia

a

Paravitrea

b

c

d

e

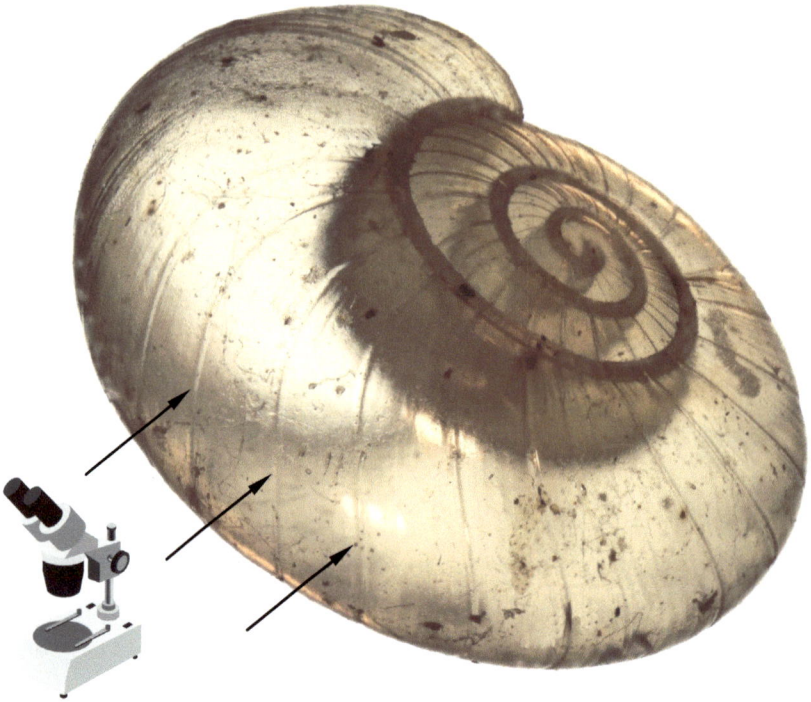

Glyphyalinia rhoadsi illustrating the remarkable indented striae

Grouping 1: shells that are imperforate, pg 123

Glyphyalinia solida

Grouping 2: shells that are perforate or rimate, pg 124-130

Glyphyalinia picea (rare)
Glyphyalinia caroliniensis (rare)
Glyphyalinia indentata (common)
Glyphyalinia species (undetermined) West Virginia glyph (rare)
Glyphyalinia praecox
Glyphyalinia rimula

Grouping 3: shells that are umbilicate, pg 131-138

Glyphyalinia rhoadsi
Glyphyalinia wheatleyi (common)
Glyphyalinia cumberlandiana
Glyphyalinia specus (a cave species)
Glyphyalinia lewisiana
Glyphyalinia raderi (rare)
Glyphyalinia virginica

Imperforate glyph

Glyphyalinia solida (H. B. Baker, 1930)

Diameter: 7.5 mm, height 4 mm

Description: Depressed heliciform; lip simple; shell with 5.25 loosely coiled whorls; shell fragile, color corneous to fulvous to almost chestnut, glossy and semi-transparent; imperforate, the umbilicus completely covered in all stages of growth (a); no teeth present; indented transverse striae are well developed, closely and nearly equally spaced, continuing to the base; the spiral striae may be weakly defined but are a constant feature under a strong lens.

Similar Species: *Glyphyalinia solida* is typically larger, heavier shelled, and has a stronger sculpture than *G. cryptomphala*; other *Glyphyalinia* species are not imperforate, the umbilicus remaining at least partially open.

Habitat: A species that can be found under moist leaf litter in mixed upland hardwood forest and in talus near rock ledges.

Status: G5/SH; Common; a widespread species in West Virginia but not encountered during WVDNR surveys

Specimen: Tennessee, Marion County, Clear Spring Cave, 2 miles NE of Jasper (author's collection).

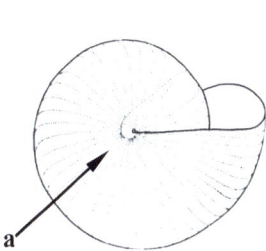

a

Pilsbry, 1946

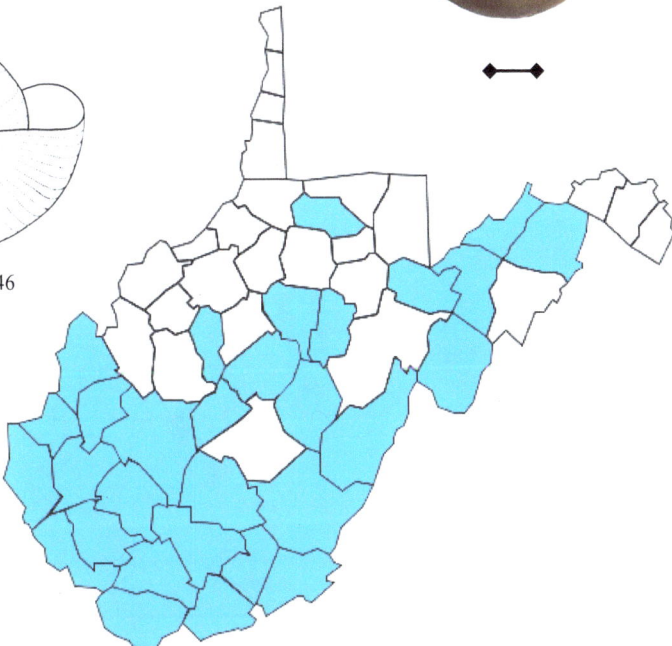

Rust glyph

Zonitidae

Glyphyalinia picea Hubricht, 1976

Diameter: 7.1 mm, height 3.0 mm

Description: Depressed heliciform; lip simple; shell with 5.5 loosely coiled whorls; shell pale reddish-brown; perforate or rimate (a); indented transverse striae are moderately and almost evenly spaced, continuing nearly as strongly onto the base of the shell; the spiral striae (b) are crowded and strongly developed across the entire shell surface including the bottom; periphery well rounded.

Similar Species: *Glyphyalinia solida* has a completely closed umbilicus; *G. indentata* is generally a smaller snail with less distinctive sculpture and a species of lower elevations.

Habitat: A species found under moist leaf litter in mixed higher elevation hardwood forest (Hubricht 1985); the type locality is located on the east side of Spruce Knob in Pendleton County, WV.

Status: G3/S2; Rare; in West Virginia, this infrequent land snail is reported from only four counties.

Specimen: West Virginia, Pendleton County, Spruce Knob, 4500 feet (FM 24177 Paratype).

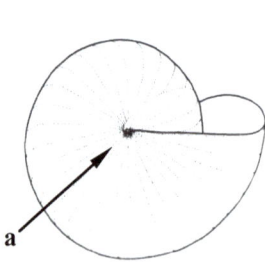

a

Pilsbry, 1946

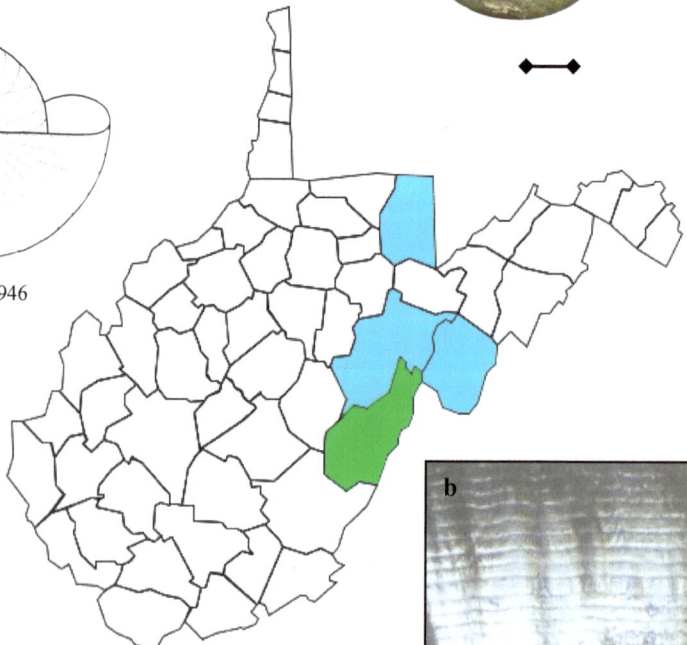

b

Spiral mountain glyph

Zonitidae

Glyphyalinia caroliniensis (Cockerell, 1890)

Diameter: 5-12 mm, height 3.4-5.5 mm

Description: Depressed heliciform; lip simple; shell with 4.5-5.5 loosely coiled whorls; perforate or rimate; shell fragile, pale-horn, shiny and semi-transparent; indented transverse striae are well developed, moderately and nearly equally spaced; a hand lens of 10X or scope will reveal numerous, closely spaced, clear-cut spiral engraved striae which, although weaker, typically continue onto the base.

Similar Species: This species is most similar to *Glyphyalinia indentata,* but is larger and has more conspicuous spiral striae; *G. praecox* is bronze-colored; *G. picea* is smaller and a species of higher elevations.

Habitat: Mixed hardwood forests under moist leaf litter along river bluffs and also a species of mountainsides.

Status: G4/SNR; although widely reported in West Virginia by MacMillan, Hubricht's (1985) records do not show *G. caroliniensis* in the state but in Virginia counties that border West Virginia; all specimens examined from the Carnegie Museum labeled as *G. caroliniensis* refer to *G. indentata;* in addition, recent and thorough land snail surveys across West Virginia by WVDNR failed to produce valid specimens of *G. caroliniensis*; therefore, it seems doubtful that the species occurs in the state and most, if not all records likely represent misidentifications of the smaller, but similar looking and very common, *G indentata;* (pg 126).

Specimen: Tennessee, Blount County, Chestnut Top Trail GSMNP (author's collection).

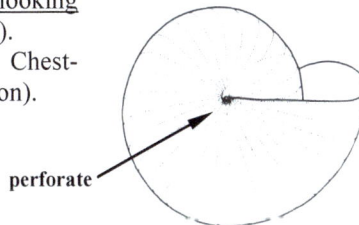

perforate

G. caroliniensis

Pilsbry, 1946

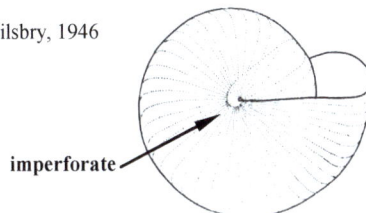

imperforate

G. solida

125

Carved glyph

Zonitidae

Glyphyalinia indentata (Say, 1823)

Diameter: 4.7-7.1 mm, height 3 mm

Description: Depressed heliciform; lip simple; shell with 4.5-5 loosely coiled whorls; perforate or rimate; shell fragile, pellucid, highly polished and translucent; indented transverse striae are well developed, rather widely but nearly equally spaced, about 28 on the last whorl; under a strong lens and good lighting, the spiral striae can usually be seen on some part of the last whorl, either rather weakly developed or only as traces (Pilsbry 1946); periphery well rounded; deceased and dried animal seen in all three views, making portions of the shell darker in places.

Similar Species: This species is most similar to *Glyphyalinia caroliniensis* but is fully 4 to 5 mm smaller when collected from the same region; *G. rhoadsi* has a wider umbilicus and more strongly developed indented transverse striae.

Habitat: Found in a variety of mixed hardwood forests under leaf litter but also occasionally living along roadsides and in urban areas.

Status: G5/S5; Common; the most common *Glyphyalinia* in the state.

Specimen: North Carolina, Macon County, Fatback Timber Sale, Nantahala National Forest (author's collection).

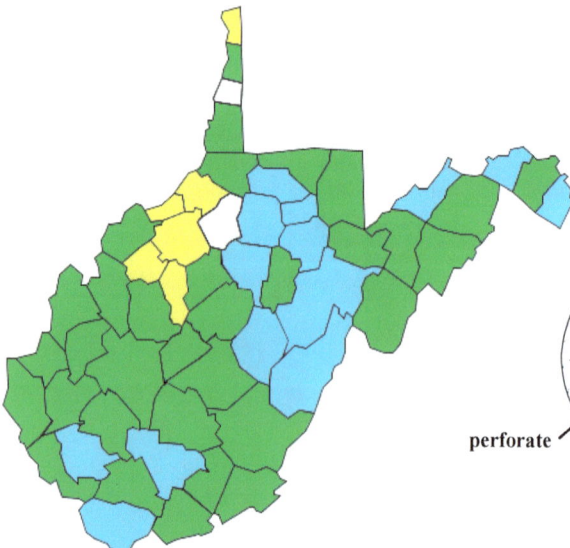

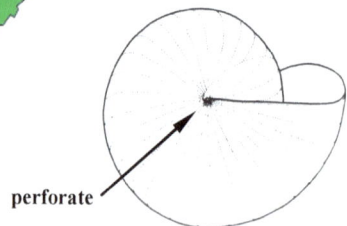

perforate

Pilsbry, 1946

126

West Virginia glyph

Glyphyalinia species (undetermined)

Diameter: 8.2 mm, height 4.2 mm

Description: Depressed heliciform; lip simple; shell with 6 loosely coiled whorls; perforate, the minute umbilical opening may be partly covered; shell slightly thicker than other *Glyphyalinia* species, pale horn-colored, glossy and translucent; indented transverse striae are well developed, equally and closely spaced; without notable spiral striae.

Similar Species: Most closely resembles *G. sculptilis* of more southern mountains, but is smaller, a different color, and most importantly, has half the number of transverse striae in the same expanse of shell, this being the single best character for their separation.

Habitat: Nothing is known of the habitat, but most *Glyphyalinia* are found living among moist leaf in ravines and on hillsides.

Status: G1/S1; Rare; Endemic to West Virginia; the specimen illustrated here is from the Carnegie Museum and was labeled as *G. sculptilis,* but differs in several important ways discussed above; precious little is known of this endemic gastropod to West Virginia and additional specimens from both sites (see below) will be required to make a final determination.

Specimen: West Virginia, Pendleton County, around Judy Gap, North Fork Mountain (CM 62.36760). Another specimen (not illustrated) from the Carnegie Museum labeled as *G. sculptilis* that agrees with the above undetermined *Glyphyalinia* was collected from near the Cheat Bridge and White Top, Randolph County, West Virginia (CM 62.3676).

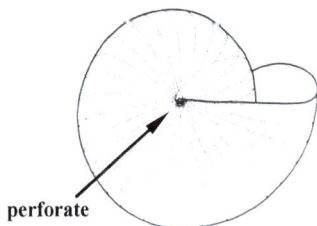

perforate

Pilsbry, 1946

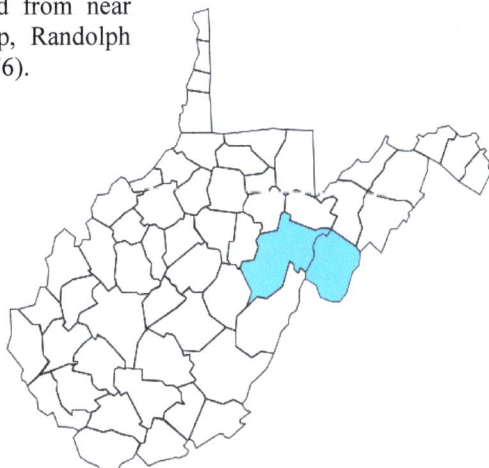

2 mm

10 striae

West Virginia glyph, *Glyphyalinia* species (undetermined),
Pendleton County, West Virginia

22 striae

Suborb glyph, *Glyphyalinia sculptilis* (Bland, 1858),
Nantahala NF, Graham County, North Carolina

Brilliant glyph

Zonitidae

Glyphyalinia praecox (H.B. Baker, 1930)

Diameter: 6.2-6.3 mm, height 2.8 mm

Description: Depressed heliciform; lip simple; shell with 4.5-5 loosely coiled whorls; perforate, the minute umbilical opening is partly covered by an expansion of the columellar lip; shell fragile, brilliant bronze-colored, glossy and transparent; indented transverse striae are well developed, moderately and nearly equally spaced and continue to the base; the spiral striae are bold and closely spaced.

Similar Species: *Glyphyalinia indentata* has a slightly higher profile, the apical whorls are less tightly coiled, the last whorl is more capacious and is corneous, not bronze colored.

Habitat: Found under moist leaf litter in floodplains and on talus slopes in mixed hardwood forests.

Status: G4/SH; Uncommon; reported from only a few scattered West Virginia counties.

Specimen: Tennessee, Polk County, CNF (author's collection).

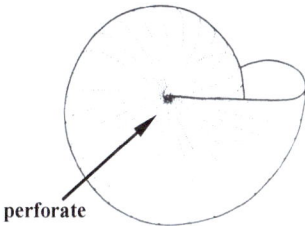

perforate

Pilsbry, 1946

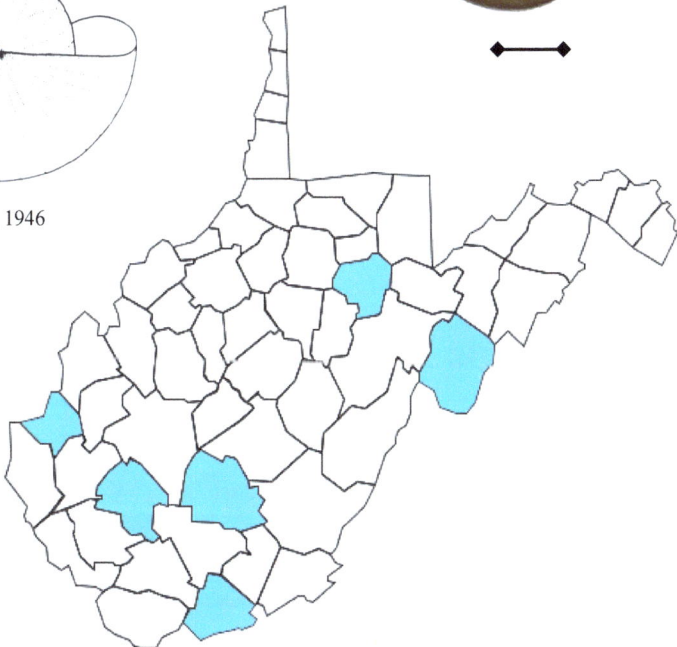

Tongued glyph

Zonitidae

Glyphyalinia rimula Hubricht, 1968

Diameter: 7.7 mm, height 4 mm

Description: Depressed heliciform; lip simple; shell with 5 loosely coiled whorls; shell fragile, pale coppery (when fresh), glossy and semi-transparent; rimate, a tongue-like callus at the columellar end which partially covers the umbilicus (a); no teeth present; indented transverse striae are well developed and widely spaced; the spiral striae may be weakly defined but generally a constant feature under a strong lens; live animal pale-gray.

Similar Species: This species is most similar to *Glyphyalinia solida,* but the tongue-like callus does not completely cover the umbilicus (a).

Habitat: Most often a snail of mixed wooded hillsides and ravines occurring under leaf litter; sometimes found in caves; also found in ever-increasing and unyielding kudzu tangles.

Status: G3/SH; Rare; in West Virginia reported from only one southwestern county.

Specimen: Kentucky, Magoffin County, 1 mile ENE of Falcon (FM 241632).

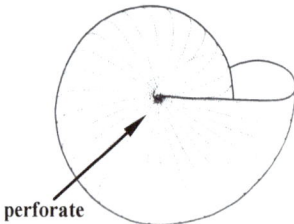

perforate

Pilsbry, 1946

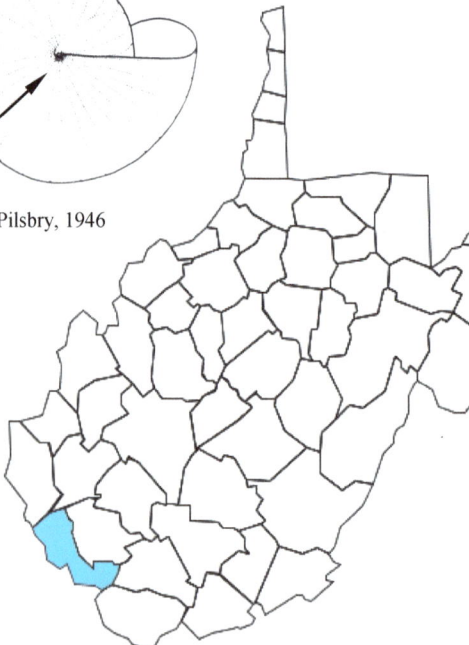

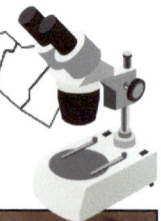

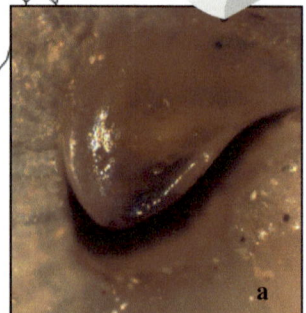

a

130

Sculpted glyph

Glyphyalinia rhoadsi (Pilsbry, 1899)

Diameter: 4.5-5.3 mm, height 2.5 mm

Description: Depressed heliciform; lip simple; shell with 4-5 loosely coiled whorls; umbilicate, umbilicus 1/12 the shell diameter or around 0.5 mm wide; shell fragile, corneous, glossy and semi-transparent; indented transverse striae are well developed and nearly equally spaced continuing to the base; the extremely weak spiral striae may be seen near the suture and umbilicus, but this feature is absent in some specimens (Pilsbry 1946).

Similar Species: This species is most similar to *Glyphyalinia indentata,* but is slightly smaller, has better developed transverse striae, and a notably wider umbilicus.

Habitat: Found in a variety of mixed upland hardwood forests under moist leaf litter.

Status: G5/S5; Common; throughout most of West Virginia.

Specimen: Virginia, Dickenson County, Garden Hole, rock talus above river (author's collection).

Zonitidae

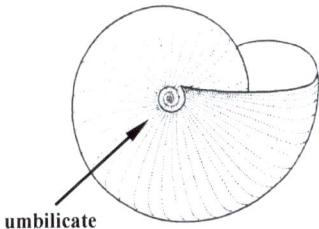

umbilicate

Pilsbry, 1946

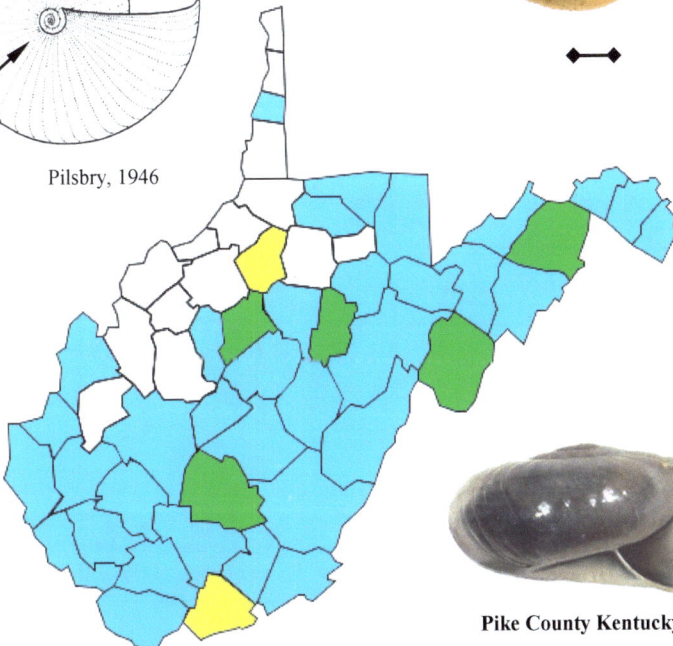

Pike County Kentucky

131

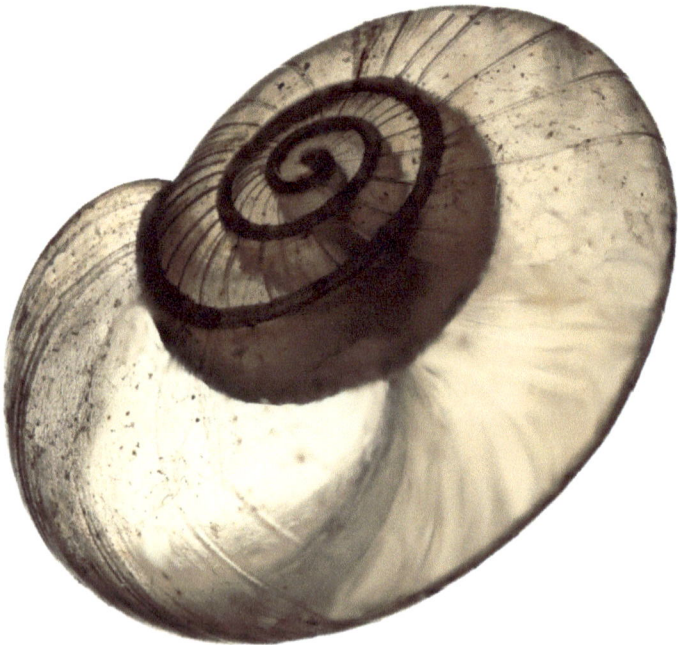

The sculpted glyph, *Glyphyalinia rhoadsi,* showing the extraordinary internal and external architecture of the shell

Bright glyph

Glyphyalinia wheatleyi (Bland, 1883)

Diameter: 5-6 mm, height 1.9 mm

Description: Depressed heliciform; lip simple; shell with 5-5.5 loosely coiled whorls; umbilicate (size varying); shell glossy and translucent; pellucid; brownish horn; indented transverse striae (radiating lines) are moderately developed and irregularly spaced; spiral striae may be absent or with only mere traces in some populations; images illustrate well the degree of translucently seen in shells.

Similar Species: *Glyphyalinia cumberlandiana* is smaller, has a thinner, more fragile shell, and is a light horn-colored; *G. rhoadsi* has a notably smaller umbilicus, a different color, and more strongly developed transverse striae.

Habitat: Found in a wide range of habitats including beneath moist leaf litter and deep detritus deposits located from valleys to mountaintops in mixed hardwood forests.

Status: G5/S5; Common; one of the most common *Glyphyalinia* in the state.

Specimen: West Virginia, Pendleton County, Seneca Rocks (author's collection).

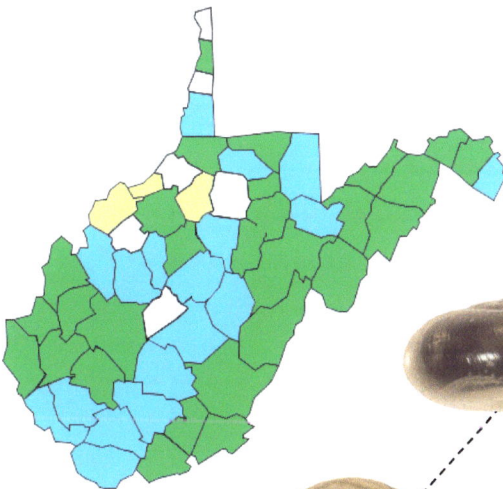

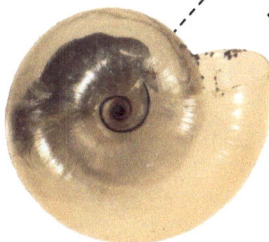

Fresh-dead *G. wheatleyi* showing translucent shell, Ravens Run Nature Reserve, Fayette County, Kentucky

Hill glyph

Zonitidae

Glyphyalinia cumberlandiana (G. H. Clapp, 1919)

Diameter: 2.5-3.5 mm, height 1.3 mm

Description: Depressed heliciform; lip simple; shell with 4-4.5 loosely coiled whorls; umbilicate; shell exceedingly fragile, glossy, light horn-colored and semi-transparent; indented transverse striae are faint and irregularly spaced; under high magnification there is merely the faintest trace of spiral sculpture (Pilsbry 1946); deceased and dried animal seen through the translucent top view of the shell, giving the first few whorls a darker color.

Similar Species: *Glyphyalinia wheatleyi* is similar in form, but is fully 2 mm larger.

Habitat: Reported as a calciphile of rocky limestone areas in shaded mixed hardwood forests, but also a snail of acidic habitats especially in and around rock talus.

Status: G4/S4; Uncommon; reported from scattered locations across the state.

Specimen: Kentucky, Powell County, Furnace Mountain (author's collection).

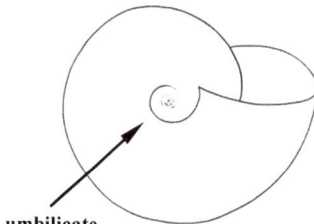

umbilicate

Pilsbry, 1946

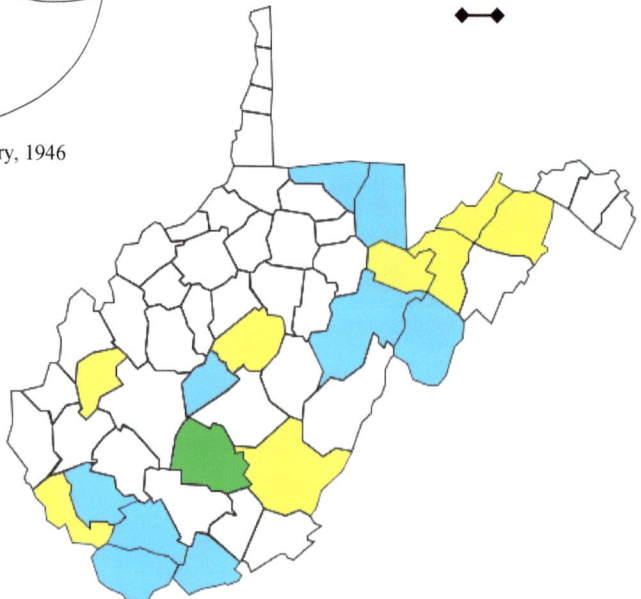

Hollow glyph

Zonitidae

Glyphyalinia specus Hubricht, 1965

Diameter: 4.8 mm, height 1.3 mm

Description: Depressed heliciform; lip simple; shell with 4.5 whorls, shell subhyaline, shiny; widely umbilicate; no teeth present; the indented transverse striae are rather well developed and very closely spaced (a); the spiral striae are only weakly defined, but often obsolete in many specimens or populations and in aged shells (dissecting scope required to view this fine feature); periphery is well rounded; the live animal is white and without functional eyes (Hubricht 1965).

Similar Species: *Glyphyalinia specus* is most closely related to *G. lewisiana*, differing principally in its larger size (Hubricht 1965).

Habitat: Troglobitic; known only from the total darkness of caves where it is reported to feed on cricket scat and possibly bat guano.

Status: G4/SH; Rare; in West Virginia, this troglobitic species is known from a single site, McClung Zenith Cave in Monroe County.

Specimen: Tennessee, White County, Ghost River Cave (author's collection).

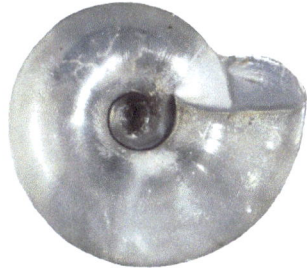

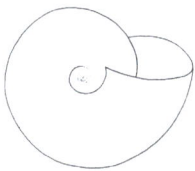

Pilsbry, 1946

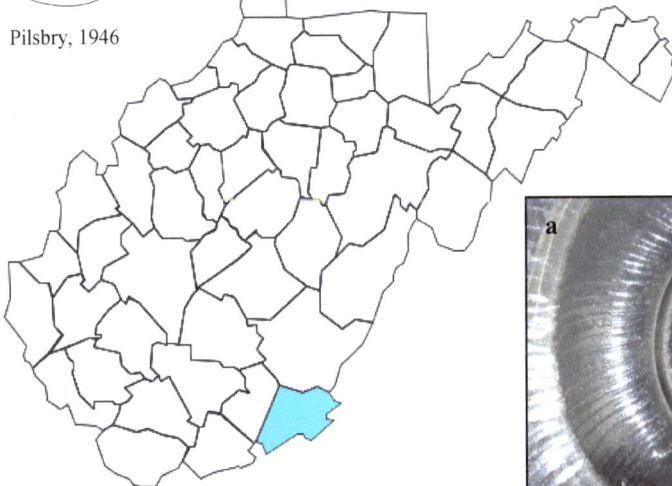

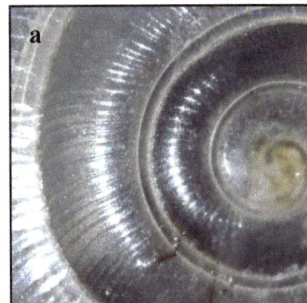

a

135

Pale glyph

Zonitidae

Glyphyalinia lewisiana (G. H. Clapp, 1908)

Diameter: 2.8-3.5 mm, height 1.5 mm

Description: Depressed heliciform; lip simple; shell with 3.5-4 loosely coiled whorls, the last whorl greatly expanded; shell very glossy, a yellowish-white; widely umbilicate, about one-fifth to one-sixth the shell diameter; no teeth present; the delicate indented transverse striae are rather well developed and very closely spaced (a) on top but mostly absent from the base; the spiral striae may be weakly defined, but usually a constant feature under a strong lens; periphery is well-rounded.

Similar Species: *Glyphyalinia specus* is larger by around 1.5 millimeters and is reported to be a strict obligate to limestone caves.

Habitat: A calciphile species that burrows; usually found at the base of limestone outcrops and under rocks that provide the snail ample space to crawl; occasionally carried into limestone caves during storm events.

Status: G4/S2; Uncommon; reported from scattered locations across West Virginia.

Specimen: Kentucky, Edmonson County, Mammoth Cave NP (FM 240450).

Pilsbry, 1946

a

Maryland glyph

Glyphyalinia raderi (Dall, 1898)

Diameter: 3.9-4 mm, height 1.5 mm

Description: Depressed heliciform; lip simple; shell with 4.5 loosely coiled whorls; shell pale waxen white; widely umbilicate (one of the widest of any West Virginia *Glyphyalinia*), about one-forth the shell diameter; aperture turning downward but evidently this feature varies among specimens; no teeth present; <u>indented transverse striae are weakly developed, regularly and very closely spaced</u> (a); periphery is rounded.

Similar Species: The aperture of *G. lewisiana* does not curve downward as it does in *G. raderi* (in front view), has a notably smaller umbilicus and more strongly developed transverse striae; *G. wheatleyi* is larger and has more widely spaced indented striae.

Habitat: A calciphile and possibly a burrower, usually amongst rocks (Hubricht 1985).

Status: G2/SH; Rare; although reported in West Virginia from nine counties, Hubricht (1985) has no records of it in WV; other extant populations are known only from single locations in Kentucky, Virginia and Maryland; WV specimens remain highly questionable and are in need of closer examination.

Specimen: Virginia, Alleghany County, nine miles NNE of Covington (FM 240955).

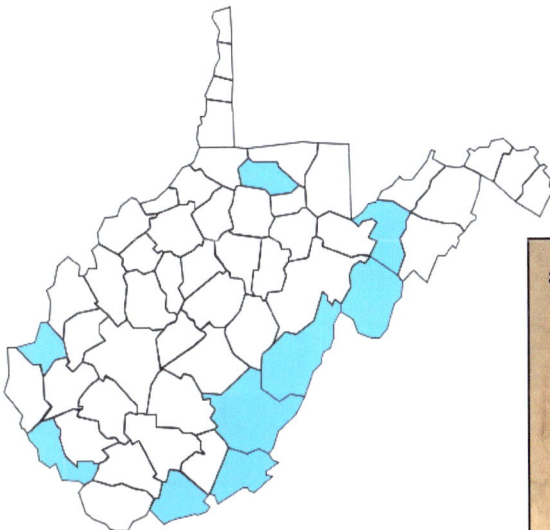

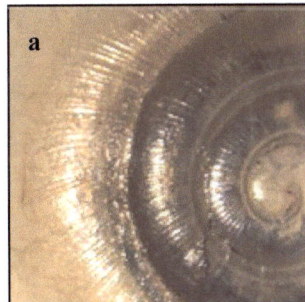

Zonitidae

a

Depressed glyph

Zonitidae

Glyphyalinia virginica (Morrison, 1937)

Diameter: 4.6-5.3 mm, height 2.1 mm

Description: Depressed heliciform; the spire lower than any other *Glyphyalinia* species in West Virginia, in some specimens, approaching a plane; lip simple; shell with 5-6 loosely coiled whorls; shell pinkish-horn colored; in figure (a) the deceased and dried animal can be seen through the translucent shell; adult shells are extremely fragile and easily broken during handling; widely umbilicate (about one-quarter the shell diameter); no teeth present; indented transverse striae are rather closely, but irregularly spaced (a), with minute spiral striae above and below; periphery is rounded.

Similar Species: This is the flattest *Glyphyalinia* species in West Virginia; similar to *G. wheatleyi,* but slightly smaller, having a finer sculpture and flatter build.

Habitat: Found in pockets of deep leaf litter on mountainsides (Hubricht 1985).

Status: G3/SH; uncommon; in West Virginia the species is reported from nine eastern counties.

Specimen: West Virginia, Pendleton County, marsh, NE Franklin (FM 240503).

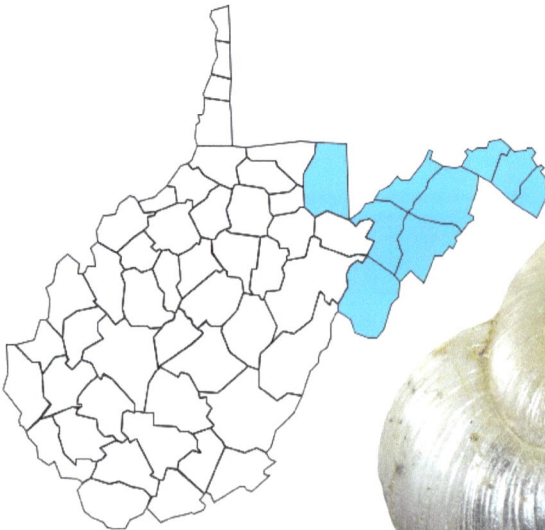

a

Shells 2-8 mm, teeth typically present, simple lip, the whorls are tightly coiled (*Paravitrea*)

Paravitrea species are undoubtedly one of the most difficult groups of land snails to differentiate. Nevertheless, most are under 6 mm (a few slightly larger), have more than 5 or 6 tightly coiled whorls, the last whorl not greatly expanded (a) and most species at some stage of development are with internal armature; *Glyphyalinia* species have loosely coiled whorls, the last whorl greatly expanded (b) and are without teeth. *Paravitrea* have either teeth (c and d) or lamellae (e). Without a good series of *Paravitrea* shells at various stages of growth, taxon recognition is difficult. Immature specimens are essential for the identification of species in the genus. If adult *Paravitrea* species are encountered during collections, it is imperative that leaf litter samples (always a good idea anyway) also be secured at the same locations. This will likely insure that young *Paravitrea* specimens are included during the survey. Many *Paravitrea* snails contain teeth at the juvenile shell-stage that are absorbed (for the calcium content) in the adult shell-stage (see next page, figures f, g, h and i). *Paravitrea* shells in this book are arranged in three groups: (1) juvenile and adult shells without teeth or lamella barriers, (2) juvenile shells with teeth or lamella barriers but as adults are usually without these protective structures and (3) both juvenile and adult shells contain teeth or lamella barriers.

Grouping 1: juvenile & adult shells without teeth or lamella barriers, pg 141-143

Paravitrea ceres (rare)
Paravitrea capsella (common)

Grouping 2: juvenile shells with teeth or lamella, adults without, pg 144-145

Paravitrea bellona (rare)
Paravitrea pontis

Grouping 3: juvenile and adult shells with teeth or lamella barriers, pg 146-149

Paravitrea reesei
Paravitrea subtilis
Paravitrea multidentata (very common)
Paravitrea seradens

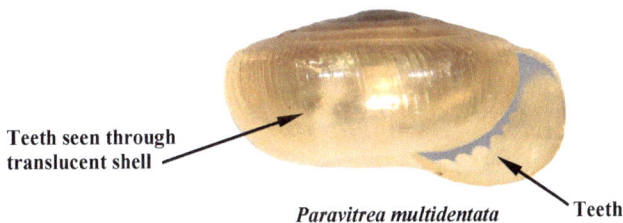

Teeth seen through translucent shell

Paravitrea multidentata   Teeth

Shell Morphology & Armature of *Paravitrea* Species

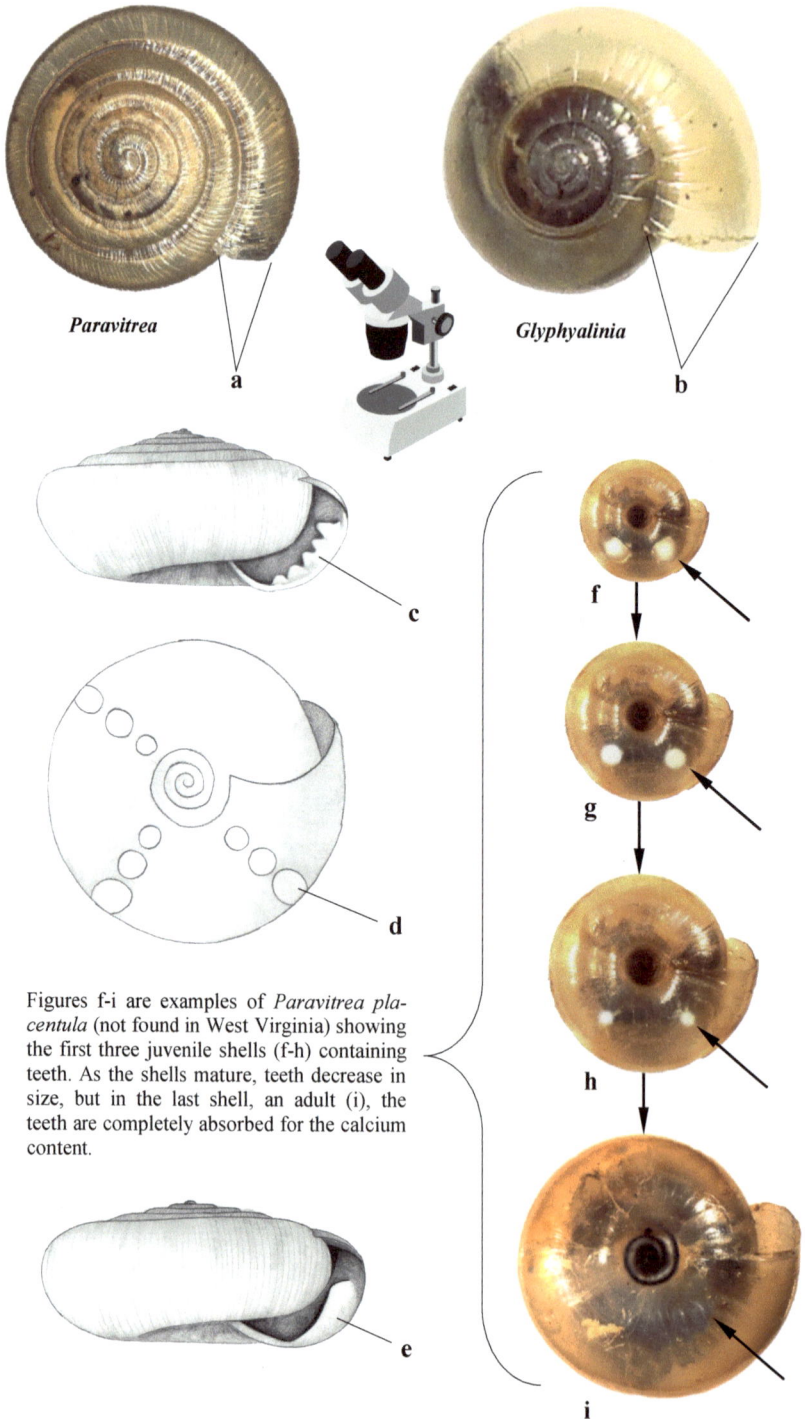

Paravitrea

a

Glyphyalinia

b

c

d

Figures f-i are examples of *Paravitrea placentula* (not found in West Virginia) showing
the first three juvenile shells (f-h) containing
teeth. As the shells mature, teeth decrease in
size, but in the last shell, an adult (i), the
teeth are completely absorbed for the calcium
content.

e

f

g

h

i

Sidelong supercoil

Zonitidae

Paravitrea ceres Hubricht, 1978

Diameter: 4.3 mm, height 2 mm

Description: Depressed heliciform, spire very low, and in some specimens, nearly discoidal; lip simple; shell with 6 tightly coiled whorls, the last whorl notably expanded (a); shell pale brownish, subhyaline, shiny; live animal with a bluish-gray foot; umbilicate; umbilicus diameter 0.8 mm; transverse striae closely but irregularly spaced, slightly wider on the last whorl, distinct above, but becoming obsolete below; adults without teeth; evidently young shells of this rare species have never been observed, all shells contained in the type collection by Hubricht have no less than 5 whorls (Hubricht 1978).

Similar Species: *Paravitrea pontis* is larger, has a wider umbilicus, more translucent shell, and has internal armature in very young shells of less than 3 whorls; *P. petrophila* has a more elevated shell and wider umbilicus.

Habitat: Found in pockets of moist leaf litter on a wooded hillside (Hubricht, 1985).

Status: G1/SH; Rare; Endemic to West Virginia; precious little is known of this highly restricted land snail, specifically if young shells carry internal barriers.

Specimen: West Virginia, Pocahontas County, Buckeye (FM 249226 Paratype).

? See page 142

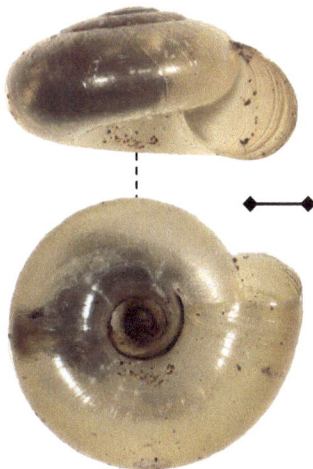

Paravitrea petrophila, Tate Gap, Macon County, North Carolina (authors collection).

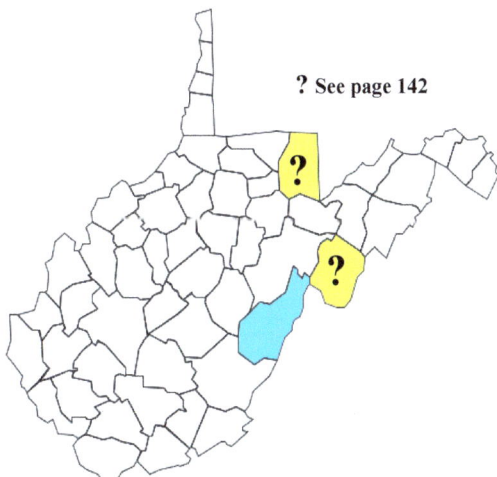

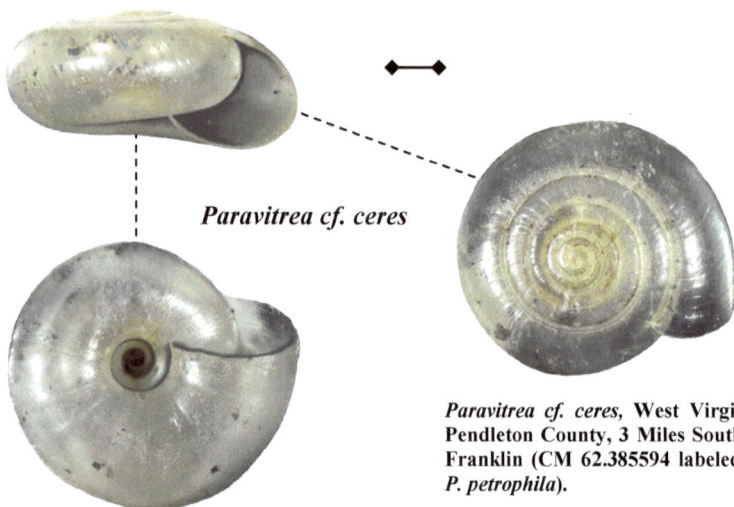

Paravitrea cf. ceres

Paravitrea cf. ceres, **West Virginia, Pendleton County, 3 Miles South of Franklin (CM 62.385594 labeled as P. petrophila).**

The specimen above from CMNH labeled as *P. petrophila* (MacMillan 1940) and the specimens below from Camp Dawson stand closer in build and size to *P. ceres* than to *P. petrophila*. Not known to MacMillan at the time, *P. ceres* was described by Hubricht in 1978. Moreover, *P. petrophila* is a species of the Cumberland Plateau and mountains of Kentucky, Tennessee and North Carolina, with a disjunct population in Oklahoma and western Arkansas, (Hubricht 1985). Specimens illustrated on this page (including *P. ceres*, opposite page), differ from *P. petrophila* (page 141, bottom two figures) by being slightly smaller, flatter, with one more whorl, having traverse striae more closely set, a different color and a notably smaller umbilicus. It is therefore the opinion of the author that shell collections in West Virginia labeled as *P. petrophila* are likely *P. ceres*, but further studies are warranted for a final deliberation. Furthermore, juvenile shells of 3 whorls (1.75 mm) from Camp Dawson are without internal armature, which, if *P. ceres*, suggests that the species is without internal armature in early stages of growth..

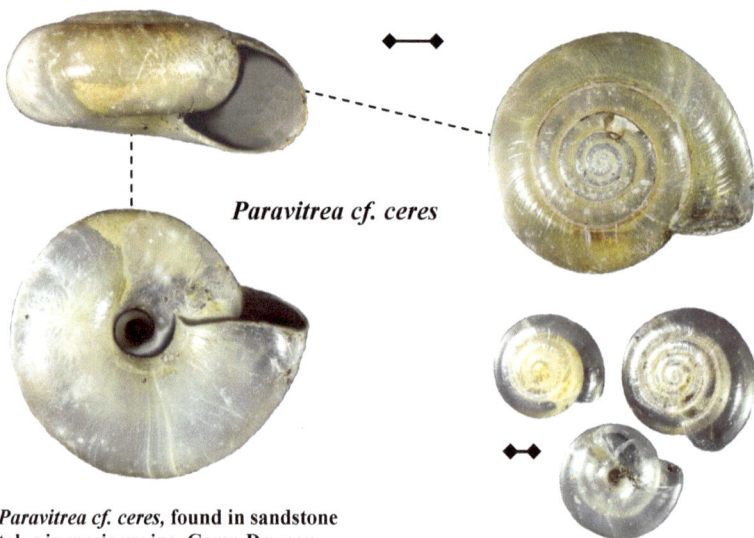

Paravitrea cf. ceres

Paravitrea cf. ceres, **found in sandstone talus in mesic ravine, Camp Dawson, Preston County, West Virginia (author's collection)**

Juvenile shells are without internal armature, Camp Dawson

Dimple supercoil

Paravitrea capsella (Gould, 1851)

Diameter: 4.8-6.2 mm, height 3 mm

Description: Depressed heliciform; lip simple, roundish; shell with 6.5-7.5 tightly coiled whorls; umbilicate; aperture roundish; amber color; live animal pale-slate colored; glossy and translucent; transverse striae moderately developed on top of shell, but nearly smooth on the base; without teeth in adults; paired teeth very rarely seen in juvenile shells; deceased and dried animal seen through the translucent bottom and top views of the shell giving it a darker color in places (a).

Similar Species: *Paravitrea placentula* (Part II, page 367) is larger and has paired teeth in young shells; *P. ceres* has a notably wider umbilicus and a flatter build; *P. bellona* has paired teeth in juvenile shells and a more restricted range in West Virginia.

Habitat: A habitat generalist found in a wide range of mixed hardwood forests.

Status: G4/S4; Common; one of the most frequent and widespread *Paravitrea* species in the state.

Specimen: North Carolina, Graham County, Nantahala NF (author's collection).

Zonitidae

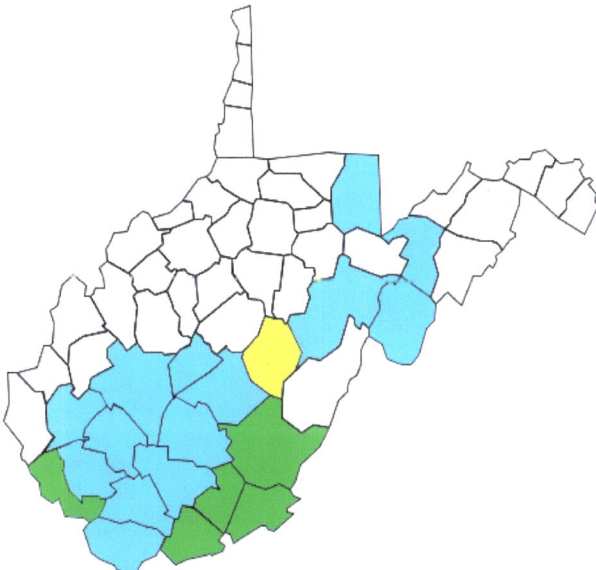

Club supercoil

Zonitidae

Paravitrea bellona Hubricht, 1978

Diameter: 5.8 mm, height 2.1 mm

Description: Depressed heliciform; lip simple; shell with 6-7 tightly coiled whorls; shell thin, glossy and translucent in fresh dead and live shells; pale brownish above; umbilicate; umbilicus diameter 1.3 mm; transverse striae closely but irregularly spaced, distinct above but becoming obsolete below; two or three pairs of teeth are present in immature shells of less than 5 whorls (a), but are absent in adults, these teeth seen through the bottom of fresh shells; live animal with a bluish-gray foot.

Similar Species: *Paravitrea bellona* is most like *P. capsella* differing in having at least one pair of teeth in juvenile shells and having a smaller umbilicus, it differs from *P. seradens* by the absence of teeth in adult shells (or shells under 5 whorls).

Habitat: Found in pockets of moist leaf litter on wooded river bluffs and in ravines.

Status: G1/S1; Rare; Endemic to West Virginia; although specimens from Clay, Randolph and Greenbrier counties strongly refer to *P. bellona*, further collections and a closer examination are recommended in making a final determination.

Specimen: West Virginia, Logan County, Horsepen Mountain 4.5 miles south of Sarah Ann (FM 249562).

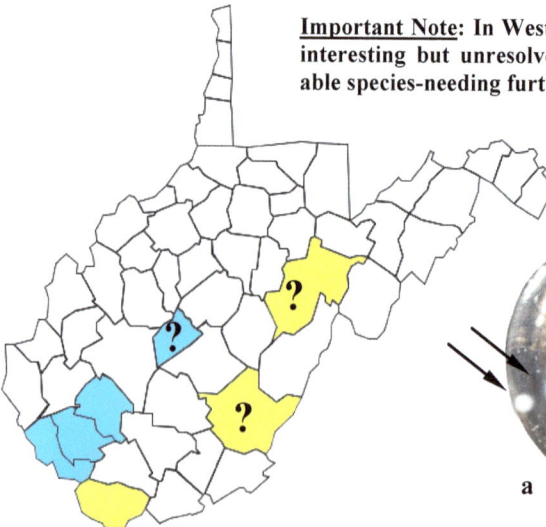

Important Note: In West Virginia *Paravitrea* are an interesting but unresolved complex of highly variable species-needing further study!

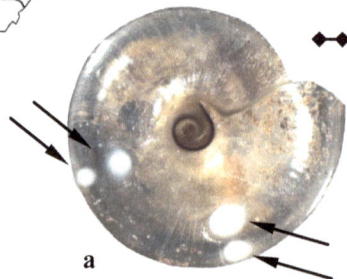

a

Natural Bridge supercoil

Paravitrea pontis Baker,1931

Diameter: 5.12-6.5 mm, height 2.8 mm

Description: Depressed heliciform; lip simple; shell with 6-7 tightly coiled whorls, except for the last which is the widest and quite suddenly expanded; shell thin, hyaline; umbilicate; umbilicus wide, deep and well-like exhibiting all the whorls; transverse striae numerous and closely but irregularly spaced, distinct above but becoming weaker below; the internal armature (only present in shells of 2-3 whorls and diameter of 1.5 mm) consisting of smooth, radial barriers extending through the middle half of the periphery in final whorl (a & b); deceased and dried animal visible through shell.

Similar Species: *Paravitrea ceres* has a similar build, but is smaller by 1.5-2 mm, has a narrower umbilicus, and is believed to be without internal armature in all stages of growth.

Habitat: A calciphile found in pockets of moist leaf litter in wooded ravines and on hillsides; the type locality was in limestone talus in a canyon just above the base of Natural Bridge, Rockbridge County, Virginia.

Status: G3/S2; Uncommon; a species with scattered locations in West Virginia.

Specimen: West Virginia, Pendleton County, 3.4 miles NW Onego (FM 249591).

Canaan Valley, Tucker Co, WV

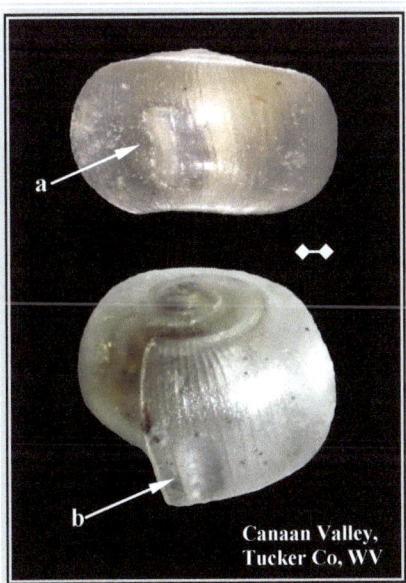

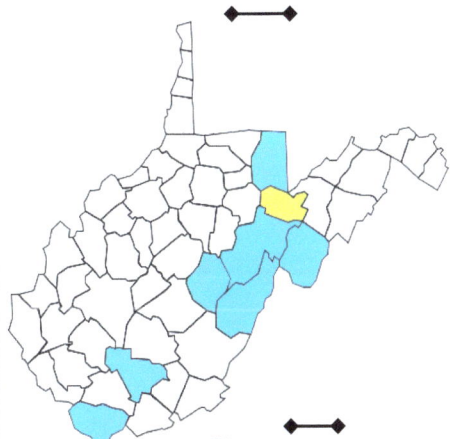

A nearly adult shell, Canaan Valley, Tucker County, WV

Round supercoil

Zonitidae

Paravitrea reesei Morrison, 1937

Diameter: 3.5-4.7 mm, height 1.6 mm

Description: Depressed heliciform; lip simple; shell with 5-6.5 tightly coiled whorls, the last whorl not expanded; umbilicate, the umbilicus deep and well-like; shell pale-amber; glossy; aperture lunate; indented transverse striae are closely but irregularly spaced; spiral sculpture indistinct above and below; in adult shells there are rows of 3 rather large teeth (a), these rows of teeth regularly spaced and visible through the base of fresh shells (Burch 1962); young shells of 2-2.5 whorls have 1 or 2 sets of paired conical teeth (b), the third tooth appearing as shells mature (c); teeth are usually without the callus bridge seen in *P. mira*, a species of northwestern Virginia and eastern Kentucky (page 363).

Similar Species: *Paravitrea capsella* is larger and is typically without teeth at all stages of growth.

Habitat: Found under moist leaf litter, detritus, and rocks on wooded hillsides, ravines, and river bluffs of mixed hardwood forests.

Status: G3/S2; Uncommon; in West Virginia this species is reported from scattered locations.

Specimen: Virginia, Pulaski County, opposite Radford (FM 249118).

Tooth being absorbed

a

b

c

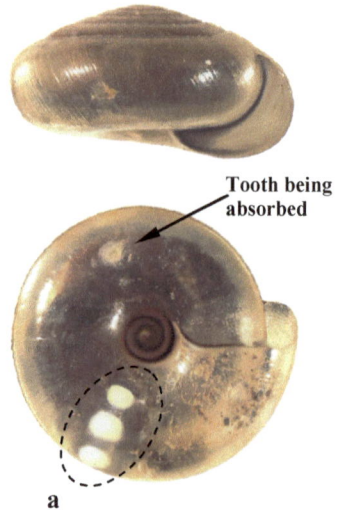

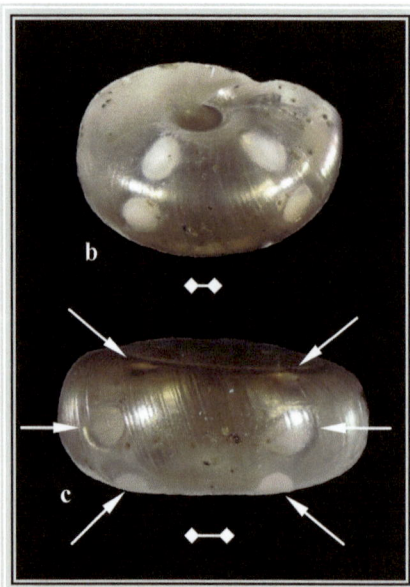

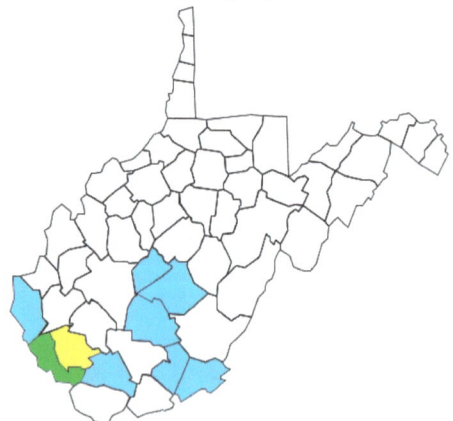

Figure (b) a juvenile shell showing the paired teeth and figure (c) a slightly older shell with the top-developing third tooth, Hatfield Cemetery, Logan County, West Virginia

146

Slender supercoil

Zonitidae

Paravitrea subtilis Hubricht, 1978

Diameter: 2.9 mm, height 1.4 mm

Description: Depressed heliciform; lip simple; shell with 6.3 tightly coiled whorls; pale reddish-brown; glossy; umbilicate; aperture squared; indented transverse striae are closely but somewhat irregularly spaced, becoming more widely spaced and irregular on the last whorl; 1 to 2 radial rows of 4 to 6 teeth in the last whorl which are coalesced in some specimens to form a lamella, found in both juvenile and adult shells (Hubricht 1978).

Similar Species: *Paravitrea multidentata* is most similar, but has indented transverse striae more closely set and evenly spaced, especially on the last 2 whorls (a); *P. dentilla* of nearby states (page 366), is 3-4 mm larger and adult shells are without teeth.

Habitat: Found under moist leaf litter and detritus on wooded hillsides and in ravines of mixed hardwood forests.

Status: G2/S1; Rare; in West Virginia this scarce species is reported from two southern counties.

Specimen: Kentucky, Harlan County, Pine Mountain north side of mountain 2 miles Southeast of Bledsoe (FM 248709).

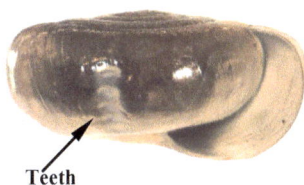

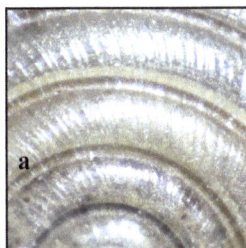

Teeth

Paravitrea multidentata
(top view)

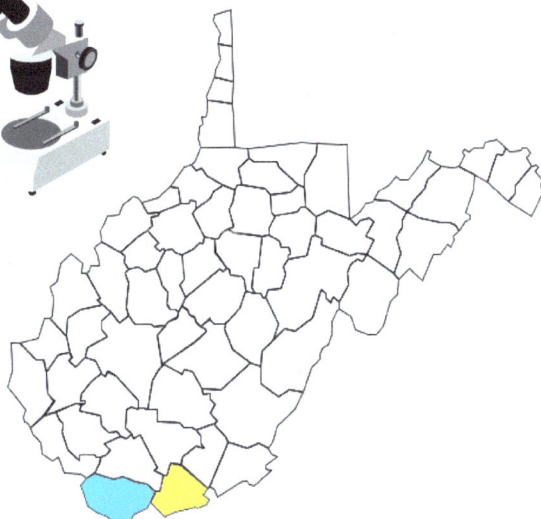

147

Dentate supercoil

Zonitidae

Paravitrea multidentata (A. Binney, 1840)

Diameter: 2.5-3 mm, height 1.3 mm

Description: Depressed heliciform; lip simple, shell with 6 tightly coiled whorls, fully adult shells are umbilicate, while young shells are perforate; last whorl not expanding; shell thin, glossy and translucent; transverse striae are poorly developed, barely visible, <u>very closely and evenly spaced</u>; without spiral striae; two to four radial rows of usually five teeth that can be seen through the bottom of fresh shells of both juveniles and adults; on occasion will have rows of lamellae, figure (a); fully adult shells have a faint callus (b) connecting the end of the lip and a slight reflection in the peristome (c).

Similar Species: *P. lamellidens* of nearby states is slightly larger, has lamellae instead of teeth, a notably smaller umbilicus, and more well developed transverse striae (see page 364).

Habitat: Found in pockets of moist leaf litter on hillsides in mixed hardwood forests; especially common in wet leaf litter surrounding small headwater seeps.

Status: G5/S5; Common; one of the most widespread *Paravitrea* species in the state.

Specimen: West Virginia, Upshur County, Stonecoal WMA (author's collection).

Teeth

Teeth

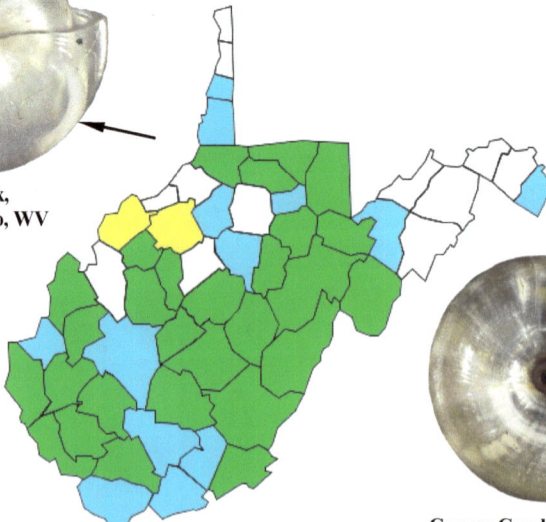

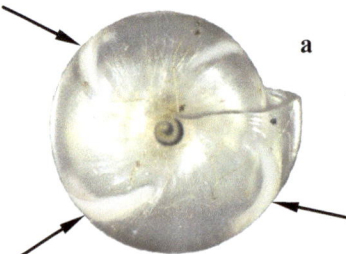

a

Shavers Fork,
Randolph Co, WV

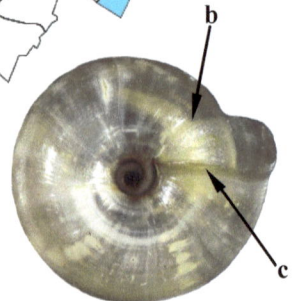

b

c

Groves Creek, Clay County, WV

Barred supercoil

Zonitidae

Paravitrea seradens Hubricht, 1972

Diameter: 5.5 mm, 2.8 mm

Description: Depressed heliciform; lip simple; shell with 6-7 tightly coiled whorls; shell thin, glossy and translucent, sometimes with a hint of yellow in fresh and live shells; umbilicate, umbilicus deep and well-like, exhibiting all the whorls; transverse striae numerous and irregularly spaced; whorls slowly expanding, last whorl expanding more rapidly and deflecting downward to the aperture; internal armature of one to three pairs of rather large teeth on the outer and basal walls (a) at all stages of growth, however these teeth occasionally missing in very old shells; these teeth can be seen through the bottom of fresh shells of both juveniles and adults (Hubricht 1972).

Similar Species: *Paravitrea seradens* is most like *P. capsella* differing in having at least one pair of teeth at all stages of growth and having a larger umbilicus in immature shells.

Habitat: Like many *Paravitrea* species, found in pockets of moist leaf litter located in mixed hardwood forests on hillsides and in ravines.

Status: G3/S2; Uncommon; most of the records of this infrequent land snail are from West Virginia.

Specimen: West Virginia, Mingo County, Guyandotte River, opposite Justice (FM 249218).

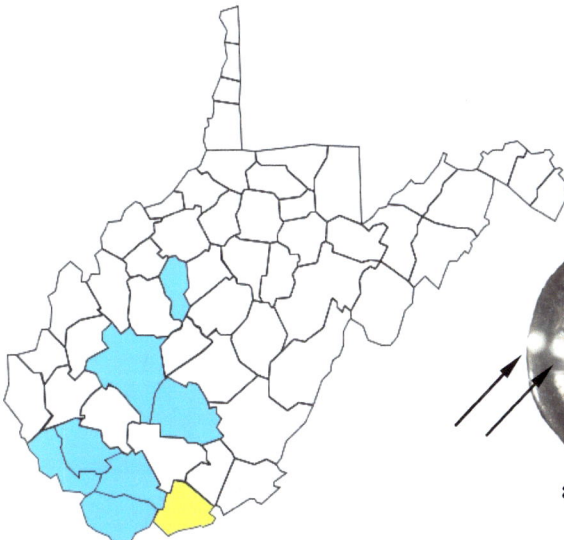

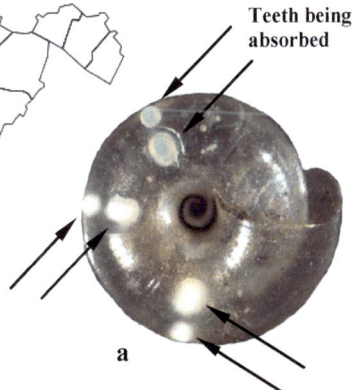

Teeth being absorbed

a

Shells 6-8 mm, simple lip, shell finely ribbed, <u>without teeth</u>, defense mucus fluorescent under UV light (*Discus*)

Discus species are depressed heliciform, less tightly coiled than *Helicodiscus* and do not feature color bands or blotches on shells as in *Anguispira*. *Discus* snails are usually without internal teeth, although *D. patulus* has a small callus tooth a short distance within its shell. *Discus* species have an entirely different shell sculpture than *Helicodiscus*, having transverse riblets (a), instead of spiral sculpture or fringed ornamentation (b). Like *Anguispira,* the defensive mucus or slime of *Discus* is fluorescent under UV light (Dourson 2013). *Discus* are generally found associated with rotting hardwood in advanced stages of decay and are sometimes found in large numbers.

Shells 6-8 mm, ribbed usually without teeth, pg 151-153

Discus patulus
Discus catskillensis
Discus whitneyi

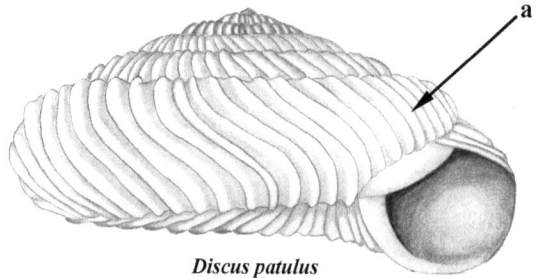

Discus patulus

Shells 3-5 mm, simple lip, raised spiral sculpture, <u>with teeth</u>, defense mucus fluorescent under UV light (*Helicodiscus*)

Helicodiscus are flattened or discoidal and under a hand lens of 10X, have raised spiral striae, fringes or hairs. All species in West Virginia have internal teeth (c) which can usually be seen in frontal view or through the bottom of fresh dead or live shells. Only the defense mucus is florescent, not the crawling slime. *Helicodiscus* species have some of the most extraordinary micro-sculpture of any land snails found in North America (see page 372).

Shells 3-5 mm, spiral striae strongly developed, always with teeth, pg 154-161

Helicodiscus shimeki
Helicodiscus notius
Helicodiscus parallelus
Helicodiscus villosus, new species (rare)
Helicodiscus triodus (rare)

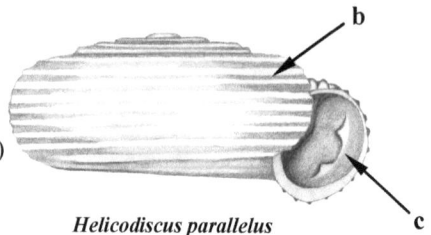

Helicodiscus parallelus

Domed disc

Discidae

Discus patulus (Deshayes, 1830)

Diameter: 7-8 mm, height 3.3-4 mm

Description: Depressed heliciform; lip simple; aperture roundish to oval shape; shell with 5.5 whorls; widely umbilicate; there is a small rounded tubercle or callous tooth a short distance within the aperture (a) but this may be a wanting feature in some specimens; transverse striae are well developed, rib-like (3-4 ribs per mm) and are distinct on top, sides and base of shell; periphery rounded.

Similar Species: Other *Discus* species in West Virginia are without an internal parietal tooth; *D. nigrimontanus* found in nearby states (page 376) has a smaller aperture, is flatter in profile, and has an angular periphery.

Habitat: A species of upper slopes in mixed hardwood under leaf litter; it is most common on or under the exfoliating bark of rotting hardwood logs in advanced stages of decay.

Status: G5/S5; Common; the most common *Discus* species in West Virginia and likely found in every single county.

Specimen: North Carolina, Macon County, Nantahala National Forest (author's collection).

a

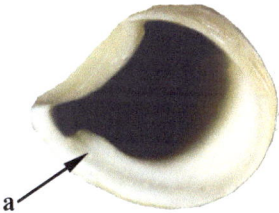

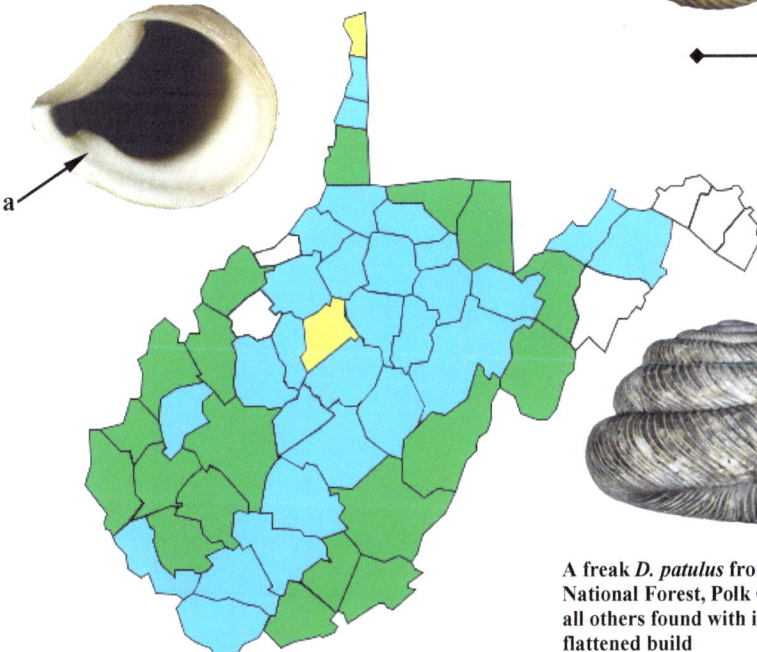

A freak *D. patulus* from the Cherokee National Forest, Polk County Tennessee; all others found with it had the typical flattened build

Angular disc

Discidae

Discus catskillensis (Pilsbry, 1896)

Diameter: 5 mm, height 2.5 mm

Description: Depressed heliciform; lip simple; shell thin with 4 whorls; umbilicate; no teeth present; transverse striae are well-developed and are rib-like (a) and are distinct on the entire shell including the umbilicus region; periphery bluntly angular.

Similar Species: *Discus patulus* is larger, has a small rounded tubercle (figure a, pg. 151) and generally a wider umbilicus; *D. whitneyi* is larger and has a more rounded aperture; *D. nigrimontanus* (page 376) has a smaller aperture, is flatter in profile and has an angular periphery (not bluntly angular as in *D. catskillensis*).

Habitat: In New York, it often occurs in low ground in company with *D. whitneyi*. A species of upper slopes in mixed hardwood under leaf litter and rotting hardwood logs in advanced stages of decay.

Status: G5/S2; Rare; in West Virginia reported from only two counties.

Specimen: West Virginia, Tucker County, Canaan Valley (author's collection).

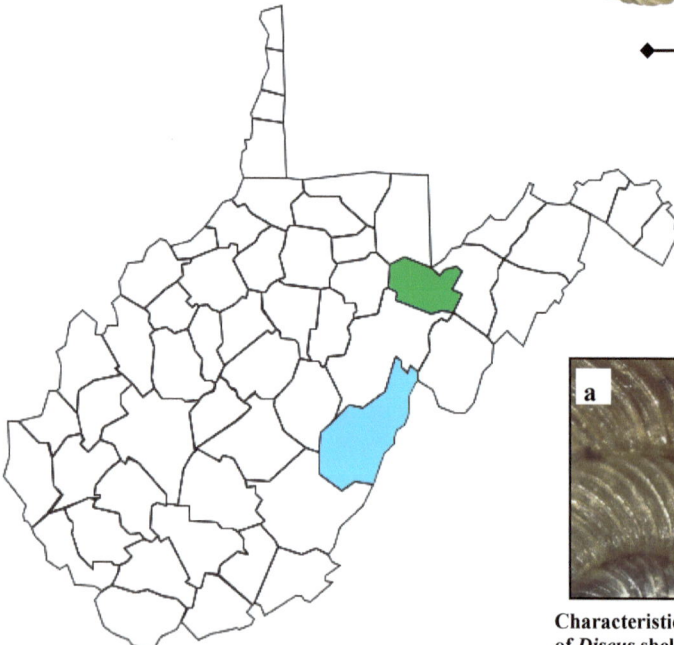

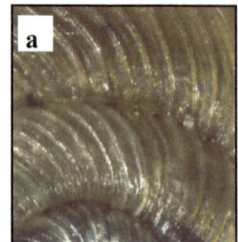

Characteristic rib sculpture of *Discus* shells

152

Forest disc

Discidae

Discus whitneyi (Newcomb, 1864)

Diameter: 5-7 mm, height 3 mm

Description: Depressed heliciform; lip simple; shell thin with 3.5-4.5 whorls; umbilicate; no teeth present; transverse striae are well developed, rib-like and nearly as well defined on base of shell around the umbilicus; periphery usually well rounded.

Similar Species: *Discus patulus* is slightly larger, has a thicker shell, egg shaped aperture, and has a small callous tooth; *D. nigrimontanus* (page 376) has a smaller aperture, is less compact, being flatter in profile, and has an angular periphery (not rounded).

Habitat: A species of rather low damp soils around open grassy places, roadsides, and edges of wetlands; also a frequent snail in untidy urban areas.

Status: G5/S2; Uncommon; this mostly northern species is known from scattered counties across West Virginia; previously known as *D. cronkhitei,* but Roth (1988) demonstrated that *D. whitneyi* is a senior synonym of *D. cronkhitei.*

Specimen: Kentucky, Estill County, along railroad tracks in the town of Irvine (author's collection).

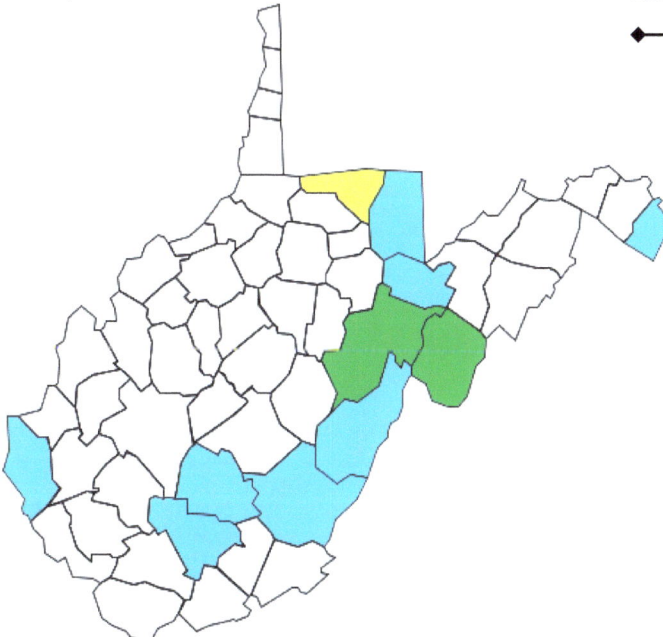

Temperate coil

Helicodiscidae

Helicodiscus shimeki Hubricht, 1962

Diameter: 4.2 mm, height 1 mm

Description: Depressed heliciform-discoidal; lip simple; shell with 5-6 well rounded whorls; fresh shells somewhat shiny and translucent; widely umbilicate; embryonic whorl smooth; in the last whorl there are usually three pairs of small conical teeth on the outer and basal walls (seen in bottom view of juvenile shells below, figures a and b); spiral striae clearly present as in most *Helicodiscus* species on all whorls (dissecting scope required).

Similar Species: Both *Helicodiscus parallelus* and *H. notius* have a narrower and deeper umbilicus.

Habitat: A land snail found in upland mixed hardwood forests living among the leaf litter on acid soils, also found in low places and in rock talus.

Status: G4/S3; Uncommon; in West Virginia reported from eight counties.

Specimen: West Virginia, Monongalia County, Snake Hill WMA (author's collection).

Teeth

Teeth

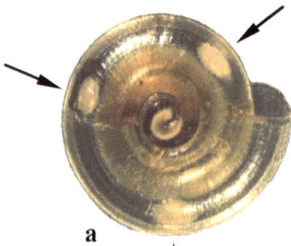

a

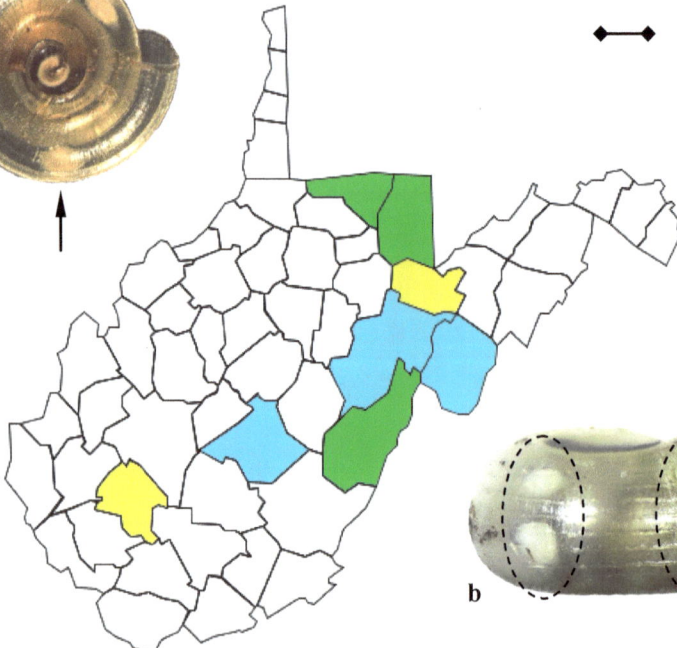

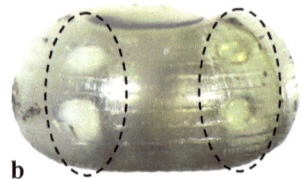

b

Tight coil   Helicodiscidae

Helicodiscus notius Hubricht, 1962

Diameter: 3.6 mm, height 1.2 mm

Description: Depressed heliciform-discoidal; lip simple; shell with 5-5.5 whorls; widely umbilicate; within the last whorl of juvenile and adult shells there are 2 to 3 pairs of conical teeth which can be seen through the bottom of fresh shells; the raised spiral striae are well developed including the ones on the embryonic whorl (a).

Similar Species: *Helicodiscus parallelus* has poorly developed spiral striae on the embryonic whorl; adult *H. notius specus,* a troglobitic land snail reported from Kentucky and eastern Tennessee cave systems close to West Virginia, is without internal armature.

Habitat: A species found on forested hillsides, in ravines and occasionally washed into caves; usually found in drier and higher habitats than *H. parallelus*.

Status: G5/S5; Common; in West Virginia known from scattered locations across the state.

Specimen: Tennessee, Warren County, Storage Cave (author's collection).

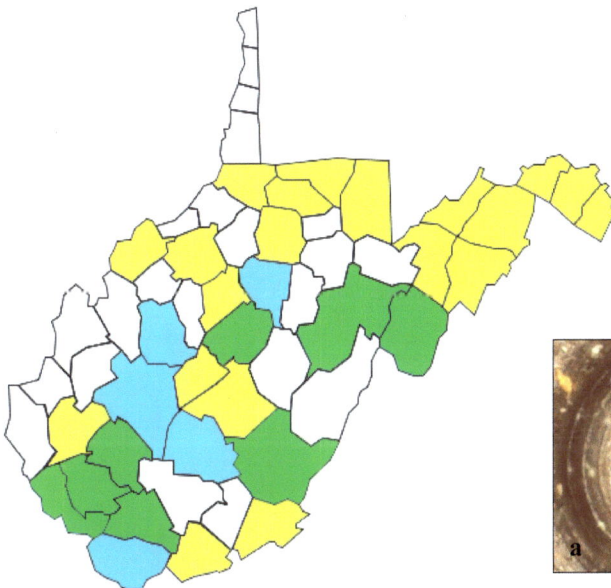

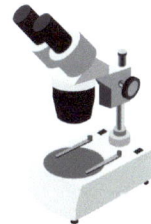

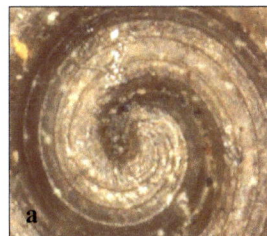

Top view

155

Compound coil

Helicodiscidae

Helicodiscus parallelus (Say, 1817)

Diameter: 3.2-3.5 mm, height 1.2 mm

Description: Depressed heliciform-discoidal; lip simple; shell with 4-4.5 whorls; widely umbilicate; paired teeth present usually deep within the aperture, can be seen in fresh shells (a); spiral striae are well developed except on the embryonic whorl where there is only a hint of spiral striae (b); periphery rounded.

Similar Species: *Helicodiscus notius* is larger by one whorl, has a slightly wider umbilicus, and most importantly, the spiral striae on the embryonic whorl are more strongly developed, this being a variable characteristic, however.

Habitat: Found in floodplains and upland mixed hardwoods under leaf litter and around rock structures; also in urban areas under wood debris and around vacant lots.

Status: G5/S5; Common; the most common *Helicodiscus* in the state with many records.

Specimen: Kentucky, Powell County, Star Gap, Red River Gorge (authors collection).

a

Shavers Fork, Randolph County, West Virginia

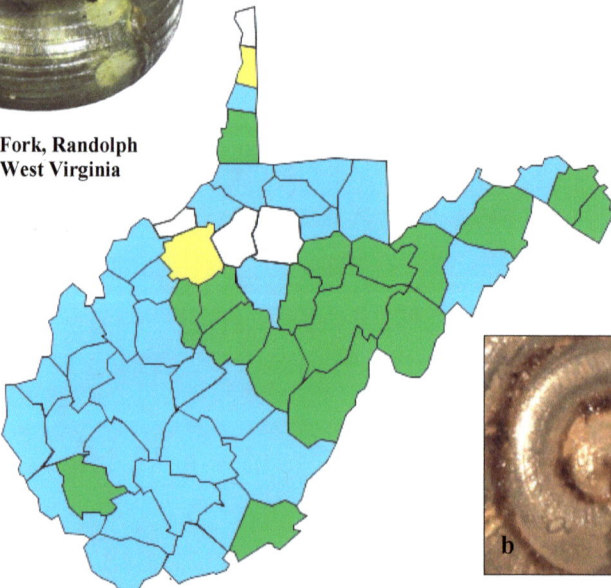

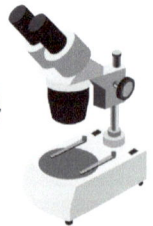

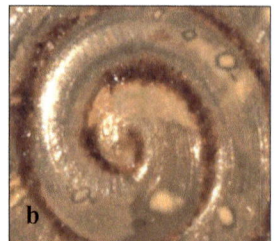

b

Top view

156

Helicodiscus villosus, new species
Figures a, b & c (page 158)

Shell. Mature specimens are approximately 4 mm in diameter and 1.2 mm in height. Shell depressed heliciform–discoidal. Lip simple. Shell with 4.5 whorls, widely umbilicate. Starting at the embryonic whorl, there are raised spiral striae, which in later whorls, appear as fringes from which long curving hairs extend (about 30 striae in the final whorl). While the hairs may be lost in aging shells, at least some remnant fringes are retained. Within the last whorl, a pair of small knob-like teeth form on the basal and outer wall and are nearly equal in size, but these teeth may be absent. The cupped parietal tooth is positioned close by and in front of the paired teeth and around the same size or slightly larger. However, this description is based exclusively on four shells; two adults and two juveniles.

Type Locality. A talus slope below a limestone cliff face and above a spring, close to the Greenbrier River, Greenbrier County, West Virginia. Holotype: (FMNH 344363); Paratype: (FMNH 344364), same data as the Holotype. Collected 25 July, 2013 by Jeff Hajenga, John Slapcinsky, and the author.

Distribution. *Helicodiscus villosus* is known only from the type locality, around a spring above the Greenbrier River. The species is endemic to West Virginia.

Remarks. *Helicodiscus villosus* is most similar to *H. triodus* in size and build, but differs in having only one set of teeth in the last whorl; these teeth may be absent however. Most importantly, there is a generous covering of fine hairs on the shell surface; *H. triodus* is without hairs (Hubricht 1958). *Helicodiscus diadema,* an endemic to Rockbridge and Allegheny Counties, Virginia (page 160, bottom figure and page 374), has larger teeth and fewer spiral striae on the final whorl. Other small snails found with *H. villosus* included *Paravitrea multidentata, Carychium nannodes, Gastrocopta contracta, Glyphyalinia wheatleyi, Guppya sterkii, Punctum minutissimum, Striatura meridionalis, Vertigo gouldii, Strobilops aeneus,* and *Euconulus dentatus. Helicodiscus villosus* appears to be a burrower of deep limestone talus, making it a difficult species to harvest and little is known of its biology. One of the rarest and habitat restricted land snails in West Virginia.

Etymology. Villosus means hairy in Latin.

Specimen	Diameter	Height	ApH	ApW	Whorls	Umbilicus W
Holotype (FMNH 344363)	4	1.2	.75	1.0	4.5	2.0
Paratype (FMNH 344364)	3.5	1.0	0.5	.75	4.5	1.75

Holotype Specimen and Shell Details

a) **Helicodiscus villosus,** Holotype (FMNH 344363), (4 mm) above a spring, Greenbrier County, West Virginia; b) **H. villosus** showing the fine hairs which are lost in aging shells; c) a cutaway of **H. villosus,** illustrating the spatial arrangement of the cupped parietal tooth and opposite basal and palatal teeth, a feature not seen in the frontal view of the Holotype specimen.

Greenbrier coil

Helicodiscus villosus new species

Diameter: 4 mm, height 1.2 mm

Description: Depressed heliciform-discoidal; lip simple; shell with 4.5 whorls; shell dull, ivory with a yellow tinge; widely umbilicate; starting at the embryonic whorl there are raised spiral striae, which in later whorls; appear as fringes and curving hairs (about 30 in the final whorl); hairs may be lost in older shells; within the last quarter whorl, a pair of small knob-like teeth form on the basal and outer wall and are nearly equal in size; one parietal tooth is positioned close by and in front of the paired teeth; the cupped parietal tooth is around the same size or slightly larger; this description based on two adult and two juvenile shells.

Similar Species: Most similar to *H. triodus,* but differs in having hairs and only one set of teeth in the last whorl; *H. diadema* of Virginia (page 374 and bottom figure next page) has larger teeth and fewer spiral striae.

Habitat: A burrower in limestone talus above a spring just below the main cliffline outcrops.

Status: G1/S1; Rare; Endemic to West Virginia; currently known only from the type locality, above a spring, Greenbrier Co, WV.

Specimen: West Virginia, Greenbrier County, (FM 344364 Paratype).

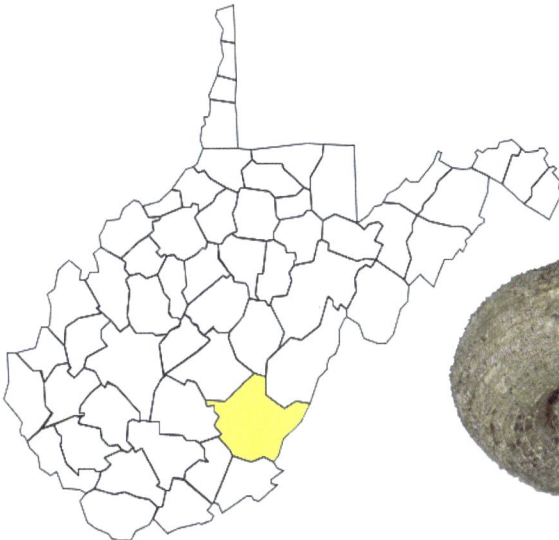

Helicodiscus Species Containing 3 Teeth Compared

Helicodiscus villosus new species, with hairs, small teeth

Helicodiscus triodus, without hairs, small teeth, page 161

Helicodiscus lirellus, without hairs, large teeth, page 373

Helicodiscus multidens, without hairs, large teeth, page 371

Helicodiscus diadema, with hairs, large teeth, page 374

Reported From West Virginia

Reported From Virginia

Talus coil

Helicodiscus triodus Hubricht, 1958

Diameter: 4-4.2 mm, height 1.5 mm

Description: Depressed heliciform-discoidal; lip simple; shell with 5-6 well rounded whorls; fresh shells somewhat shiny and translucent, dull pale corneous; sculptured with numerous spiral threads; inside the shell and occurring at irregular intervals are pairs of small conical teeth on the outer and basal walls, on the parietal wall in front of each pair of teeth is a broad flat-topped tooth, buttressed behind, a cutaway shell below showing two of these teeth (a), the cupped parietal teeth are nearly twice as broad as high, the ends turned forward; there are usually three or four groups of teeth in the last whorl (Hubricht 1958); as in all *Helicodiscus* species, the teeth in the preceding whorls are absorbed as the shell grows.

Similar Species: *Helicodiscus parallelus* and *H. notius* are without the parietal teeth seen in *Helicodiscus triodus.*

Habitat: A calciphile, found under leaves and in limestone rubble on wooded hillsides; also washed into caves (Hubricht 1985).

Status: G2/SH; Rare; All locations of this species in West Virginia are from three counties.

Specimen: West Virginia, Fayette County (FM 239104 Paratype)

H. triodus, Nekola photo collection

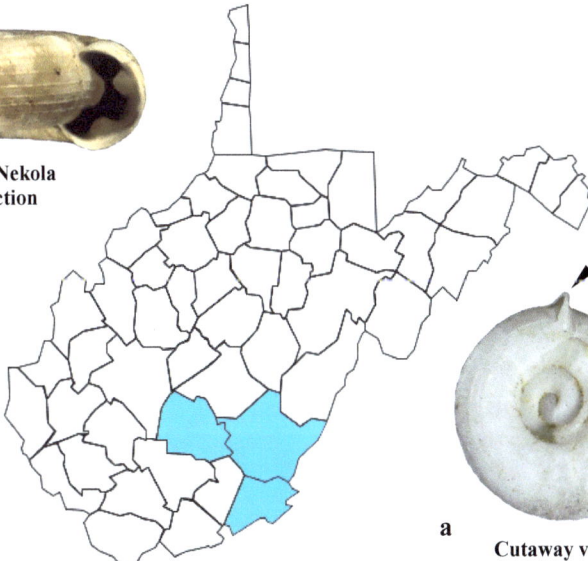

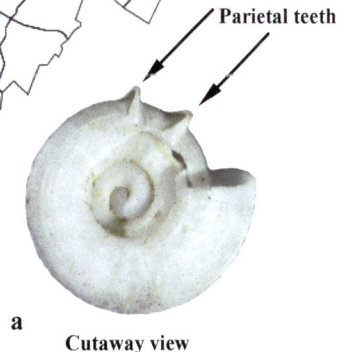

Parietal teeth

a

Cutaway view

An iridescent, snail-hunting *Cychrine* beetle dines on the flesh of an immature highlands slitmouth, *Stenotrema altispira,* Roan Mountain, Mitchell County, North Carolina.

Shells 6-15 mm, with or without teeth, simple lip (*Ventridens*)

Ventridens superficially look like miniature *Mesomphix*, but a closer look reveals that *Ventridens* have tightly-coiled shells (a), the last whorl not expanded like seen in *Mesomphix*. *Ventridens* are also more compact in form and average 2-3 more whorls than *Mesomphix*. *Ventridens* often have teeth or lamellae (b) while *Mesomphix* are without these defensive features. Like *Mesomphix* however, *Ventridens* have a thickening or thin callus, usually a whitish color, just inside the aperture's bottom (c). The shell is perforate (never completely closed) to umbilicate and in many *Ventridens* the umbilicus is proportionately wider in the youngest shells. The bidentate species of the genus *Ventridens* are characterized by the presence of 2 apertural lamellae in the pre-adult shells (Hubricht 1964). *Ventridens* in this group can be divided readily into two basic groups by the color of the live animal and the presence or absence of internal armature. In the **V. gularis** group, the live animal is dark, olive or bluish-gray (opposite page). In the **V. pilsbryi** group, the live animal is pale with a hint of yellow and perhaps some grayish flecking along the back; opposite page (Hubricht 1964). Multiple species of the *V. gularis* group are frequently found with species of the *V. pilsbryi* group. Species of the *V. gularis* group have not been found living together. Species of the *V. pilsbryi* group have not been found together with the exception of *V. lasmodon* (these observations of co-existence are according to Hubricht 1964). In the third and last group of *Ventridens*, the adult shells are without teeth. It is therefore important that the color of live animals be noted in the field (before preservation) as well as the presence of any teeth in shells collected.

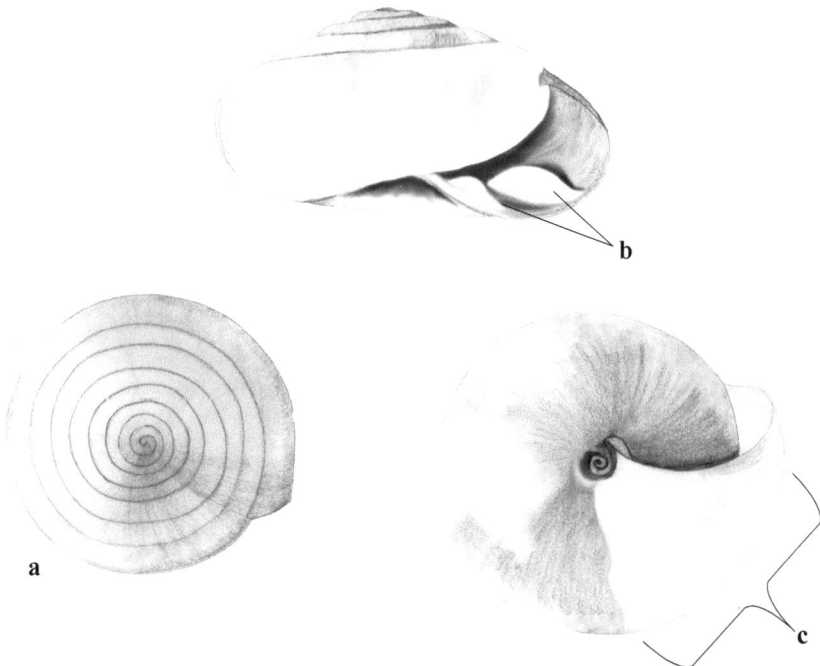

a

b

c

Grouping 1: *Ventridens gularis* group, pg 165-167: Live animal dark, olive or bluish-gray. Two or more apertural lamellae in the juvenile shell stage (see below). One, both or all of these lamellae may be reduced or wanting in mature shells of this group and shells rarely have more than 6.5 whorls. The aperture is proportionally larger than in the *V. pilsbryi* group of shells.

*Ventridens gularis
*Ventridens suppressus
*Ventridens virginicus

Animal dark

Juvenile shell

Grouping 2: *Ventridens pilsbryi* group, pg. 168-171: Animal pale, (may have a hint of yellow) with some grayish flecking along the back and sides. Like the *V. gularis* group, there are two or more apertural lamellae in the juvenile shell stage and one, both or all of these lamellae may be reduced or wanting in mature shells of this group. The diameter of the whorls usually increases more slowly and the whorls are more flattened than seen in the *V. gularis* group. *V. pilsbryi* shells may have as many as 9 whorls.

*Ventridens coelaxis (rare)
*Ventridens lasmodon (rare)
*Ventridens lawae (rare)
*Ventridens collisella

Animal light

Grouping 3: Other *Ventridens*, no teeth in the adult shells, pg 172-177

*Ventridens theloides
*Ventridens demissus
*Ventridens ligera
*Ventridens acerra (uncommon)
*Ventridens arcellus
*Ventridens intertextus

No internal armature in adult shells

Throaty dome

Gastrodontidae

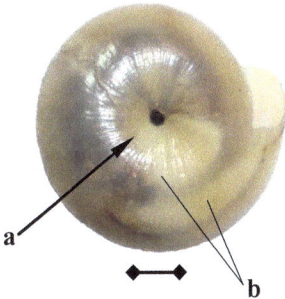

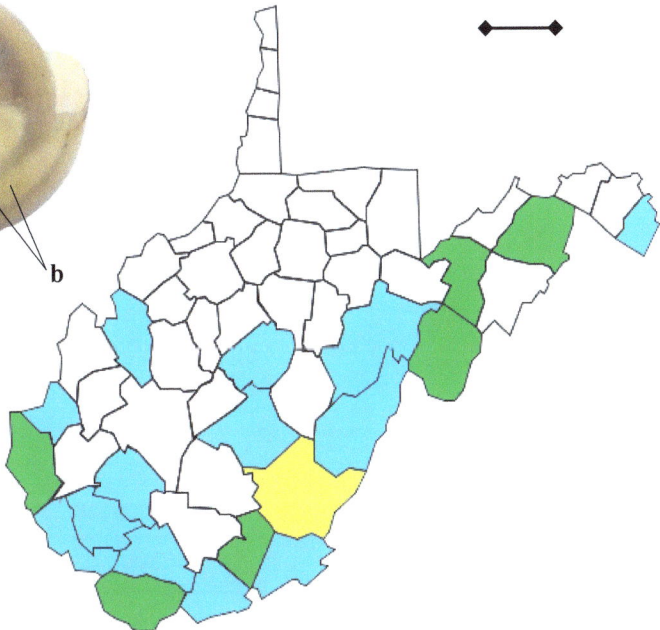

Ventridens gularis (Say, 1822)

Diameter: 7.5-9 mm, height 5-6.3 mm

Description: Depressed heliciform to heliciform, young shells are more depressed; lip simple; shell with 6-7 tightly coiled whorls; perforate, usually widest in young shells (a) but in some adults, the perforation is as wide as half a millimeter; live animal nearly black (an important feature); shell translucent and glossy; transverse striae are distinct on top, but weaker on the sides and bottom; spiral striae are indistinct; within the aperture of both young and adult shells there are two lamellae which can be seen through the bottom of fresh shells (b); a member of the *V. gularis* group.

Similar Species: Both *V. suppressus* and *V. virginicus* are smaller with a notably wider umbilicus; *Ventridens pilsbryi* (page 377) has 7-8 whorls and the live animal is pale yellow.

Habitat: Found on wooded hillsides, ravines, under leaf litter and around logs.

Status: G5/S4; Common; a common snail across the state of West Virginia.

Specimen: Tennessee, Monroe County, Towhee Falls Baptist Church (author's collection).

Flat dome

Gastrodontidae

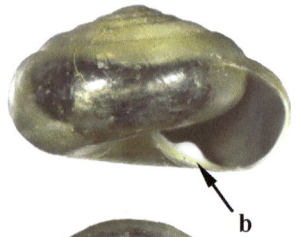

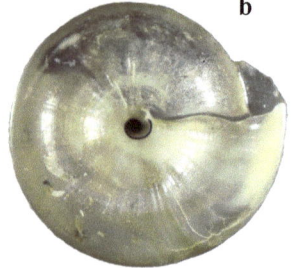

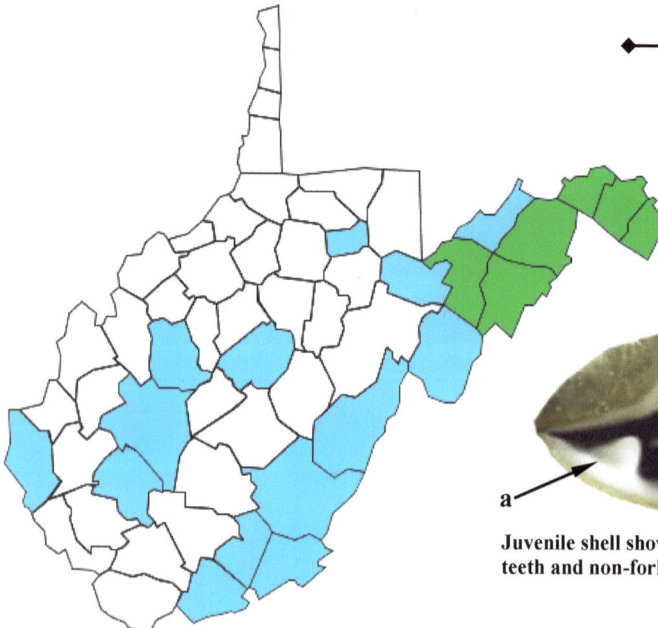

Ventridens suppressus (Say, 1829)

Diameter: 5.4-7.8 mm, height 3.5-4 mm

Description: Depressed heliciform; lip simple; shell with 6 moderately coiled whorls; perforate to umbilicate; live animal dark, olive or bluish-gray; shell thin, pale horn, glossy and semi-transparent; transverse striae (sometimes called growth wrinkles) are poorly developed; spiral striae present but a weak feature; in early shell stages multiple teeth or lamella are present (a) but in fully mature shells 6–6.5 mm with more than 6 whorls, there is usually only one basal lamella (b), or very occasionally, no lamella at all; a member of the *V. gularis* group (Hubricht 1964).

Similar Species: *Ventridens virginicus* is smaller and contains an outer wall tooth in adult shells.

Habitat: Found in mixed hardwood forests on hillsides and in ravines under leaf litter, often times in rocky areas.

Status: G5/S3; Common; found in scattered locations across the state, mostly along the Virginia border.

Specimen: Virginia, Craig County, Millers Cove (author's collection).

Juvenile shell showing multiple teeth and non-forked basal tooth

Split-tooth dome

Gastrodontidae

Ventridens virginicus (Vanatta, 1936)

Diameter: 5.7 mm, height 3.8 mm

Description: Depressed heliciform; lip simple; shell with 6 whorls; perforate to umbilicate; live animal dark, olive or bluish-gray; shell glossy and semi-transparent; transverse striae are poorly developed; spiral striae present but a weak feature; aperture with basal lamellae in all stages of growth, in old, adult shells the outer-wall tooth is often absorbed; in the juvenile stage (occasionally adults) there is a distinctive double columellar lamella (a); a member of the *V. gularis* group.

Similar Species: *Ventridens suppressus* is slightly larger and contains no outer wall lamellae in the adult shell.

Habitat: Found in the same habitat as *V. suppressus,* but not with it, according to surveys done by Hubricht 1985.

Status: G4/S3; Relatively Common; most records of this *Ventridens* are along the border with Virginia.

Specimen: West Virginia, Mineral County, Patterson Creek Road (author's collection).

North Fork Mountain, Pendleton County, West Virginia (FM 246975)

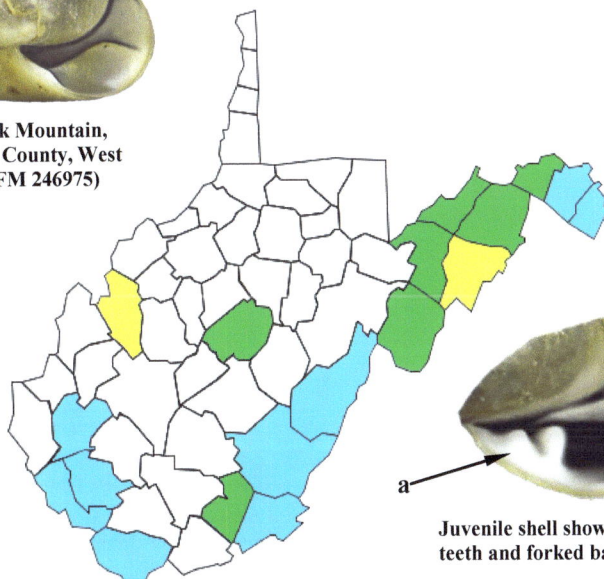

a

Juvenile shell showing multiple teeth and forked basal tooth

Bidentate dome

Gastrodontidae

Ventridens coelaxis (Pilsbry, 1899)

Diameter: 6.5-6.7 mm, height 3 mm

Description: Depressed heliciform; lip simple; shell with 6.5-7 tightly coiled whorls; umbilicate; shell yellow-corneous, delicate for this group and glossy; transverse striae (growth wrinkles) are relatively distinctive on top but almost smooth on the sides and base of shell; only traces of spiral striae; within the aperture there are usually two elongated lamellae or teeth that are present at every stage of growth; live animal pale, not dark as seen in species of the *V. gularis* assemblage; a member of the *V. pilsbryi* group.

Similar Species: *V. lawae* is larger, has a thicker shell and a higher dome shape. *V. lasmodon* has a wider umbilicus.

Habitat: Found in wooded ravines of higher elevation mountainsides under leaf litter and around log structure.

Status: G3/S1; Rare; little is known of this unexpected resident of West Virginia; it is a delicate species of lofty peaks, that centers around Roan Mountain North Carolina to Mount Rogers, Virginia.

Specimen: Virginia, Washington County (all figures in author's collection).

Umbilicus wide

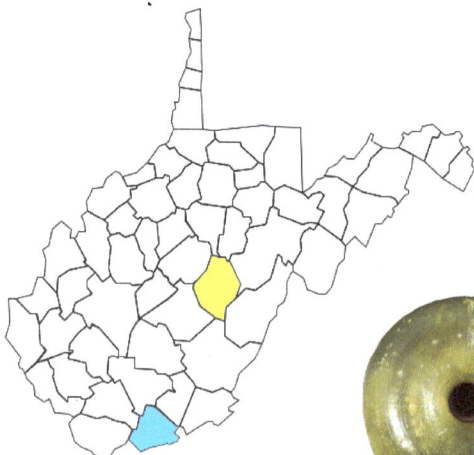

Ventridens coelaxis, Holly River State Park, Webster County, West Virginia

Hollow dome

Gastrodontidae

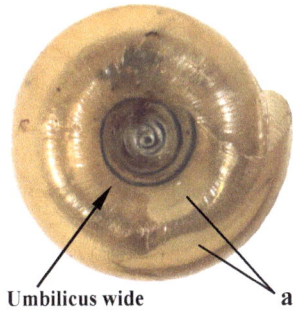

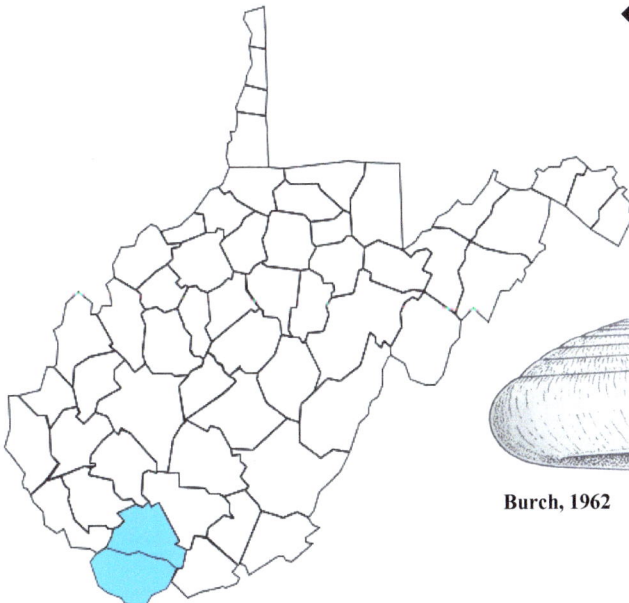

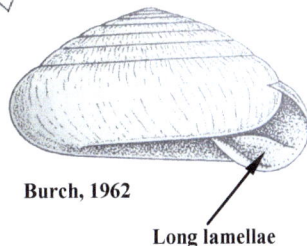

Ventridens lasmodon (Phillips, 1841)

Diameter: 7.5-7.8 mm, height 3.8 mm

Description: Depressed heliciform; lip simple; shell with 7.5-8 tightly coiled whorls; umbilicate (the widest of any *Ventridens*), the umbilicus is funnel-shaped and nearly one third the diameter of the shell; live animal pale; shell translucent, thin and glossy; finely developed transverse striae are distinct on top, but grow weaker on the sides and base of shell; spiral striae not well defined; within the aperture are two elongated lamellae (a) or teeth which are present at every stage of growth, but as in *V. lawae,* sometimes out of sight in frontal view; a member of the *V. pilsbryi* group.

Similar Species: *Ventridens lawae* has a higher shell profile, a thicker shell, and a notably smaller umbilicus.

Habitat: Found on wooded hillsides and in ravines under leaf litter on both limestone and sandstone sites.

Status: G4/SH; Rare; reported from only two southern counties in the state.

Specimen: Tennessee, Monroe County, Cherokee National Forest (author's collection).

Umbilicus wide   a

Burch, 1962

Long lamellae

169

Rounded dome

Gastrodontidae

Ventridens lawae (W. G. Binney, 1892)

Diameter: 7.8-9 mm, height 4 mm

Description: Depressed heliciform; lip simple; shell with 7.5-8.5 tightly coiled whorls; umbilicate, the umbilicus is deep with nearly parallel sides (well-like); in adults the umbilicus may vary from 0.4 mm to 2 mm in size (Hubricht 1964) ; live animal pale; shell light horn to yellow and glossy; finely developed transverse striae are distinct on top, but dwindle on the sides and base of shell; young shells with lamellae (a); adults with or without a basal lamella; a member of the *V. pilsbryi* group.

Similar Species: *V. theloides* is similar in build, but with a smaller umbilicus and typically without adult lamellae.

Habitat: Found in mixed hardwood forests on hillsides and in ravines under leaf litter and around logs, more common on limestone, but also occurs on sandstone sites (Hubricht 1985).

Status: G3/SH; Rare; in West Virginia this species is reported from only two southwestern counties.

Specimen: North Carolina, Macon County, Nantahala National Forest (author's collection).

Umbilicus wide

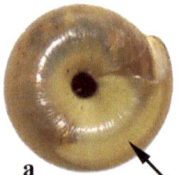

a

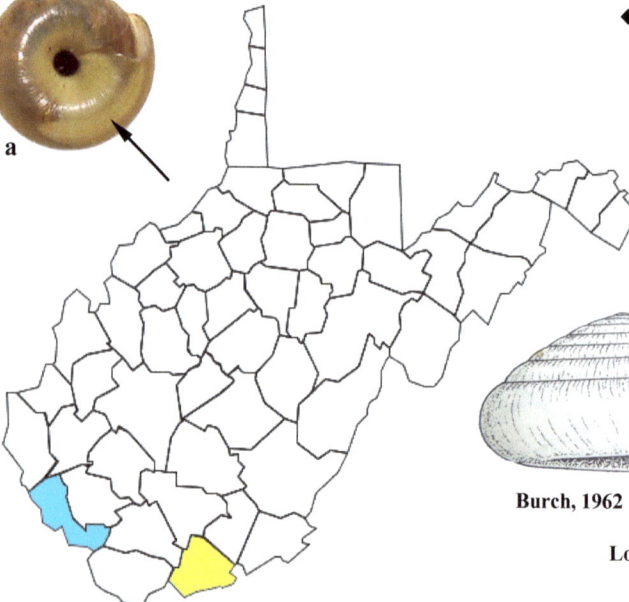

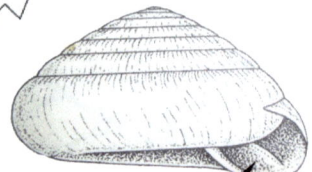

Burch, 1962

Long lamellae

Sculptured dome

Ventridens collisella (Pilsbry, 1896)

Diameter: 8.4-9.6 mm, height 6-8.1 mm

Description: Heliciform; lip simple; shell with 7.5-8 tightly coiled whorls; perforate, young shells are umbilicate, the umbilicus becoming smaller as shells mature (a); live animal pale; shell pale yellowish horn color and dull glossy; transverse striae are well-defined and boldest near the sutures; without any notable spiral striae; usually with two entering lamella, in adult shells, the lamellae becoming lower in stature, although some specimens will be without this armature; a member of the *V. pilsbryi* group (Hubricht 1964).

Similar Species: *Ventridens ligera* is larger and does not contain teeth at any stage of growth.

Habitat: A calciphile species occurring on wooded hillsides in hardwood forests.

Status: G4G5/S3; Relatively Common; known from scattered locations in West Virginia.

Specimen: Virginia, Tazewell County, (author's collection).

Gastrodontidae

a

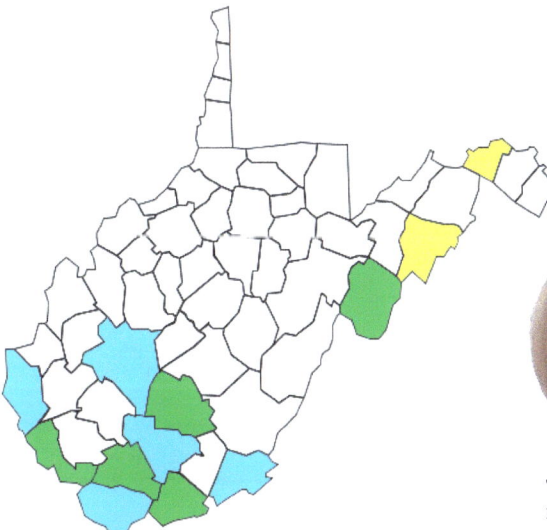

Translucent shell showing internal lamellae

Copper dome

Gastrodontidae

Ventridens theloides (Walker & Pilsbry 1902)
Diameter: 7.5-8 mm, height 4.5 mm
Description: Heliciform and dome shaped; lip simple or with a slight basal deflection (a); 7.5-9 tightly coiled whorls; perforate but open, the area around umbilicus well excavated and funnel-like (b); live animal pale; shell yellowish or coppery, the base very glossy; transverse striae are moderately well developed on top, but are diminished on the base; young shells are with basal lamellae, adult shells typically without teeth; a member of the *V. pilsbryi* group.
Similar Species: *Ventridens gularis* has well-developed basal lamellae and a smaller umbilicus; *V. lawae* has a much wider umbilicus and is more compressed.
Habitat: Found in upland hardwood forests, in ravines, under leaf litter and around logs; generally a species of limestone but also occurs on sandstone sites.
Status: G5/SH; Uncommon; another *Ventridens* which in West Virginia is restricted to southern counties.
Specimen: North Carolina, Swain County, Nantahala NF (author's collection).

a

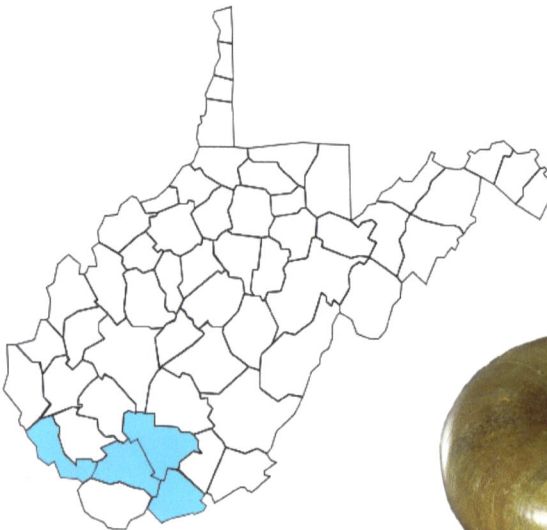

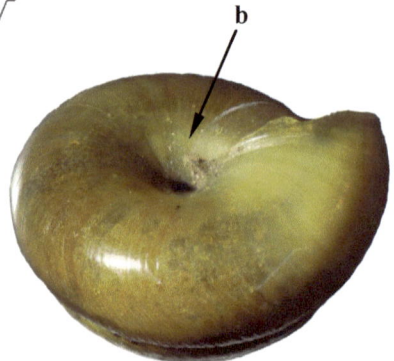

b

172

Perforate dome

Gastrodontidae

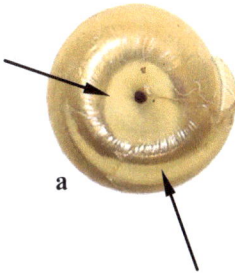

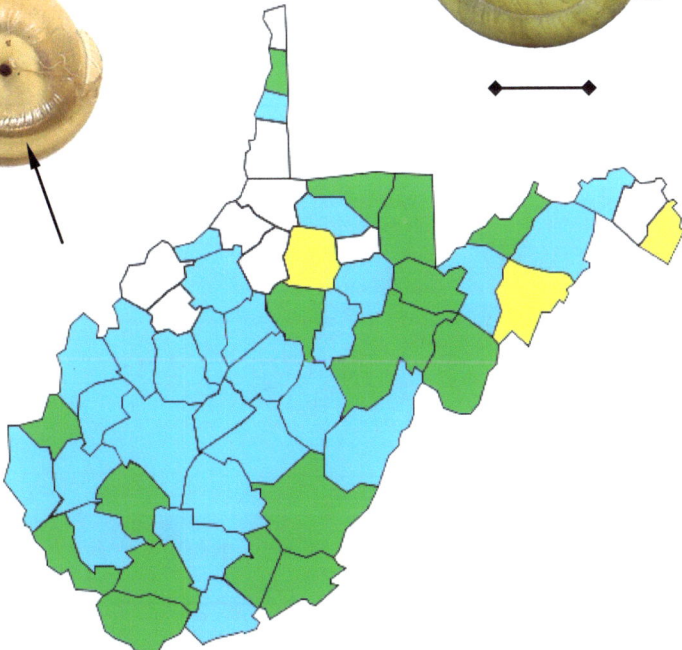

Ventridens demissus (A. Binney, 1843)

Diameter: 7.5-11 mm, height 4.8-6.8 mm

Description: Heliciform; lip simple; shell with 6-7.5 tightly coiled whorls; perforate; shell horn-yellow and glossy; transverse striae are moderately developed on top of shell, but less so on the sides and bottom; aperture typically without basal lamella in adult shells, but young shells (a) are typically with this diagnostic feature.

Similar Species: *Ventridens ligera* is larger, has a slightly smaller umbilicus, no teeth in juvenile shells and in general a higher shell than *V. demissus*; *V. collisella* is only slightly smaller, has better developed transverse striae and teeth in the adult stage.

Habitat: Found on wooded hillsides, in ravines and floodplains under leaf litter; also a snail found frequently in urban areas.

Status: G5/S5; Common; one of the most common and widespread *Ventridens* in West Virginia.

Specimen: Kentucky, Letcher County, Bad Branch Nature Reserve (author's collection).

a

Globose dome

Gastrodontidae

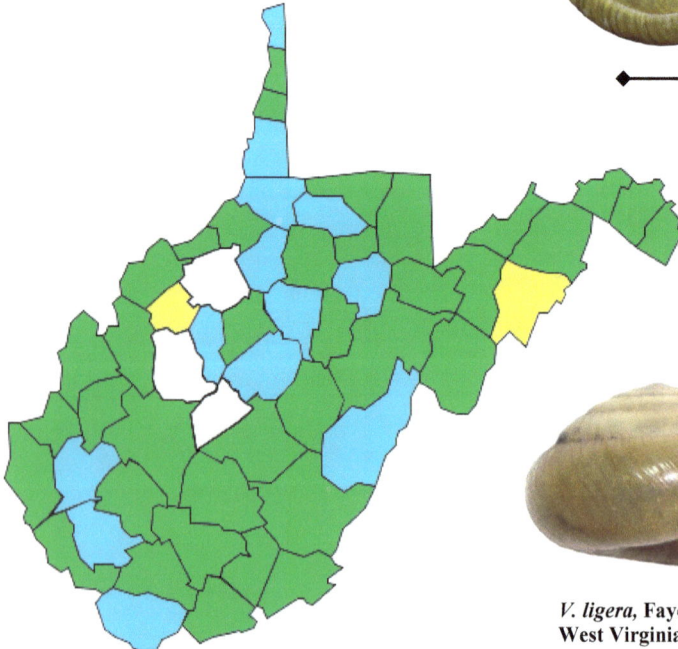

Ventridens ligera (Say, 1821)

Diameter: 11-15.6 mm, height 8-12.2 mm

Description: Heliciform, the height of this species will vary considerably (page 178); lip simple; shell with 6-7 moderately coiled whorls; perforate; shell pale yellowish-horn and somewhat glossy; transverse striae are comparatively well developed on top of shell but much less defined on the sides and bottom; aperture without a basal lamella in any stages of growth.

Similar Species: *Ventridens intertextus* has a similar form but has better developed transverse striae, a slightly larger umbilicus, often, a weakly angular periphery, and on occasions, a light color band on its periphery; *V. demissus* is smaller and young shells are armed with internal lamellae.

Habitat: Found in a variety of open, weedy and mixed hardwood forests in floodplains and other wet low-lying areas; also along roadsides.

Status: G5/S5; Common; one of the most common and widespread *Ventridens* in West Virginia and likely occurs in every county.

Specimen: Kentucky, Estill County, along the Kentucky River around Irvine (author's collection).

V. ligera, Fayette County, West Virginia

Glossy dome

Gastrodontidae

Ventridens acerra (J. Lewis, 1870)

Diameter: 12.6-19 mm, height 11-12 mm

Description: Heliciform; lip simple; shell with 7-8 moderately coiled whorls, the last whorl only slightly expanded; perforate; shell light yellowish-olive and very glossy; transverse striae are poorly developed, strongest on the top but fade away on the sides and base; without notable spiral striae; no teeth at any stage of growth; periphery rounded.

Similar Species: *Ventridens arcellus* is smaller, has a larger umbilicus, is more tightly wound (the last whorl not expanding greatly), and is usually found above 1200 m; *V. ligera* is less glossy and found in low places.

Habitat: A habitat generalist found in a variety of mixed hardwood forests on hillsides under leaf litter usually under 1200 m in elevation.

Status: G4/S2; Uncommon; reported from only seven West Virginia counties; the species becomes increasingly common south of the state.

Specimen: North Carolina, Swain County, Nantahala National Forest (author's collection).

V. acerra, Graham County, North Carolina

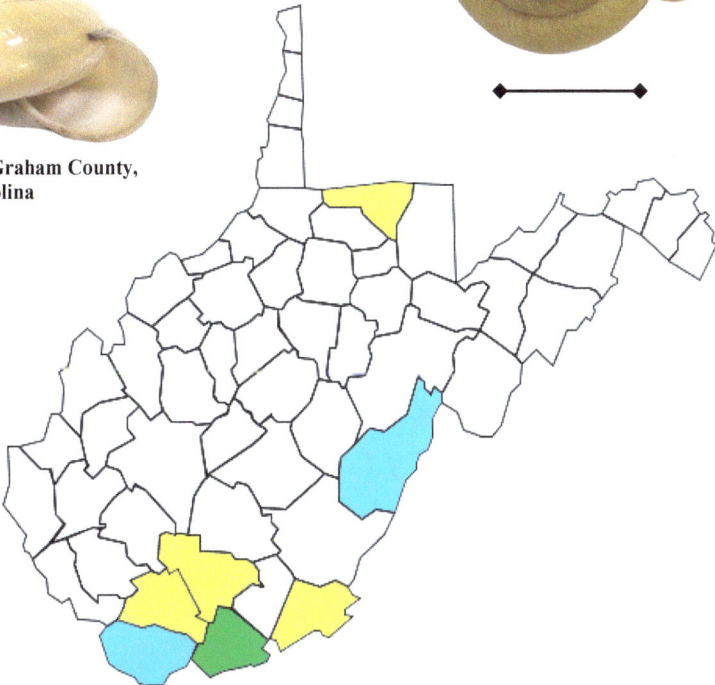

175

Golden dome

Ventridens arcellus Hubricht, 1976

Gastrodontidae

Diameter: 12.9-13.8 mm, height 9.2-11.1 mm

Description: Heliciform; lip simple; shell with 6-7.5 moderately coiled whorls, the last whorl only slightly expanded; perforate but open (a); shell light yellowish-olive and glossy; transverse striae are moderately developed, strongest on the top but fading away on the sides and base; without notable spiral striae; no teeth at any stage of growth.

Similar Species: *Ventridens acerra* is larger, slightly more compressed, has a smaller umbilicus, is less tightly coiled, and is a species of lower elevations; *V. ligera* is less glossy and found in low places.

Habitat: A habitat generalist found in a variety of mixed hardwood and northern hardwood forests on hillsides and mountaintops under leaf litter, generally above 1200 m elevation.

Status: G4/S3; Fairly Common; in West Virginia this species is restricted mostly to the higher elevation counties of the state.

Specimen: West Virginia, Tucker County, Canaan Valley NWR (author's collection).

a

V. arcellus, Tennessee, Swain County, around Clingman's Dome, GSMNP

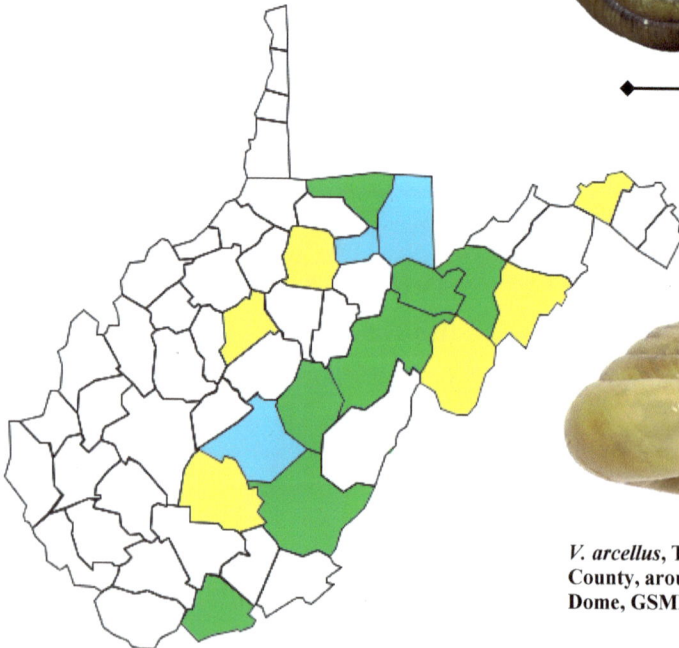

Pyramid dome

Gastrodontidae

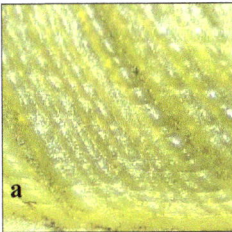

Ventridens intertextus (A. Binney, 1841)

Diameter: 8-20 mm, height 10-13.3 mm

Description: Heliciform; lip simple; shell with 5.5-6 moderately coiled whorls; perforate; shell yellowish-horn to olive-buff and dull; broken transverse striae (a) are well-developed and are the single best feature for identifying this species; spiral striae are well developed; without teeth or basal lamellae in all stages of growth; shell periphery can be round or slightly angular (b) and occasionally with a faint color band.

Similar Species: *V. ligera* has a similar form, but does not have the unique broken striae found in *V. intertextus*.

Habitat: Found in mixed hardwood forests on hillsides, ravines, and acidic ridgetops, under leaf litter.

Status: G5/S5; Common; a frequent and widespread land snail in West Virginia.

Specimen: West Virginia, Mineral County, Potomac River Gorge (author's collection).

Valley Falls State Park, Taylor County, West Virginia

177

West Virginia *Ventridens* Compared (shells proportionate)

Adult Shells with Teeth

V. collisella, Tazewell Co, VA

V. pilsbryi, Monroe Co, TN

V. gularis, Haywood Co, NC

V. lawae, Macon Co, NC

V. lasmodon, Monroe Co, TN

V. suppressus,
Craig Co, VA

10 mm

V. coelaxis,
Washington Co, VA

V. virginicus,
Mineral Co, WV

Adult Shells without Teeth

10 mm

V. acerra, Swain Co, NC

V. ligera, Powell Co, KY

V. ligera, Fayette Co, WV

V. arcellus, Tucker Co, WV

V. arcellus, GSMNP, TN

V. ligera, Fayette Co, KY

V. intertextus, Taylor Co, WV

V. demissus,
Letcher Co, KY

V. theloides,
Swain Co, NC

Shells 15-36 mm, without teeth, simple lip (*Mesomphix*)

Mesomphix species have medium to large shells (15-35 mm) that are either depressed heliciform or heliciform. The shells are loosely coiled, the last whorl expanded (a), while *Ventridens* species have more tightly coiled whorls (b) and are typically smaller. The shell surface can be dull to glossy (some species with a glass-like surface). Lips are not reflected and are without teeth in all stages of development. The transverse striae vary from closely spaced to indistinguishable. Several species of this genus exhibit spirally arranged papillae. The shells are umbilicate to perforate; the umbilicus never completely closed. Like *Ventridens*, most *Mesomphix* have a thickening which often forms a whitish callus just within the aperture. Two or three fresh shells of *M. cupreus,* shaken in one loosely closed hand will sound remarkably like glass. Shaking other species of snails like *Mesodon* will have a duller sound. Immature *Mesodon* species are easily confused with mature *Mesomphix* snails (see below comparison). The *Mesomphix* species will be arranged according to the absence or presence of papillae. **Group 1** shells that have spiral papillae and **Group 2** shells that are generally without any spiral papillae, although some shells in this group may occasionally have scant traces of this micro-feature (see next page)

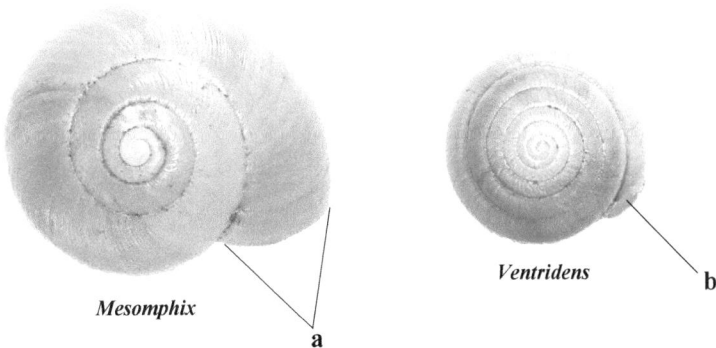

Ventridens **b**

Mesomphix

a

Important Note: Immature *Mesodon* shells are often confused with immature and mature *Mesomphix* species (see below illustrations). An immature *Mesodon* has a more rounded lip, while the mature *Mesomphix* lip is more oval in shape. *Mesodon* species will have a reflected lip only when fully mature.

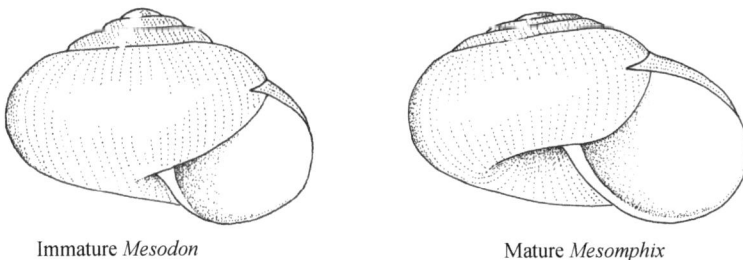

Immature *Mesodon*   Mature *Mesomphix*

179

Grouping 1: typically with spiral rows of papillae, pg 181-189

Mesomphix perlaevis
Mesomphix species (undetermined), Pygmy button (exceptionally rare)
Mesomphix capnodes (rare)
Mesomphix inornatus
Mesomphix luisant new species (rare)
Mesomphix rugeli oxycoccus (rare)

Spiral rows of papillae

Grouping 2: typically without spiral rows of papillae, pg 191-192

Mesomphix cupreus

Shell generally smooth without spiral papillae

Smooth button

Zonitidae

Mesomphix perlaevis (Pilsbry, 1900)

Diameter: 17-21 mm, height 11-12 mm

Description: Heliciform, shell height variable, even within the same populations (see below); lip simple; shell with 5 loosely coiled whorls; perforate; shell thin, but not fragile, light olive, glossy, embryonic whorl, if not worn, with notable transverse striae (can be seen with a hand lens of 10X); with or without a thin whitish callus just inside the aperture's bottom of the last whorl (a); transverse striae are strongest on the third whorl, the striations of the last whorl become notably weaker; fine spiral rows of papillae are not well developed (wanting in most places) but usually present on at least some portions of the upper shell surface.

Similar Species: *Mesomphix inornatus* is flatter in build and has well-developed spiral papillae; *M. cupreus* is larger, with a more open umbilicus and without spiral papillae.

Habitat: A snail of rich upland mixed hardwood forests usually found under and among moist leaf litters and detritus.

Status: G5/S3; Relatively Common; this species is likely more widespread than current records indicate.

Specimen: Tennessee, Cocke County, Cosby Campground, GSMNP (GSMNP collection).

M. perlaevis, both specimens from same site, Furnace Mountain, Powell County, Kentucky

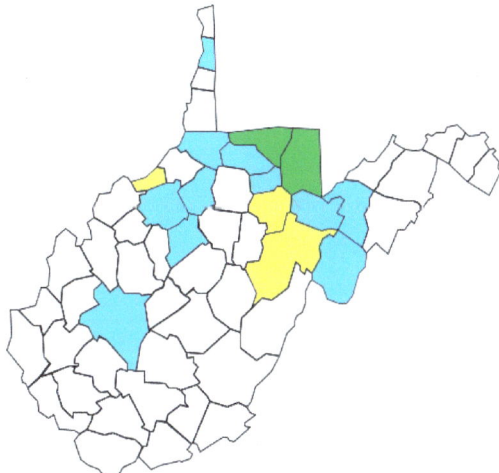

181

Pygmy button

Mesomphix species (undetermined)

Zonitidae

Diameter: 15 mm, height 11.5 mm

Description: Heliciform; lip simple; shell with 5 loosely coiled whorls; perforate, shell thin but not fragile, olivaceous with darker streaks (marking former growth stoppages), glossy; embryonic whorl worn; no teeth present; a thickening or thin whitish callus just inside the aperture's bottom of the last whorl; poorly developed transverse striae (growth wrinkles); spiral rows of papillae are present, especially strong on top of the last whorl but seen only as mere traces on the base; periphery is well rounded.

Similar Species: All other *Mesomphix* in West Virginia are larger (by around 5 mm) and are less globose in build; *Ventridens* species of the same size have a greater number of whorls, are more tightly coiled, and lack rows of papillae.

Habitat: A snail of rich, mixed hardwood forests found under moist leaf litter in a ravine below a small waterfall.

Status: G1/S1; Rare; Endemic to West Virginia; known only from a single specimen (pictured here) collected from a ravine behind Camp Dawson, Preston County.

Specimen: West Virginia, Preston County, Camp Dawson, just below a waterfall in a wet ravine behind army base buildings (author's collection).

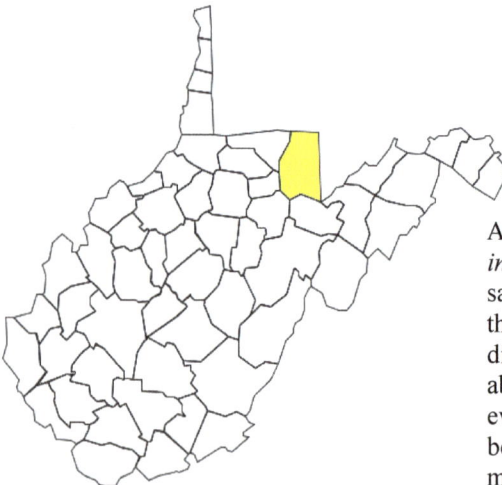

Although *M. cupreus* and *M. inornatus* are known from the same site, it seems unlikely that the Pygmy button is simply a distorted form of one of the two above mentioned species, however, additional specimens will be needed to make a final determination on this position.

182

Dusky button

Zonitidae

Mesomphix capnodes (W. G. Binney, 1857)

Diameter: 29.5-35.5 mm, height 18-26.2 mm

Description: Depressed heliciform to heliciform, in degree of elevation the shells vary widely, even within the same site, as illustrated on this page; lip simple; shell with 5 loosely coiled whorls; perforate to umbilicate; shell thin but not fragile; periostracum dark; somewhat glossy; embryonic whorl usually worn; no teeth present; with or without a thickening or thin whitish callus inside the aperture; transverse striae are poorly developed; spiral rows of papillae are present, but not well developed.

Similar Species: *Mesomphix cupreus* is typically smaller, with a slightly higher shell profile, a wider umbilicus, and very rarely displays any spiral papillae.

Habitat: A calciphile snail of upland mixed hardwood forests usually found under loose leaf litter.

Status: G5/SH; Rare; a single collection was made by Hubricht from a hillside near Lindside, Monroe County, West Virginia; all *M. capnodes* specimens from the state should be re-examined and compared to the newly described *M. luisant* (page 185).

Specimen: All figures from Tennessee, Hamilton County, Ruby Falls (author's collection).

Umbilicus wide

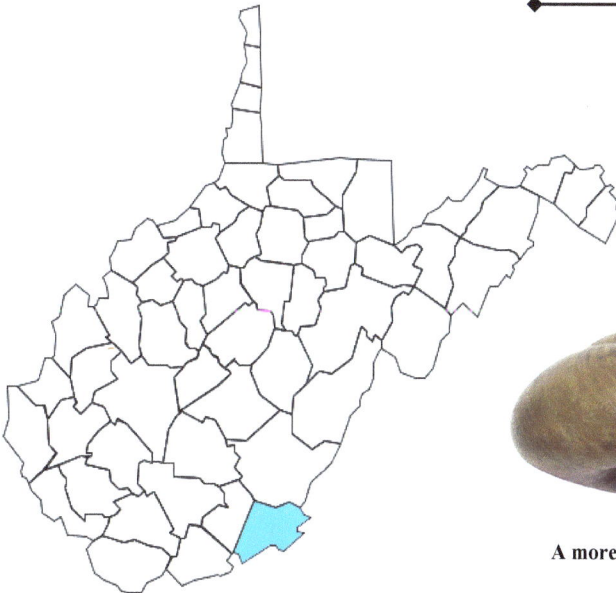

A more elevated form

Plain button

Zonitidae

Mesomphix inornatus (Say, 1821)
Diameter: 16-21 mm, height 8.4-9.8 mm
Description: Depressed heliciform; lip simple; shell with 5 loosely coiled whorls; perforate, shell thin but not fragile, olive-tan, very glossy; embryonic whorl smooth (a hand lens of 10X can be used to view this feature); no teeth present; a thickening or thin whitish callus that is typically just inside the bottom of the aperture of the last whorl; transverse striae and spiral rows of papillae are present; a carnivorous species which feeds on other live snails (Dourson pers. obs.).

Similar Species: *Mesomphix perlaevis* has a higher shell profile and the transverse striae are clearly more defined on the embryonic whorl.

Habitat: A snail of rich, upland mixed hardwood forests including beech and hemlock, usually found under or in moist leaf litter and detritus, but also a species found thriving along shale banks of road-cuts.

Status: G5/S5; Common; the most common *Mesomphix* in the state and the widest ranging in eastern North America.

Specimen: Kentucky, Powell County, Red River Gorge (author's collection).

Specimens of *M. inornatus*, from Warren Co, Pennsylvania are small, around 12-13 mm

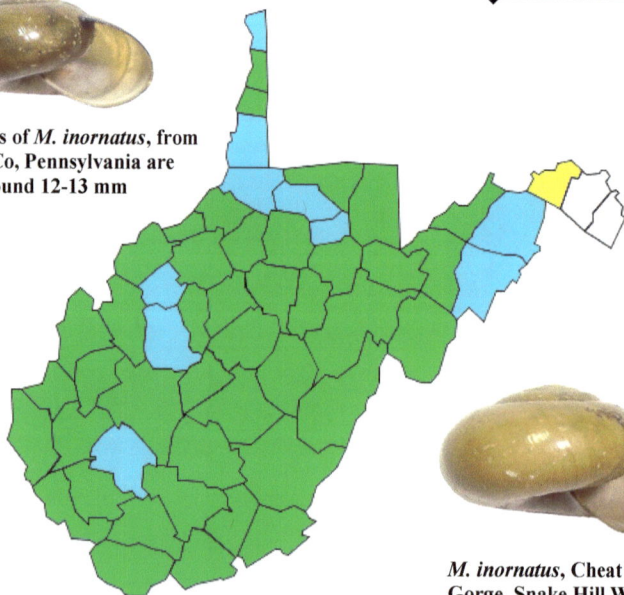

M. inornatus, Cheat River Gorge, Snake Hill WMA, West Virginia

184

Mesomphix luisant, new species
Figures a, b & c

Shell. Mature specimens 22-26 mm in diameter, 14-19 mm in height. Shell heliciform, with a great deal of variation in its elevation. Lip simple. Shell with 5 loosely coiled whorls; umbilicate, the umbilicus between 3 and 4 mm. Shell color light brown, often with darker streaks. The top of shell is dull-glossy, while the base is more polished shiny. The transverse striae are moderately developed; bolder than seen in *M. cupreus,* but not as well defined as in *M. vulgatus.* Spiral rows of papillae are a constant feature of the species and strongest on the top of the final whorl. There are some rather unique engraved spiral striae, which are strongest near the suture lines. The periphery is well rounded.

Type Locality. A hillside above a spring, near the Greenbrier River, Greenbrier County, West Virginia. Holotype: (FMNH 344365). Paratype: (FMNH 344366), Paratype from same location as above. All specimens collected 12 June, 2014 by Jeff Hajenga, John Slapcinsky and the author.

Distribution. *Mesomphix luisant* appears restricted to rich hillsides below and amid limestone bluffs along the Greenbrier River, Greenbrier County West Virginia, and possibly along Huff Creek, 2 miles east of Davin, Logan County, West Virginia (the Logan County site needing additional specimens collected for final confirmation). *Mesomphix luisant* is endemic to West Virginia.

Remarks. In terms of shell morphology, *Mesomphix luisant* is most similar to *M. capnodes* but smaller, more globose, has a slightly larger umbilicus, and has a glossier shell, especially the base. It differs from *M. cupreus* by its constant display of well-developed spiral papillae on the final whorl and smaller umbilicus. The three species have not been found together, however *M. luisant* is found in company with the much smaller and flatter *M inornatus*. It has little resemblance to other *Mesomphix* reported in West Virginia. The species may have been first collected by N.D. Richmond in 1938 on a hillside along Huff Creek, 2 miles east of Davin, Logan County, West Virginia.

Etymology. Luisant is French for glossy, shiny, or gleaming.

Specimen	Diameter	Height	ApH	ApW	Whorls	Umbilicus W
Holotype (FMNH 344365)	25	17	12	13	5	3.5
Paratype (FMNH 344366)	23	16	11	11	5	3.0
Paratype (FMNH 344366)	24	19	12	13	5	3.0
Other specimens	23	17	11	12	5	3.0
Other specimens	26	19	11	13	5	4.0

Holotype and Paratype Specimens with Shell Details

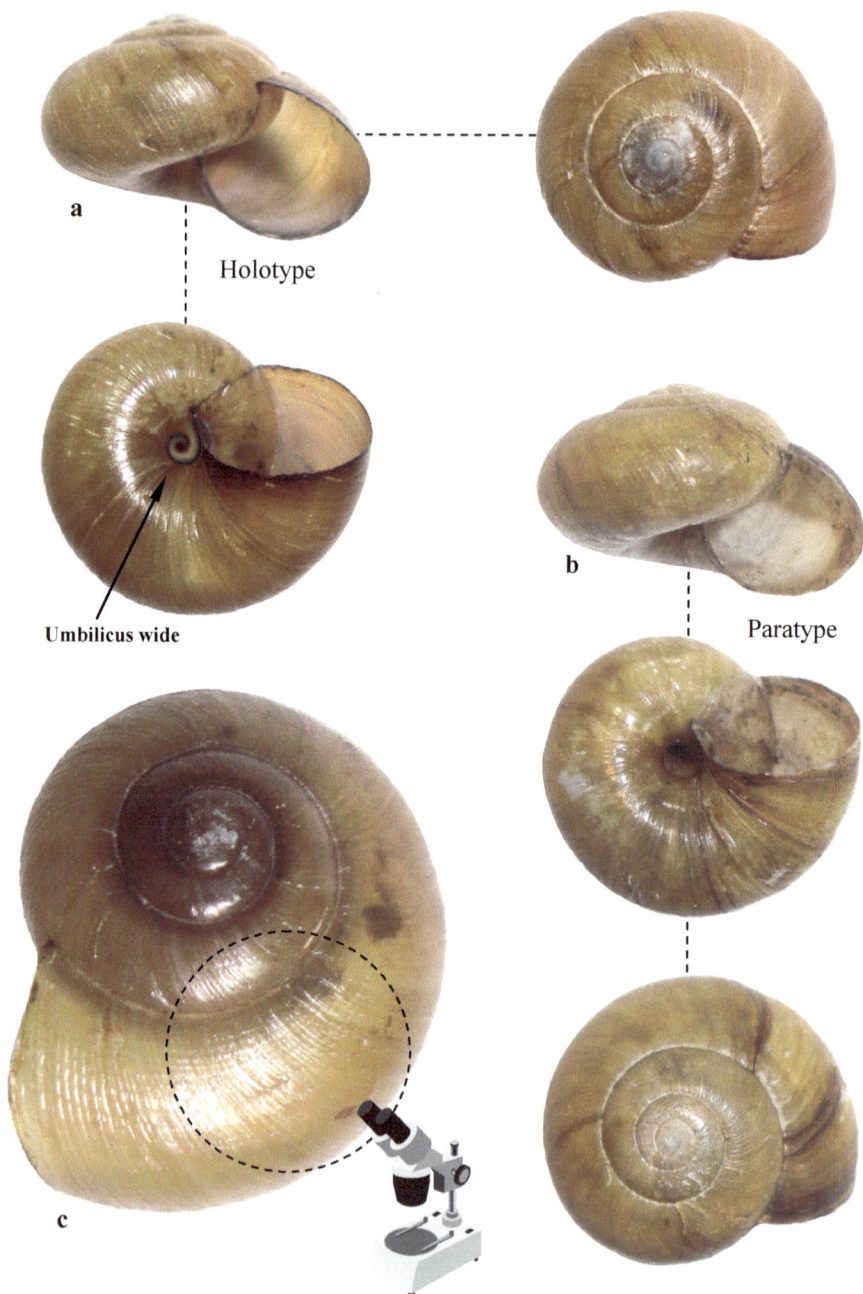

a) *Mesomphix luisant,* Holotype: (FMNH 344365), Greenbrier County, West Virginia; b) *M. luisant,* Paratype: (FMNH 344366), from same location c) *M. luisant* illustrating the rather unique (for a *Mesomphix*), engraved spiral striae, boldest near the suture lines of the final whorl.

Glossy button

Zonitidae

Mesomphix luisant new species

Diameter: 22-26 mm, height 14-19 mm

Description: Heliciform; lip simple; shell with 5 loosely-coiled whorls; umbilicate; shell thin but not fragile; light brown with darker streaks; base of shell especially glossy (see opposite page); transverse striae moderately developed; spiral rows of papillae are present and strongest on the top of the last whorl, as are some rather unique engraved spiral striae (not always prevalent), which themselves are boldest near the suture lines; periphery rounded; a globose form in figure (a).

Umbilicus wide

Similar Species: Most like *M. capnodes* but is smaller, more globose with a slightly larger umbilicus and has a glossier shell, especially the base; it differs from *M. cupreus* by its constant display of well-developed spiral papillae on the final whorl and smaller umbilicus; the two have not been found together.

Habitat: A snail of rich, mixed hardwood forests adjacent to and below limestone cliffs along the Greenbrier River.

Status: G1/S1; Rare; Endemic to West Virginia; currently confirmed from one county.

Specimen: West Virginia. Greenbrier County, (FM 344365 Holotype) and figure (a) from same location (FM 344366 Paratype).

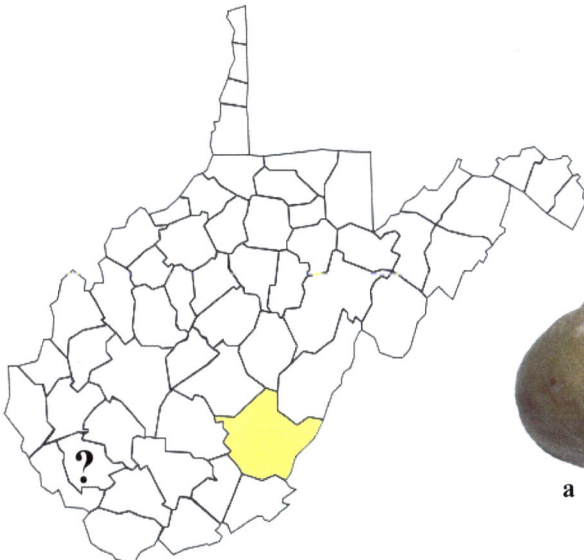

a

Dark streaks are growth stoppages in the shell

Bottom view showing the glossy surface, Greenbrier County, WV

The above specimen from the Carnegie Museum, labeled as *M. rugeli,* was found by N.D. Richmond in 1938 on a hillside along Huff Creek, 2 miles east of Davin, Logan County, West Virginia (CM 62.35779); it essentially agrees with *Mesomphix luisant* from Greenbrier County, West Virginia.

The live animal and shell of *M. luisant* Greenbrier Co, WV

Imitator button

Zonitidae

Mesomphix rugeli oxycoccus (Vanatta, 1903)

Diameter: 20-22.5 mm, height 12-15 mm

Description: Heliciform; lip simple; shell with 5.5-6 loosely coiled whorls; perforate; shell thin but not fragile, greenish-horn and glossy often with darker streaks marking former growth stoppages; occasionally has a thin whitish callus just inside the aperture; transverse striae are smooth and the shell displays finely developed spiral rows of papillae, best observed on the top last two whorls becoming nearly obsolete on the base, or seen as mere traces; periphery well rounded.

Similar Species: *Mesomphix perlaevis* is lighter in color, has better developed transverse striae and weaker spiral rows of papillae; *M. rugeli oxycoccus* differs from *M. rugeli* (page 379) by having well-developed and crowded spiral papillae and a somewhat duller shell surface.

Habitat: Low to mid-elevation hillsides and ravines in mixed hardwood forests under moist leaf litters.

Status: GNR/S3; Rare; this interesting subspecies was first described by Vanatta in 1903 and later synonymized by Hubricht in 1985, but without explanation, consequently further DNA work is needed to make a final deliberation on its status as a species; it has been reported from southern West Virginia, southeastern Kentucky, eastern Tennessee and western North Carolina.

Specimen: Tennessee, Carter County, Cherokee National Forest (author's collection).

Mesomphix rugeli, Mitchell County, North Carolina

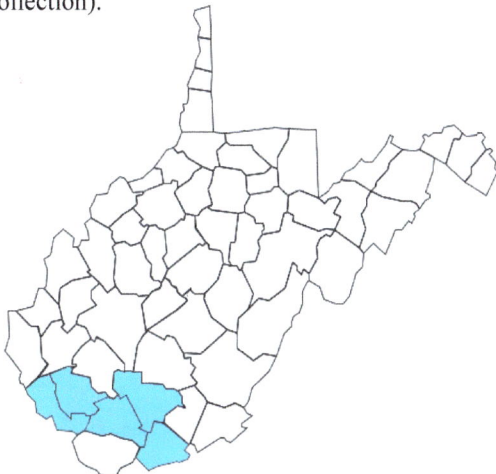

Mesomphix Shells of West Virginia Compared (proportionate)

Mesomphix with spiral papillae

M. capnodes, Ruby Falls, Hamilton Co, TN

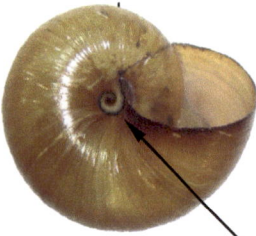

Mesomphix luisant, Holotype, Greenbrier Co, WV

Mesomphix perlaevis,
Cocke Co, TN

Mesomphix inornatus,
Powell Co, KY

Mesomphix rugeli oxycoccus,
Carter Co, TN

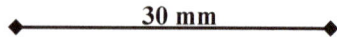

Mesomphix species (undetermined)
Pygmy button, Preston Co, WV

◆———— **30 mm** ————◆

Mesomphix without spiral papillae

M. cupreus,
Cocke Co, TN

Copper button

Zonitidae

Mesomphix cupreus (Rafinesque, 1831)

Diameter: 22-29.5 mm, height 12.4-15.4 mm

Description: Heliciform, but with a great deal of variation; lip simple; shell with 4.5-5 loosely coiled whorls; umbilicate, has the widest umbilicus of any *Mesomphix* species found in West Virginia; shell thin but not fragile; periostracum dark; shell glossy; embryonic whorl usually worn; a thickening or thin whitish callus just inside the aperture; transverse striae are poorly developed and rather smooth on some portions of the shell; without any detectable rows of spiral papillae; periphery well rounded.

Similar Species: *Mesomphix capnodes* is larger, has a slightly lower shell profile, a smaller umbilicus, and is always with spiral rows of papillae.

Habitat: A snail of both limestone and acidic upland mixed hardwood forests usually found under and among loose leaf litter and detritus.

Status: G5/S5; Common; one of the most common large *Mesomphix* in West Virginia, with records occurring in nearly every county.

Specimen: Tennessee, Cocke County, Cosby Campground, GSMNP (GSMNP collection).

Umbilicus wide

M. cupreus,
Randolph Co, WV

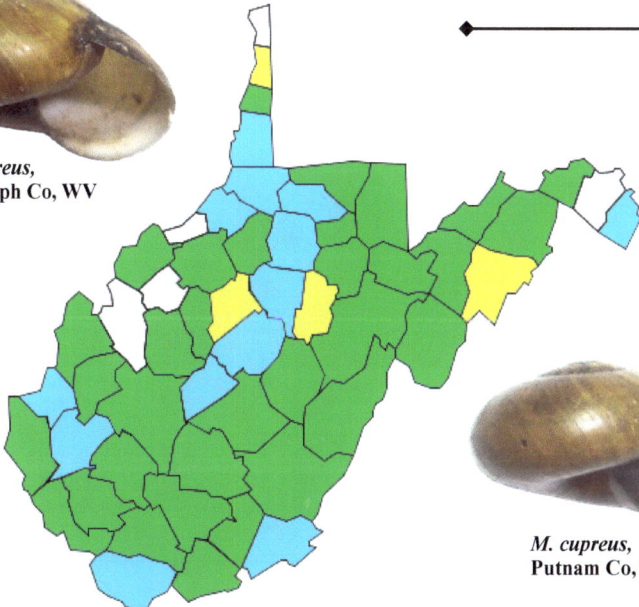

M. cupreus,
Putnam Co, WV

Mesomphix cupreus hunting a meal in Red River Gorge, Kentucky.

Extraordinary Feeding Behavior Observed in a Central America Land Snail (by entomologists Piotr Naskrecki and Kenji Nishida)

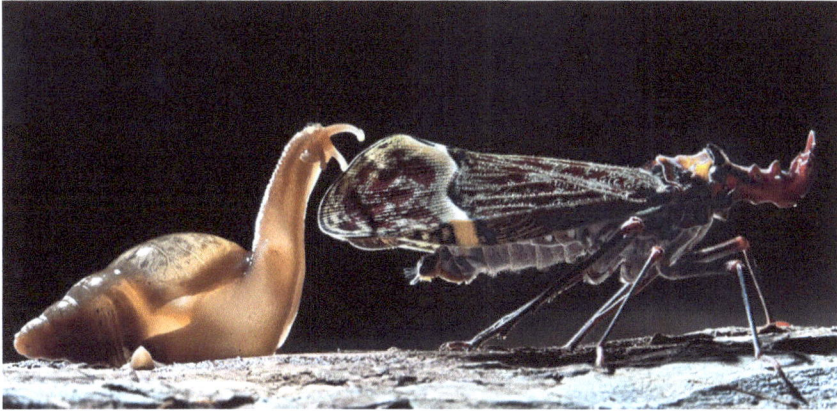

While using its tentacles, a wolf snail, *Pittieria aurantiaca,* politely taps the rear end of a lantern bug, *Phrictus quinquepartitus* in Costa Rica. This strange behavior appears uncharacteristic for a snail that normally hunts other gastropods for their flesh. The snail is not interested in eating the bug, but attracted to the honeydew which is ejected from the bugs abdomen. After tapping the bug's rear, the snail catches flying droplets of honeydew by forming a hood over the tip of the insect's abdomen with its head and foot. The honeydew accumulates on the snails ventral surface, which is than consumed. Why would a predaceous snail want honeydew in the first place? It appears that, in addition to the benefit of getting some high quality carbohydrates, the phloem (sap) of the tree species that the lantern bugs feeds on, *Simarouba amara*, contain high levels of calcium, the mineral that snails need to build shell. Carpenter ants also want in on the action but are to short and to slow to catch the honeydew that is flying out of the lantern bug's rear end at between 0.8 and 1.7 m/sec (2.6-5.6 ft/sec.) according to

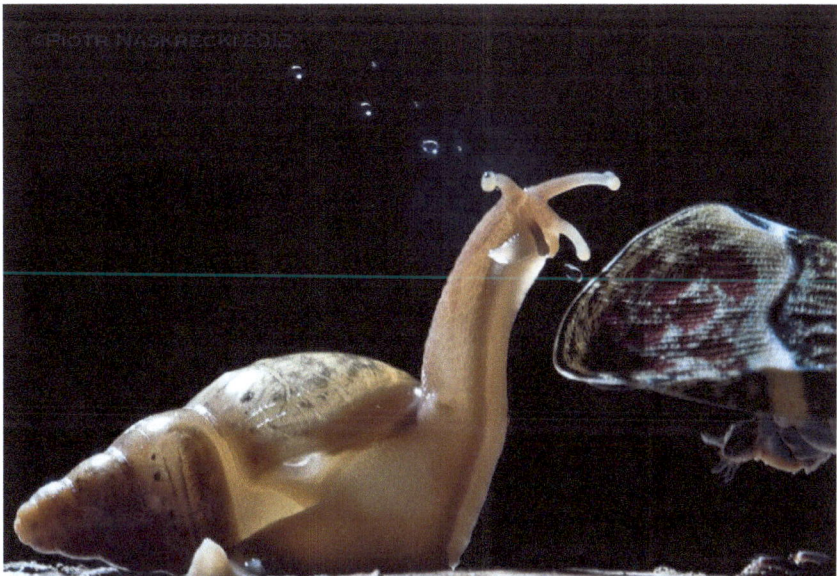

193

Naskrecki and Nishida. So the ants climb up the snail to harvest the sweet food from the surface of the gastropods head. Not surprisingly, the snail quickly become agitated with the ants that are stepping on its eyes and chemoreceptors. Unable to rid themselves of the pesky ants the snail simply crawls away. This extraordinary land snail feeding behavior between mollusk and arthropod clearly demonstrates the wide range of calcium carbonate sources in nature and that the harvesting techniques used by land snails are even more complex than we ever imagined.

Shells 11-33 mm, without teeth, simple lip, typically with bands or streaks of color, crisscrossed protoconch sculpture, defense mucus fluorescent under UV light (*Anguispira*)

Anguispira species are among the most multicolored native land snails in West Virginia. Interestingly, the defense mucus of this genus is fluorescent under black (UV) light for reasons that remain a mystery. In general, these medium size gastropods are most common around outcroppings of limestone, but also found in acidic environments. Shells of *Anguispira* are usually depressed heliciform. There are crisscrossing striae (a) on the embryonic (protoconch) whorl of most species of *Anguispira* (often a lost feature in worn shells), however not all specimens within the same population exhibit this trait; the lips of *Anguispira* shells are simple in all stages of growth and there are never any teeth present in the aperture. The transverse striae or ribbed surfaces are easily observed with a hand lens of 10X magnification. *Anguispira* shells are always umbilicate. The periphery of the body whorl is either carinate, bluntly angular or rounded. The color features are usually present but this character may be poorly defined in some species and in shells that are badly weathered; wetting the shell will usually reveal some color. The following *Anguispira* are arranged according to shape and rib counts and are as follows: **Group 1**-shells globose; **Group 2**-shells compressed that have more than 75 ribs on the last whorl and **Group 3**-shells compressed that have less than 65 ribs on the last whorl.

Grouping 1: shell globose, pg 196

Anguispira kochi (rare)

Grouping 2: shell compressed, more than 75 ribs on the last whorl of adult shells, pg 197-206

Anguispira alternata
Anguispira clarki (rare)
Anguispira stihleri, new species (rare)

Grouping 3: less than 65 ribs on the last whorl of adult shells, pg 207-208

Anguispira mordax (rare)
Anguispira strongylodes

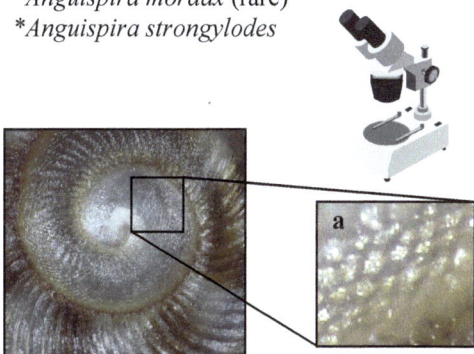

Counting Ribs

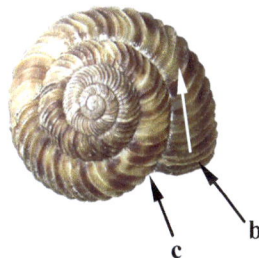

When counting ribs on the last whorl start at (b) and stop at (c); this shell having around 43 ribs

Banded globe

Discidae

Anguispira kochi (Pfeiffer, 1845)

Diameter:17-33 mm, height 14-21 mm

Description: Heliciform; lip simple; shell with 6 whorls; umbilicate; no teeth present, light color bands (can be thin or wide) always present but may be a weak feature; old snails or weathered dead shells lose their brown periostracum and weather to dull white (as seen below); shell solid (thick); embryonic whorl smooth; transverse striae distinct; no spiral striae present; periphery is well rounded.

Similar Species: The look of this *Anguispira* is like no other native species in West Virginia.

Habitat: A snail of river bluffs, but also found in upland woods at or near the base of limestone outcrops, among the loose leaf litter; a species indicative of high quality habitat.

Status: G5/S1; Rare; in West Virginia known only from three northern counties; extant populations of *A. kochi* are becoming increasingly scarce (Dourson 2010); loss of native habitat is likely the biggest threat.

Specimen: All figures illustrated, except for the Cincinnati specimen are from Furnace Mountain, Powell County, Kentucky (author's collection).

Cincinnati, Hamilton County, Ohio

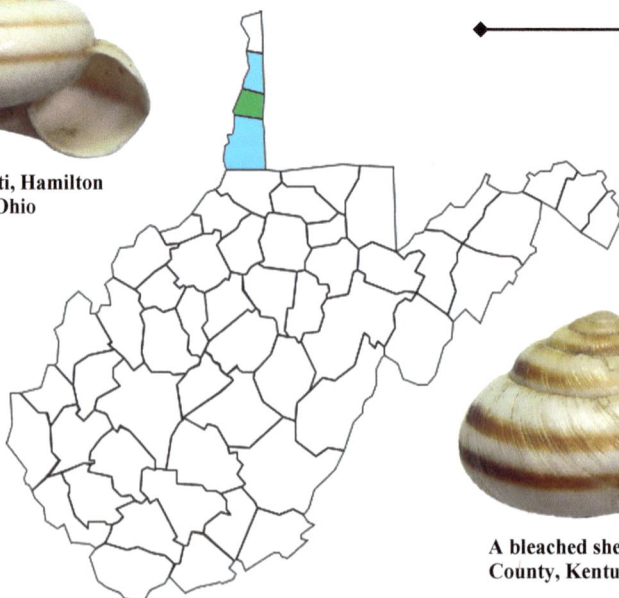

A bleached shell, Powell County, Kentucky

Flamed tigersnail

Anguispira alternata (Say, 1816)

Diameter: 15-30 mm, height 10.5-14.5 mm

Description: Depressed heliciform; lip simple; shell with 5-6 whorls; umbilicate; color features usually strong on entire shell surface; in some populations there are crisscrossing striae on the embryonic whorl; no teeth present; transverse striae are closely spaced and fairly well developed on top becoming more diminished on the base, adult shells with at least 5 whorls have a count of over 80 ribs in the last whorl; periphery can be angular or rounded but very rarely carinate.

Similar Species: *Anguispira mordax* is smaller having fewer than 50 ribs on the last whorl; *A. clarki* is substantially smaller and has a relatively smaller umbilicus.

Habitat: A species found in a wide range of lower elevation wooded and open habitats including rocky limestone regions, glades, and along forested hillsides adjacent to rivers.

Status: G5/S5; Common; the most widespread *Anguispira* in the state.

Specimen: West Virginia, Greenbrier County, around Davis Spring (author's collection).

Preston Co, WV

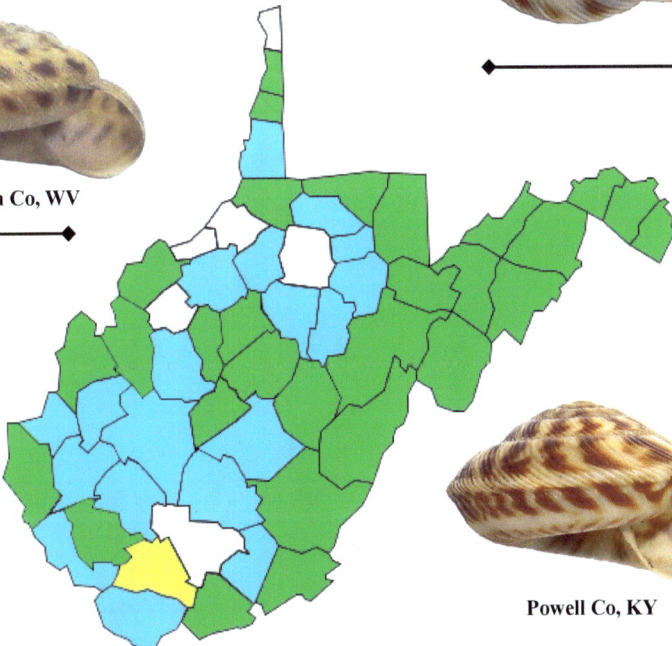

Powell Co, KY

197

Fluorescence in Land Snails

Above image of the Flamed tigersnail, *Anguispira alternata,* in typical defense posture; the live animal retracted inside its shell with its aperture blocked by a bubbly orange mucus. Below the same specimen under UV light. The function of fluorescence in land snails remains a mystery, and some scientists speculate it is nothing more than a random act of evolution. But the glow might serve as an early warning sign of distasteful compounds found in the mucus, which may repel predators and help conserve body moisture. Red River Gorge, Kentucky.

Elfin tigersnail

Discidae

Anguispira clarki Vanatta, 1924

Diameter: 11-13.5 mm, height 7-8 mm

Description: Depressed heliciform; lip simple; shell with 5 whorls; umbilicate, the umbilicus smaller than other *Anguispira* species; color features vary from light spots to relatively bold streaking; no teeth present; transverse striae are well developed on top and remain nearly as strong onto the base; adult shells with at least 5 whorls have a count of around 70-80 ribs in the last whorl; periphery rounded.

Similar Species: This is the smallest *Anguispira* species in West Virginia and one of the smallest in North America; *A. alternata* is larger, flatter and has a much wider umbilicus.

Habitat: A calciphile, in West Virginia, known only from hillsides around or in the vicinity of Cave Mountain Cave.

Status: G3/S1; Rare; this small *Anguispira* was first described by Vanatta, in 1924 and later synonymized by Hubricht in 1985, without explanation.

Specimen: All figures from West Virginia, Pendleton County, figure (a) showing a more compressed form and figure (b) showing three color morphs (author's collection).

a

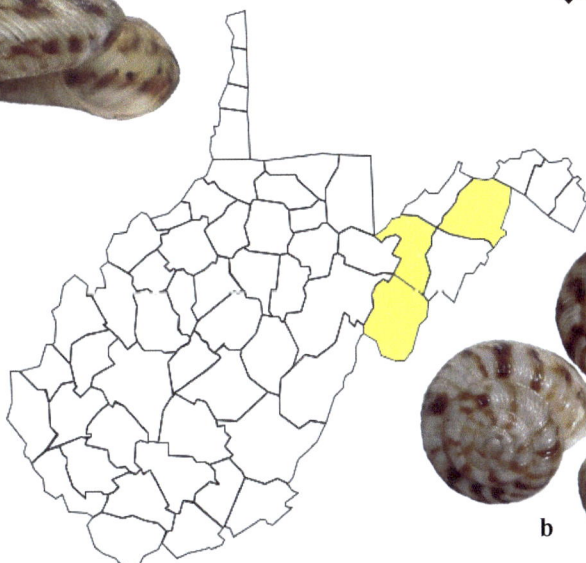

b

The Greenbrier tigersnail, *Anguispira stihleri*

Anguispira stihleri, new species
Figure a (page 202)

Shell. Mature specimens 15-19 mm in diameter, 6-10 mm in height. Shell depressed heliciform, some shells exceedingly so. Lip simple. Shell with 5-6 whorls, umbilicate, the umbilicus between 4-5 mm. Base color of shell, whitish, stained by bold rusty-brown blotches, some forming chevrons, strongest on top and periphery but seen only as faded streaks on the base. Embryonic whorl with crisscrossing striae, later whorls sculptured with transverse ribs that are strongest on the first three whorls becoming weakest on the last whorl. On the base striae grow weaker yet, or in some shells, rather smooth. Adult shells with at least 5 whorls have a count of approximately 90 ribs (striae) in the last whorl. Spiral striae typically a rather faint attribute. Periphery strongly carinate, well pinched in most specimens examined.

Type Locality. A limestone Cliff face near Manns Tunnel above Greenbrier River, Greenbrier County, West Virginia. Holotype: (UF 481164), collected on 25 July 2013 by Craig Stihler, Jeff Hajenga and Brian Streets. Paratype: (UF481165), Greenbrier County, West Virginia, collected by John Slapcinsky, Jeff Hajenga and the author.

Distribution. Restricted to limestone bluffs above the Greenbrier River, Greenbrier County, West Virginia.

Remarks. Evidently, this rare and range restricted species went undetected by early collectors in West Virginia. Not until February 23, 1993 was it discovered by Craig Stihler. Specimens lay on Stihler's desktop for 21 years before finding their way to the author. In terms of shell morphology, *A. stihleri* is most similar to *A. cumberlandiana* (figure c) and *A. alabama* (fig d & e, pg 202), but is smaller, not as compressed, and has a proportionally smaller umbilicus. It differs from *A. picta* (fig c, pg 202) by being smaller, less inflated lens shaped and having a smaller umbilicus. It has little resemblance to *A. cumberlandiana columba* (figure f). Further evidence of *A. stihleri* validity as a species is its isolation from other lens-shaped *Anguispira* (fig b-f, pg 202) tied to limestone outcrops, none of which are closer than 600 kilometers and all separated by a non-continuous, limestone geology. Found in company with *A. stihleri* are the much larger angular forms of *A. alternata*.

Etymology. Named in honor of Craig Stihler who has devoted his life protecting the more obscure creatures of West Virginia, including land snails.

Specimen	Diameter	Height	ApH	ApW	Whorls	Umbilicus W
Holotype UF 481164	17	9	6	8	5	4.2
Paratype UF 481165	16	8	6	7.5	5	4.5
Paratype UF 481165	17	9	6	7.5	5	4.0
Paratype UF 481165	15.5	8	5.5	7	5	4.0
Paratype UF 481165	16	7.5	6	7.5	5	4.0
Paratype UF 481165	16	8	5.5	7.5	5	4.0

Similar *Anguispira* Shells Compared (proportionate)

a) ***Anguispira stihleri***, Holotype: (UF 481164), (17 mm), Manns Tunnel, Greenbrier County, West Virginia b); *A. **cumberlandiana***, (19 mm) Estill Fork, Franklin County, Tennessee (author's collection); c) ***A. picta,*** (19 mm) Franklin County, Tennessee (author's collection); d) ***A. alabama***, (16 mm) Burwell Mountain, east of Jeff, Madison County, Alabama (FMNH 238001); e) ***A. alabama***, (18 mm) White County, Tennessee (FMNH 238002); f) ***A. cumberlandiana columba***, (16 mm) east side of Battle Creek, 6 miles north of Kimball, Marion County, Tennessee (authors collection).

Greenbrier tigersnail

Discidae

Anguispira stihleri new species

Diameter:15-18 mm, height 6-10 mm

Description: Depressed heliciform; lip simple; shell with 5-6 whorls; umbilicate; no teeth present, bold color bands always present but will generally fade in older specimens; shell solid; embryonic whorl with crisscrossing striae; transverse striae strongest on the first three whorls, becoming more diminished on the last; on the base, these striae are much weaker and in some shells rather smooth; adult shells with at least 5 whorls have a count of around 90 ribs in the last whorl; spiral striae sometimes present but typically a weak feature; periphery strongly carinate.

Similar Species: In West Virginia, *A. stihleri* is most similar to angular forms of *A. alternata* but has a smaller umbilicus and a carinate periphery; both species are found together and are easily separated by size alone.

Habitat: A snail restricted to crevices of limestone outcroppings and glades along the Greenbrier River; also occasionally found on trees near limestone clifflines.

Status: G1/S1; Rare; Endemic to West Virginia; this beautiful land snail is restricted to a few outcropping bluffs and glades along the Greenbrier River.

Specimen: West Virginia, Greenbrier County, near Manns Tunnel (UF 481164 Holotype).

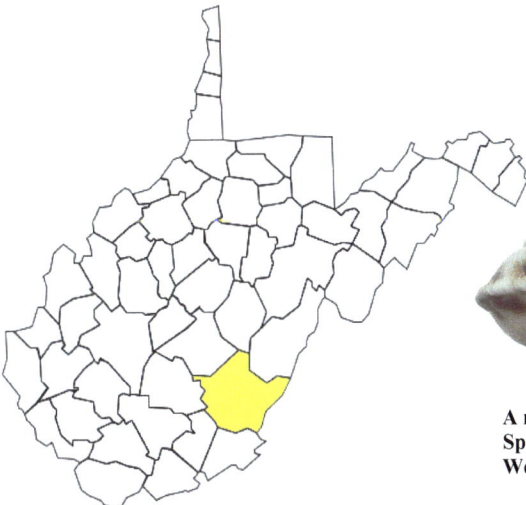

A more elevated form, Davis Spring, Greenbrier County, West Virginia

Habitat of the Greenbrier Tigersnail

The Greenbrier tigersnail is known from only a few limestone bluffs & glades along several miles of the Greenbrier River in West Virginia (above image). Below a cedar glade; one of several natural habitats of the species.

Habitat of the Greenbrier Tigersnail

Live individuals of this charming land snail are restricted to the cliffline surface and, very occasionally, trees located near rock structure. In clifflines, they
may hide in crevices (fig b, pg 206) or in the case of cedar glades, take cover
under large boulders. The species seems especially fond of dry, open cliff faces
(fig a, pg 206) where little vegetation grows. The carinate, flattened form of the
shell is believed to be an evolutionary response to its compressed environment;
the lens-shape allowing the snails to move more freely in tight quarters. Although purely speculation, the species likely feeds on a variety of cliffline fod-
der such as lichens, algae, sooty molds, cricket scat and decomposing fern
fronds. The snail has been observed feeding on decaying leaves and stems of
unknown species (fig c, pg 206).

Live snail

Habitat of the Greenbrier Tigersnail

Figure (a)

Jeff Hajenga

Figure (b)

Jeff Hajenga

Figure (c)

Appalachian tigersnail

Discidae

Anguispira mordax (Shuttleworth, 1852)

Diameter: 15-18 mm, height 6 mm

Description: Depressed heliciform; lip simple; shell with 5-6 whorls; umbilicate; color streaks can be either a strong or weakly developed feature but always present; transverse striae are modified into distinct and widely spaced ribs, having a count of 35 to 45 ribs in the last whorl and remaining bold unto the base of the shell; periphery usually bluntly angular.

Similar Species: *Anguispira alternata* is larger in size with finer sculpture with more than 80 ribs on the last whorl; *A. strongylodes* has a higher number of ribs (55 to 65 in the last whorl). See (fig. a) below and (fig. a), pg 208 comparing the two species, both at 18mm.

Habitat: Found in a variety of forested habitats, especially around limestone outcrops.

Status: G4/S2; Rare, West Virginia specimens examined from the Carnegie Museum labeled as *A. mordax* have rib counts that fall between *A. strongylodes* and *A. mordax* and therefore should be more closely scrutinized (DNA). *A. mordax* appears restricted to southern counties.

Specimen: Kentucky, Bell County, Pine Mountain State Park, figure (a) enlarged to show rib spacing (author's collection).

a

7 ribs

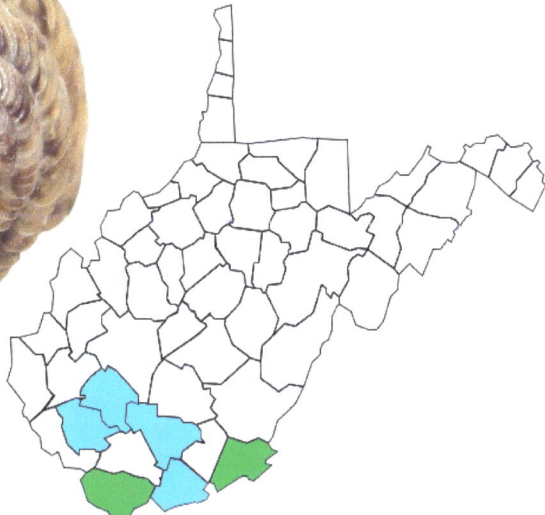

207

Southeastern tigersnail

Discidae

Anguispira strongylodes (Pfeiffer, 1854)

Diameter: 13-20 mm, height 10 mm

Description: Depressed heliciform; lip simple; shell with 5 whorls; umbilicate; color streaks and blotches can be either a strong or weakly developed feature but are always present; no teeth present; transverse striae are well developed and rib-like, having a count of 55 to 65 in the last whorl and which are well defined to the base of the shell; periphery rounded to bluntly angular, variable.

Similar Species: *Anguispira alternata* is larger, has a wider umbilicus and has transverse striae sculpture which is notably finer; *A. mordax* has more widely spaced ribs (35 to 45 in the last whorl); for additional comparisons of *Anguispira* species, review images found on page 210 of *A. strongylodes* and *A. mordax*.

Habitat: Limestone regions along forested hillsides adjacent to rivers, usually found under the leaf litter but also among loose rock formations.

Status: G5/S2; Uncommon; in West Virginia only found in scattered locations.

Specimen: All figures from West Virginia, Mercer County, around Beaver Pond Cave entrance, figure (a) enlarged to show rib spacing (author's collection).

a

14 ribs

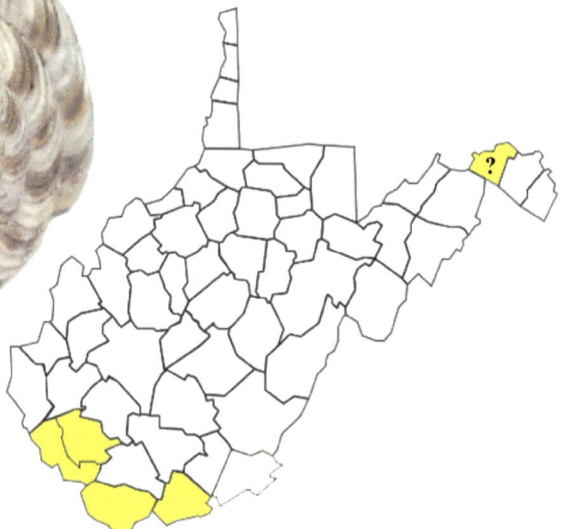

208

The newly described Greenbrier tigersnail, *Anguispira stihleri*, is a rare and endemic land snail known only from Greenbrier County, West Virginia. These exquisite watercolors of the species were painted by West Virginia resident Libby Cagle.

Illustrated below are two coarsely sculptured *Anguispira* species found in West Virginia; *A. strongylodes* and *A. mordax*, including comparative material. Shells have been cleaned to show their true color, and more importantly, the spacing of the ribs. Weathered shells are typically soiled, hiding these key features. An old toothbrush works best for cleaning the shell surface and aperture. Shells proportionate.

A. strongylodes, Beaver Pond Cave, Mercer County, West Virginia

55-65 ribs

A. strongylodes, Tennessee, Smith County, Dixon Springs (FM 238349, images by Jochen Gerber).

55-65 ribs

A. mordax, 2 miles south of Panther, McDowell County, West Virginia (CMNH 62.35086)

35-45 ribs

A. mordax, Pine Mountain State Park, Bell County, Kentucky

35-45 ribs

A. mordax, Tennessee, Campbell County, Lafollette, Well Spring (FM 238053, images by Jochen Gerber).

35-45 ribs

20 mm

The Tiger of Land Snails, *Haplotrema concavum*

Haplotrema are the tigers of land snails, eating a variety of terrestrial gastropods including *Novisuccinea ovalis, Zonitoides arboreus, Discus catskillensis* and *Strobilops labyrinthicus* (Pearce and Gaertner 1996). *Patera, Mesomphix* and *Mesodon* species are also taken by this formidable hunter (author's personal observations). To access the snail flesh, the species will typically enter through the aperture of the prey's shell where it slowly eats the snail alive. If the aperture is too small for entry, *Haplotrema* will drill (actually grind) through the side of the shell by using its radula (Pearce and Gaertner 1996). The nickel-sized gastropods have one of the widest umbilicus of any eastern land snail and are a common woodland species.

Above a Gray-foot lancetooth, Newfound Gap, Great Smoky Mountains National Park and below the same species attacking and consuming a juvenile wrinkled button, *Mesomphix rugeli* , Mitchell County, North Carolina.

Gray-foot lancetooth

Haplotrema concavum (Say, 1821)

Diameter: 16-25 mm, height 5.3-9.5 mm

Description: Depressed heliciform; lip usually simple but in fully developed shells there is a slight reflection in the lower lip (a); shell with 4.5-5.5 whorls; widely umbilicate; shell light greenish-yellow and glossy; transverse striae poorly developed; spiral striae present, but may be a wanting feature on some portion of the shell; live animal gray; a carnivorous species that feeds on other gastropods.

Similar Species: No other land snail of similar size has the wide umbilicus; the similar looking *Haplotrema kendeighi* of more southern mountains has a blue body whereas the animal of *H. concavum* is gray, their shells are indistinguishable.

Habitat: Found in mixed hardwood forests at all elevations.

Status: G5/S5; Common; one of the most common land snails in eastern North America and a widespread species in West Virginia's woodlands.

Specimen: All figures from Tennessee, Blount County, Chestnut Top Trail, GSMNP (GSMNP collection).

Umbilicus very wide

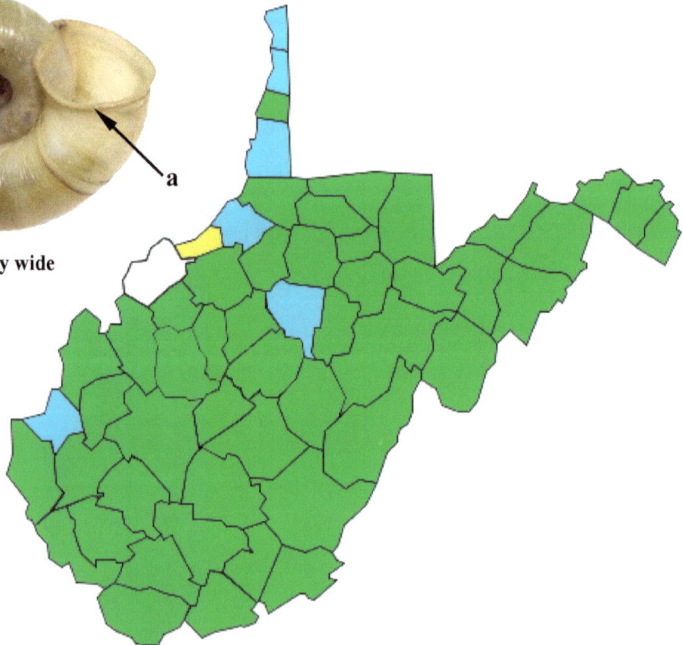

The feeding trail of an unknown species of land snail, Cheat River Gorge, Monongalia County, West Virginia.

Shells 2.4-3.4 mm, beehive shaped, no teeth, simple lip, not thickened, no operculum (*Euconulus*)

Euconulus species are small dome or beehive shaped land snails that have many tightly coiled whorls. They are generally uncommon in leaf litter of mixed hardwood forests. They differ from *Strobilops* by having a simple not reflected lip, a smoother surface (*Strobilops* having a ribbed surface), and being more dome-shaped. *Euconulus* are typically without teeth, although *E. dentatus* has internal armature in young shells. Three species, *E. dentatus, E. fulvus* and *E. polygyratus* are reliably reported from West Virginia.

Shells 2.4-3.4 mm, dome or bee-hive shape, shell surface more or less smooth (not ribbed), pg. 215-218

Euconulus dentatus
Euconulus fulvus
Euconulus polygyratus (rare)

Shells 6-8 mm, domed shaped, no teeth, simple lip and thickened, with an operculum (*Hendersonia*)

Hendersonia species are similar in shape to *Euconulus* but are much larger with a thicker shell and in live specimens, contain an operculum. *Hendersonia* differ from *Strobilops* by having a less reflected lip, a smoother surface (*Strobilops* having a ribbed surface, see below) and averages 5-6 mm larger.

Shells 6-8 mm, heliciform, transverse striae moderately developed but not ribbed, pg. 219

Shells are proportionate

Hendersonia occulta

Strobilops aeneus

Toothed hive

Euconulidae

Euconulus dentatus <small>(Sterki, 1893)</small>

Diameter: 2.4 mm, height 2.3 mm

Description: Dome shape; lip simple; shell with 6.5 tightly coiled whorls; scarcely perforate; yellowish-white and rather dull; shell thin, translucent in live snails and fresh dead; transverse striae are poorly developed (dissecting scope required); spiral striae not well defined, but always present; teeth or lamellae in young shells (a, b & c), adults usually without teeth.

Similar Species: Most closely related to *Euconulus chersinus* but differs in having armature in young shells and 2 to 3 less whorls; <u>in West Virginia, this is the only *Euconulus* that can be easily and reliably identified to species using just young shells, a result of its internal teeth</u>.

Habitat: A species found in mixed hardwood forests on hillsides and ravines living under layers of leaf litter usually in drier sites than other species of *Euconulus*; it is rarely found in large numbers.

Status: G5/S5; Fairly Common; Hubricht (1985) records indicate the species in only a few scattered counties in southern West Virginia.

Specimen: Kentucky, Hardin County, Spurrier (FM 251488).

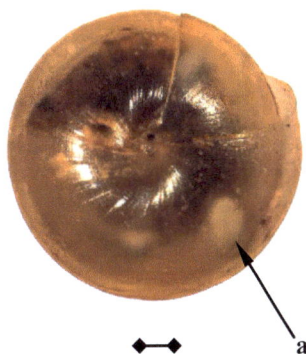

a

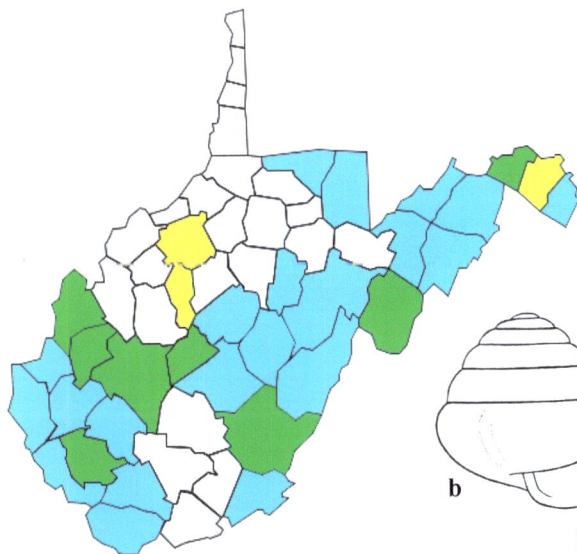

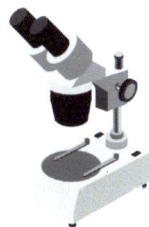

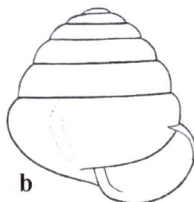

b

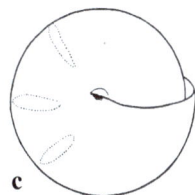

c

Pilsbry, 1946

Brown hive   Euconulidae

Euconulus fulvus (Müller, 1774)

Diameter: 3.1-3.4 mm, height 2.4 mm

Description: Dome shape; lip simple; shell with 4.5-6 tightly coiled whorls; minutely perforate or nearly closed; cinnamon or dilute tawny, the summit paler; glossy; shell thin, translucent in live snails and fresh dead; transverse striae are poorly developed (dissecting scope required); spiral striae not well defined but always present; without teeth; periphery rounded or weakly angular.

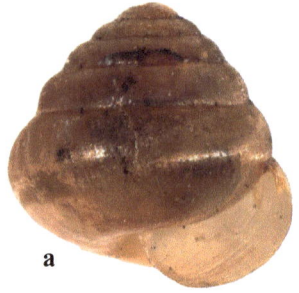

a

Similar Species: *Euconulus chersinus* (page 361) is more finely sculptured, has a more elevated shell, is less glossy and has two more whorls; *Euconulus* IDs in West Virginia should be based on having a good series of adult shells in hand; this will require that large quantities of leaf litters be collected at sites where *Euconulus* shells are encountered.

b

Habitat: A Holarctic species found in mixed hardwood forests on well-shaded hillsides and ravines living under layers of moist leaf litter; also found in creek and river drift.

Status: G5/S5; Common; this is the common *Euconulus* in West Virginia and the most likely species that will be found in leaf litter samples.

Specimen: Figure (a) from Michigan, Wayne County, Belle Isle, Detroit (FM 251565) and figure (b) from David Kirsh and Bill Frank at www.jaxshells.org.

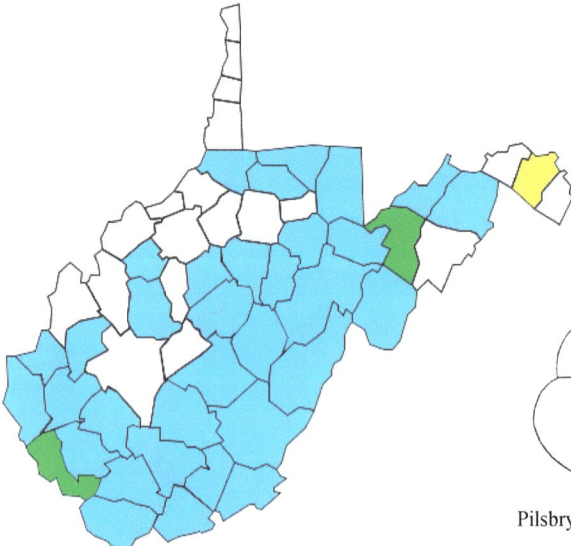

Pilsbry, 1946

Fat hive

Euconulus polygyratus (Pilsbry, 1899)

Diameter: 2.7-3.3 mm, height 2.3-2.8 mm

Description: Dome shape; lip simple; shell with 6-7 tightly coiled whorls; scarcely perforate; aperture narrow; cinnamon-buff, somewhat dull and silky except in the middle of the base which is glossy and somewhat transparent; without teeth; in mature specimens the periphery is rounded, but in immature shells the periphery is sharply angular (Pilsbry 1946); it will be difficult to determine a correct ID with only half grown *Euconulus* shells, therefore it is imperative that adult shells also be secured at collection sites.

Similar Species: Shell much like *E. chersinus*, but with a fatter build; *Euconulus fulvus* is more glossy and has 1 to 2 fewer whorls in adult shells.

Habitat: A northern species found in mixed hardwood forests on hillsides and ravines living under layers of moist leaf litter; a Holarctic land snail.

Status: G5/S1; Rare; this mostly northern *Euconulus* is probably only reliably reported from northern West Virginia; it was not reported by Hubricht (1985), his closest records being in western Maryland, along WV eastern Panhandle; like *E. chersinus*, its widespread occurrence in West Virginia remains questionable.

Specimen: Maine, Aroostock County, (FM 252035).

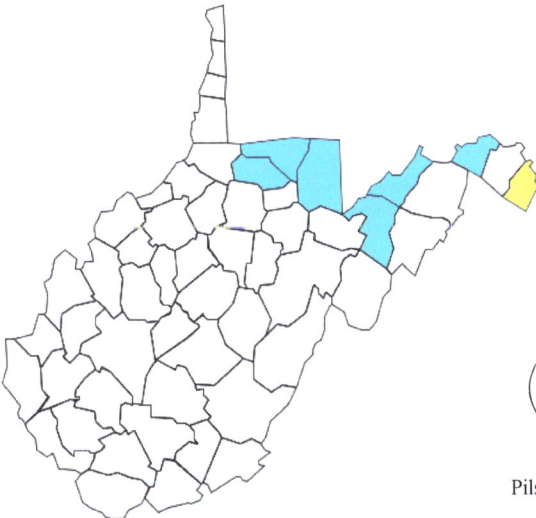

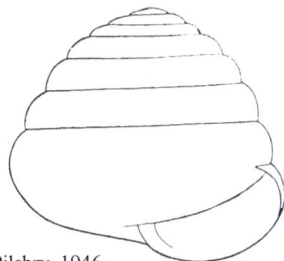

Pilsbry, 1946

Aperture narrow

Euconulus Species Compared (proportionate)

Euconulus dentatus, juvenile shell and some adult shells with teeth, Hardin County, Kentucky (see page 215)

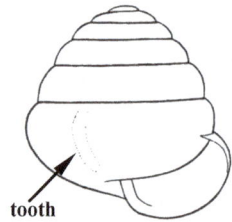

tooth

Euconulus fulvus, without teeth in young shells, Wayne County, Michigan (see page 216)

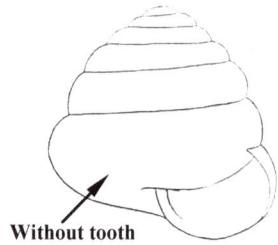

Without tooth

Euconulus polygyratus,, shells are wide (obese), or somewhat compressed for a *Euconulus,* with a narrow opening in the aperture Aroostock County, Maine (see page 217)

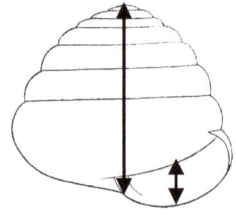

Short shell & narrow aperture

Euconulus chersinus, adult shell tall with 7-8 whorls, Swain County, North Carolina (see page 361)

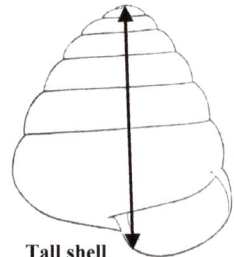

Tall shell

Euconulus trochulus, adult shell with a slightly angular periphery, Pulaski County, Kentucky (see page 362)

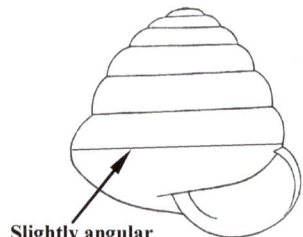

Slightly angular

218

Pilsbry, 1946

Cherrystone drop

Helicinidae

Hendersonia occulta (Say, 1831)

Diameter: 6-8 mm, height 4-6 mm

Description: Heliciform; lip thickened (a & b); shell with 4.5-5 whorls; imperforate; with an operculum in live individuals; transverse striae moderately developed; spiral striae are well developed and sometimes seen as fringes in young shells under 4 mm; periphery slightly angular and occasionally weakly keeled, particularly in juvenile shells; eyes are located at the base of the tentacles, (a trait common to a primitive and largely tropical family of gastropods); this species is unique among West Virginia land snails by being dioecious (having the male and females separate).

Similar Species: *Stenotrema* species are around the same size, but have a slit like aperture and are without an operculum.

Habitat: A calciphile snail of river bluffs, talus slopes, ravines and mountainsides.

Status: G4/S3; Uncommon; this species is reported from scattered locations across the state and will likely continue to be found in new locations as surveys for minute species increase.

Specimen: West Virginia, Randolph County, mouth of Bowden Cave (author's collection).

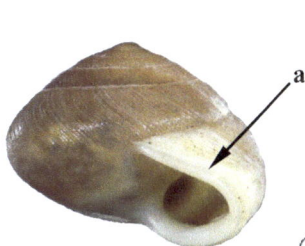

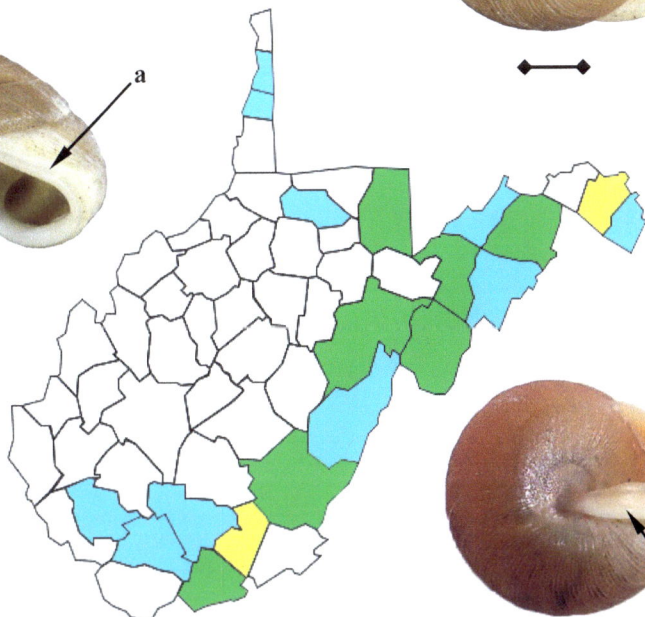

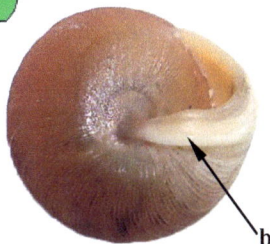

Close up of micro-sculpture

A juvenile Cherrystone drop, *Hendersonia occulta,* illustrating the stark difference between adult and young shells. Note the remarkable micro-sculpture and carination of the shell before adulthood. Specimen from "The Trough" along the South Branch of the Potomac River, Hampshire County, West Virginia.

Carinate periphery

Tiny Gems Hiding Among Rock Talus

Punctum vitreum

Helicodiscus villosus

Guppya sterkii

Carychium clappi

Vertigo gouldii

Glyphyalinia rhoadsi

Paravitrea multidentata

Cochlicopa morseana

Deep rock talus creates an exceptional hideaway for a number of minute (under 5 mm) land snails, including several gastropods that are globally rare. An uncommon micro-climate forms in talus, keeping conditions cool and humid, a land snail's best friend. The deeper the talus, the more stable the micro-climate. Upwards to 15 species can be found in just one heaping handful of leaf litter pulled from the compact spaces between the boulders.

This section includes land snails that are more or less the size and shape of pills. All species of *Stenotrema* are under 15 mm in diameter (in WV most species are between 7 and 10 mm), the majority possessing hairs (see below picture) that cover the entire shell surface. These hairs collect forest debris, thought to help keep the snails well hidden from possible predators like birds and salamanders. The aperture of *Stenotrema* species are narrow (slit like), denying easy access to predatory beetle larva. The long and wide parietal tooth of *Stenotrema* easily exceeds that of all other snails of similar size.

Genera Included:
(in order of appearance in text)

Stenotrema

Shells 4-12 mm, <u>pill shape</u>, long parietal tooth (*Stenotrema*)

Stenotrema species are usually compact and pill-like in shape, having tightly coiled shells (a), the last whorl not expanded and most are with fine or coarse hairs. All species have a long parietal (blade-like) tooth. *Stenotrema* species are best distinguished in frontal and bottom views, top views having very little diagnostic value within the genus. Examining the aperture is key (figures e through m, page 224). The aperture opens on the underside of the shell and is slit-like and narrow. The interdenticular sinus (figures l and m, page 224) is deep in some species and shallow in others. Another important trait is the basal notch (figures g, h, i, j and k, page 224), which can be nearly absent or a prominent feature of the lip. The fulcrum (below, bottom image), an internal structure and part of the center supporting column of the shell, can be different lengths and can be seen through the bottom of live and fresh shells, but is difficult to see in old and badly weathered shells. *Stenotrema* differ from *Euchemotrema* in having a more closed aperture (b), a longer and wider parietal tooth (c), a completely sealed umbilicus (d), and average 2-3 mm smaller.

<u>*Stenotrema* species of West Virginia, pg 225-232</u>

*Stenotrema stenotrema
*Stenotrema macgregori (rare)
*Stenotrema hirsutum
*Stenotrema simile (rare)
*Stenotrema edvardsi
⁺Stenotrema barbatum

a

d

Stenotrema

c

Fulcrum

b

Euchemotrema fraternum

Umbilicus slightly open

Aperture Terminology in *Stenotrema*

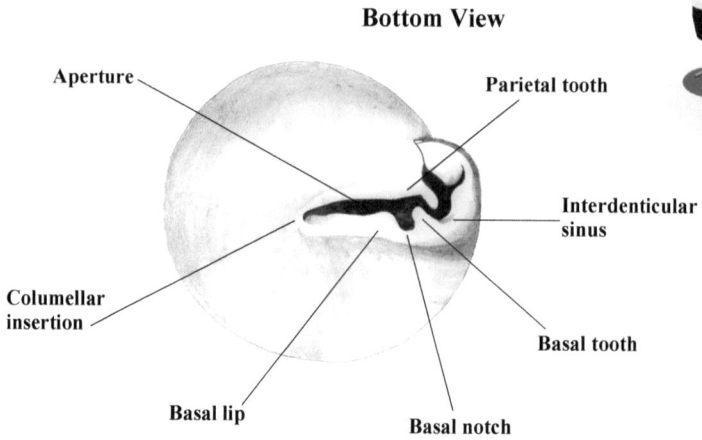

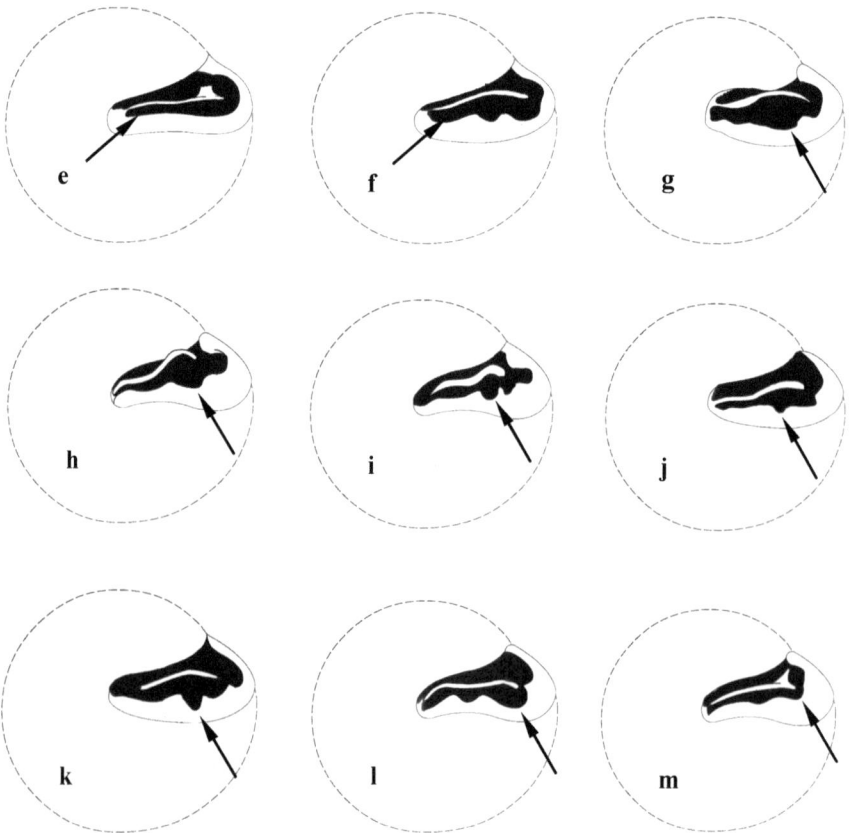

Bottom View

e) aperture narrow; f) aperture wide; g) basal notch wide; h) basal notch medium; i) basal notch narrow; j) basal notch shallow; k) basal notch deep; l) interdenticular sinus deep; m) interdenticular sinus shallow

Inland slitmouth

Polygyridae

Stenotrema stenotrema (L. Pfeiffer,1842)

Diameter: 7.8-15 mm, height 5.3-8.6 mm

Description: Pill shape; aperture narrow, especially near the columellar insertion (a); shell with 5-6 whorls; imperforate; brownish-tan or cinnamon-brown; minute hairs present on entire surface, but are usually lost in aging shells; transverse striae poorly developed; basal notch is large and deep (b); interdenticular sinus deep (c); fulcrum is well developed with a convex edge (seen through an opening made on the side of the shell); periphery well rounded; this is the largest *Stenotrema* in the state.

Similar Species: *Stenotrema edvardsi* has an angular periphery; *S. macgregori* is smaller, has a more open aperture opening, but most importantly, a smaller basal notch and indistinct interdenticular sinus.

Habitat: Usually found under leaf litter in lower elevation mixed hardwood forests.

Status: G5/S5; Common; one of the most wide ranging *Stenotrema* in eastern North America and a common resident of West Virginia, especially in the southern portions of the state.

Specimen: Tennessee, Sevier County, Green Brier Picnic Area, GSMNP (GSMNP collection).

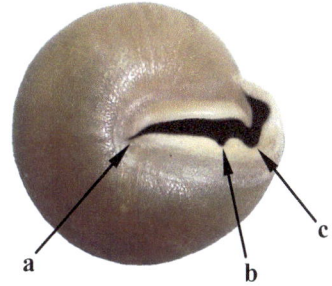

a b c

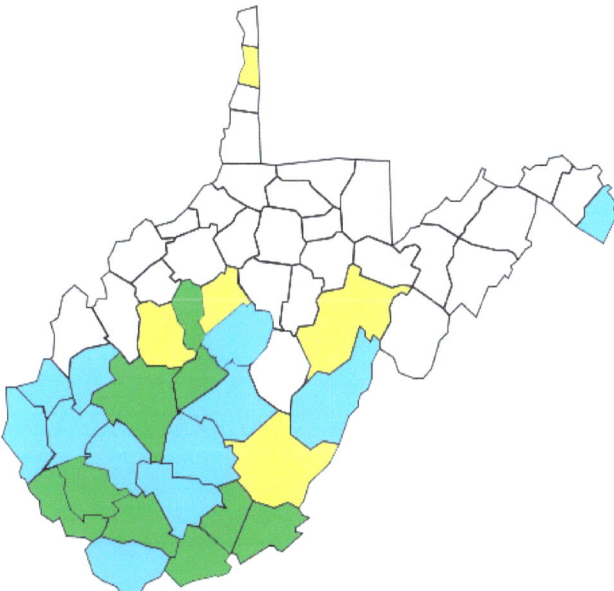

Fraudulent slitmouth

Polygyridae

Stenotrema macgregori Dourson, 2011

Diameter: 8-10 mm, height 6 mm

Description: Pill shape; lip reflected; shell with 5-6 whorls; imperforate; cinnamon-buff; short stiff hairs present on entire surface, but hairs are often lost in older shells; transverse striae poorly developed; basal notch shallow and the interdenticular sinus is indistinct in most specimens; fulcrum is well developed, but short; periphery rounded.

Similar Species: *Stenotrema stenotrema* is 2-3 mm larger, has a deeper basal notch, more-closed aperture, and the interdenticular sinus is notably excavated; the two species are occasionally found together and can readily be separated by size alone.

Habitat: Found in rich hardwood forests under forest litter, apparently absent from the drier mountain tops, dense rhododendron thickets, and Virginia pine forests.

Status: GNR/S2; Uncommon; in Kentucky this recently described species was formerly reported as *S. stenotrema*, as was likely the case in West Virginia where it is now confirmed from at least six counties.

Specimen: Kentucky, Pike County, Breaks Interstate Park (author's collection).

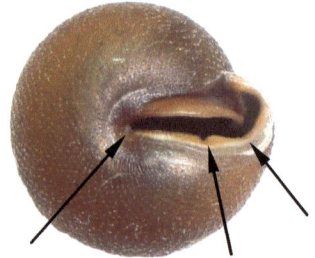

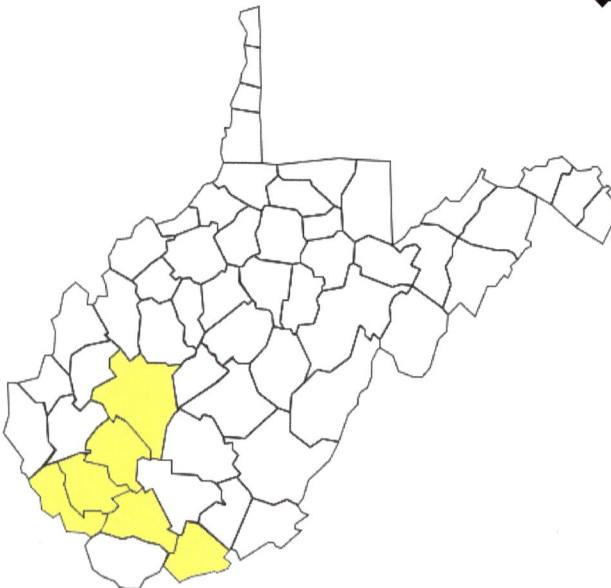

Hairy slitmouth

Polygyridae

Stenotrema hirsutum (Say, 1817)

Diameter: 6.2-9.6 mm, height 4.3–5.7 mm

Description: Pill shape; aperture width generally wider than other *Stenotrema* species; shell with 5-5.5 whorls; imperforate; cinnamon-buff to tan; short, stiff hairs present on entire surface, but hairs are generally lost in older shells; transverse striae poorly developed; basal notch and the interdenticular sinus are moderately deep; the fulcrum (a) is well developed, long (around 2 mm), and can be seen through the bottom view of fresh shells; shell periphery is boxy-shaped in frontal view.

Similar Species: Most similar to *S. simile,* but with less prominent lip features, and most importantly, a much longer fulcrum than seen in *S. simile.*

Habitat: A species of dry upland mixed hardwood forests found under a variety of forest litter; also found hiding under logs and rocks and in clearings.

Status: G5/S5; Common; this widespread and frequent land snail is reported from nearly every West Virginia county.

Specimen: West Virginia, Pendleton County, along the South Branch of the Potomac River near the town of Franklin (author's collection).

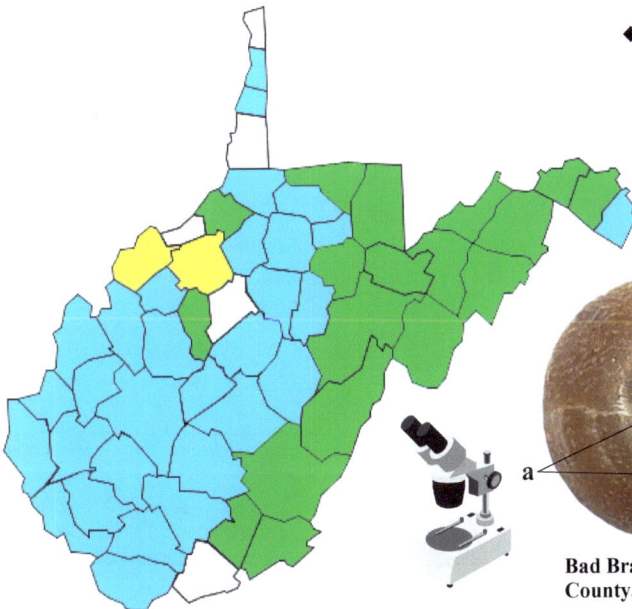

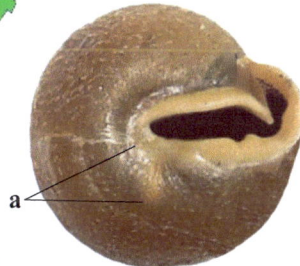

Bad Branch NR, Letcher County, Kentucky

227

Bear Creek slitmouth

Stenotrema simile Grimm, 1971

Diameter: 9.3 mm, height 6.7 mm

Description: Pill shape; lip reflected; shell with 5.5 whorls; imperforate; color variable, ranging from deep cinnamon-brown to olive buff; short stiff hairs usually present on live and fresh shells; transverse striae poorly developed; basal notch and interdenticular sinus (a) well-developed and deep; fulcrum quite short, barely projecting beyond the basal lip callus (b & c), around 1 mm long; shell periphery roundish, with a slight shoulder.

Similar Species: *Stenotrema edvardsi* has an angular periphery; *S. stenotrema* is larger and has a more narrow aperture; the interdenticular sinus and basal notch of *S. hirsutum* are less well developed, but the fulcrum is fully 2 mm long.

Habitat: A snail of cool, wet hardwood forests under leaf litter and logs of rocky hillsides.

Status: G2/S2; Uncommon; a gastropod nearly endemic to West Virginia; while this scarce snail is reported from eight counties, little is known of its basic biology.

Specimen: West Virginia, Randolph County, Mingo Knob (author's collection).

Polygyridae

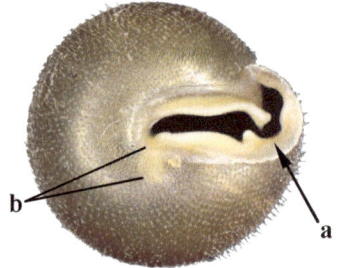

Side view

Ridge and Valley slitmouth

Polygyridae

Stenotrema edvardsi (Bland, 1856)

Diameter: 7-8 mm, height 4.8-5.4 mm

Description: Pill shape; shell with 5-5.5 whorls; imperforate; tawny-olive to light cinnamon colored; stiff and well-developed hairs, especially on young animals, but usually lost on aging shells (see opposite page); transverse striae poorly developed; basal notch small and shallow; interdenticular sinus indistinct; fulcrum short and moderately developed; shell periphery distinctly and always angular.

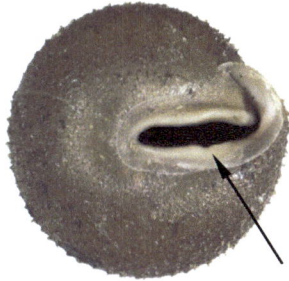

Similar Species: *Stenotrema hirsutum* is more or less the same size but differs in having a deeper interdenticular sinus, a deeper basal notch, and most notably, a boxy (not angular) periphery.

Habitat: Rocky mixed hardwood forests, around and under logs and in leaf litter; reported to be very common in hemlock dominated ravines on hillsides (Hubricht 1985).

Status: G4G5/S3; Fairly Common; in West Virginia, this long-haired gastropod is reported from scattered locations across the state.

Specimen: West Virginia, Mercer County, Camp Creek State Park (author's collection)

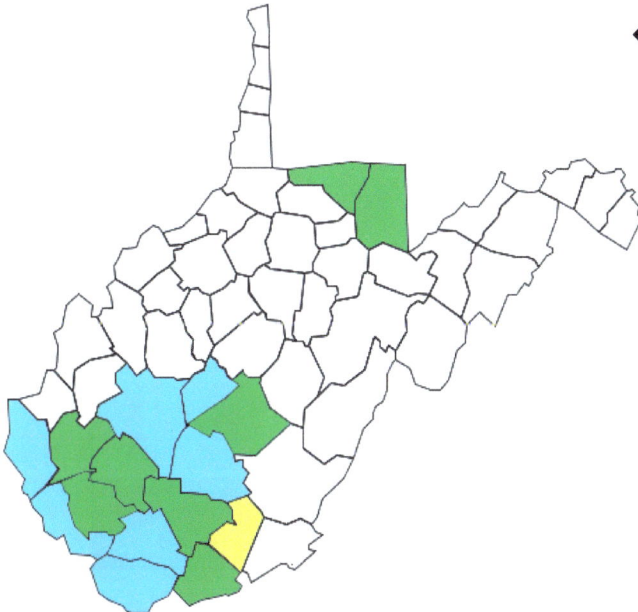

Above image of a juvenile Ridge-and-Valley slitmouth, *Stenotrema edvardsi,* crawling toward dinner; a colony of maturing slime mold growing from rotting wood. Below, observe the numerous hairs that cover the entire shell surface; the longest ones growing from the sutures. These hairs are usually lost as shells mature. Both images from Red River Gorge, Kentucky.

Bristled slitmouth

Polygyridae

Stenotrema barbatum (G. H. Clapp, 1904)

Diameter: 7.5-11 mm, height 5.5-7 mm

Description: Pill shape; lip reflected; shell with 5.5 whorls; imperforate; tawny-olive to light cinnamon colored; well-developed, stiff hairs (0.5 mm long) usually present on live and fresh shells, but lost with age; transverse striae poorly developed; basal notch shallow and the interdenticular sinus is indistinct; fulcrum long and well developed, fully 3 mm in length (figures a, b & c); shell periphery not well-rounded, with a slight shoulder.

Similar Species: Most like S. *hirsutum,* but larger and with strongly developed, stiffer hairs, 0.5 mm long; S. *edvardsi* has an angular periphery; S. *stenotrema* has a more narrow aperture opening and is less hairy.

Habitat: A snail found in mixed hardwood forests of floodplains under leaf litter and logs; sometimes an upland species.

Status: G5/S3; Uncommon; reported from scattered locations across the state.

Specimen: West Virginia, Summer County, Bluestone River (author's collection).

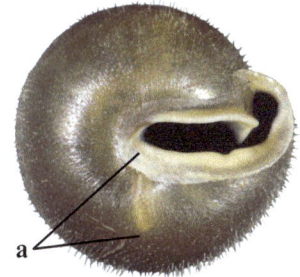

a

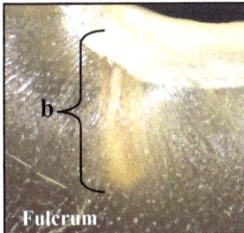

b

Fulcrum

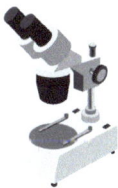

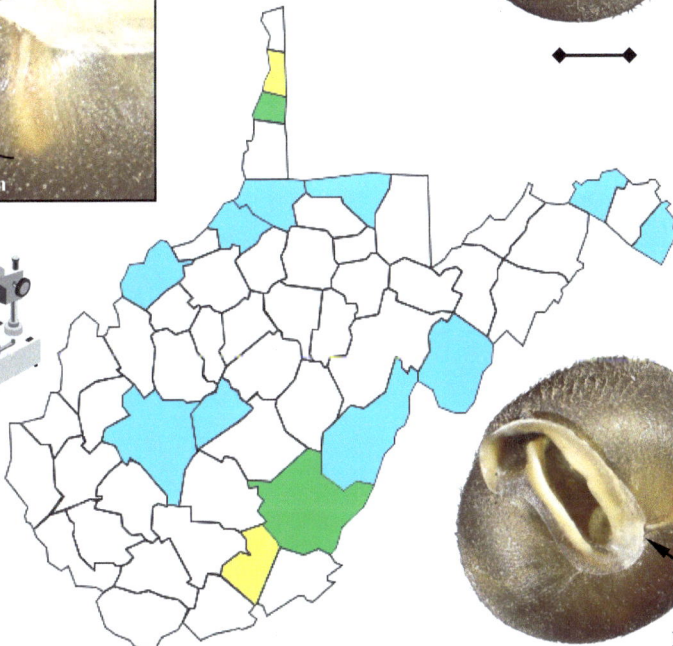

c

Fulcrum

Stenotrema and *Euchemotrema* species compared (proportionate)

10 mm

S. stenotrema, KY

S. macgregori, KY

angular

S. edvardsi, WV

Shouldered

S. barbatum, WV

Boxy

Rounded

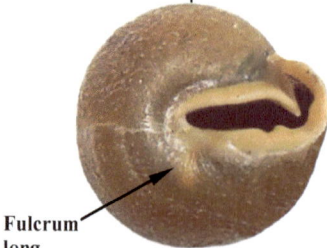

Fulcrum
long

S. hirsutum, KY

Fulcrum
short

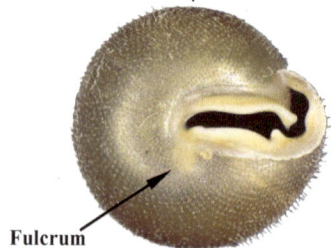

S. simile, WV

More open

Euchemotrema leai, WV

Euchemotrema fraternum, WV

The Protective Hairs of *Stenotrema*

Above image of a Highland slitmouth, *Stenotrema altispira,* covered in bits of soil and detritus. Below, a clean shell of same species showing the remarkable hairs that jacket the shell's exterior. The hairs collect material, thought to aide in crypsis, perhaps keeping the snail out of plain view of predators such as salamanders and birds. Both images from Roan Mountain, North Carolina.

This section includes land snails that possess shells which are wider than tall, heliciform or depressed heliciform, with or without teeth, and most importantly, have reflected lips. Shells in this group can be small, 5 mm or less, such as *Vallonia*, but most are among the largest land snails found in West Virginia, some reaching 45 mm in diameter. Most are unadorned in color, although some gastropods such as *Webbhelix* and *Allogona* are rather colorful.

Genera Included:
(in order of appearance in text)

Strobilops
Vallonia
Euchemotrema
Inflectarius
Patera
Mesodon
Appalachina
Neohelix
Triodopsis
Xolotrema

A grand globe, *Mesodon normalis,* strips flesh for protein and rasps bone for calcium from a decomposing raccoon carcass, Hot Springs, North Carolina.

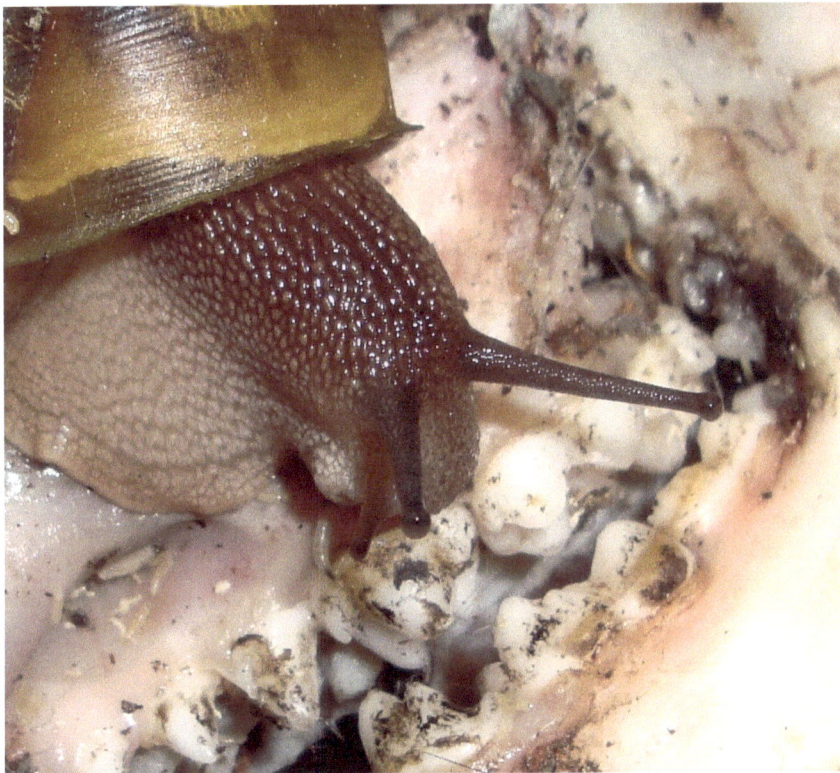

Shells under 3 mm, perforate to umbilicate, lip only slightly reflected, <u>teeth always present</u> (*Strobilops*)

Strobilops are generally uncommon gastropods in rich woods, becoming more common in glades, dry limestone outcrops and strangely enough, parking lots. They are easily separated from all snails of similar size and form by having a reflected lip; most snails of comparable size have simple lips. While *Strobilops* are dissimilar enough from *Vallonia* to make their separation trouble free, separating *Strobilops* from each other is an arduous task. Illustrated below are the two groupings to follow.

<u>Shell 2.4-2.8 mm, lip narrowly reflected, top of shell notably ribbed, internal armature present, narrowly umbilicate, pg 237-241</u>

Strobilops aeneus
Strobilops labyrinthicus
Strobilops affinis
Strobilops texasianus

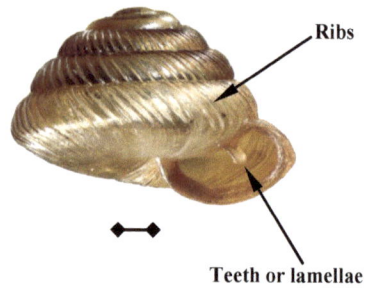

Ribs

Teeth or lamellae

Shells under 3 mm, umbilicate, reflected lip, <u>without teeth,</u> with or without paper-thin ribs (*Vallonia*)

Vallonia like *Strobilops* are generally uncommon gastropods in rich woods, becoming more common in glades, dry limestone outcrops and abandoned parking lots. They are easily separated from all snails of similar size and form by usually having a widely reflected lip, although in several species the lip is rather narrowly reflected; most other snails of comparable size containing simple lips.

<u>Shell 1.8-2.3 mm, lip widely reflected, without internal armature, widely umbilicate, pg 242-247</u>

Vallonia excentrica
Vallonia pulchella
Vallonia costata
Vallonia perspectiva

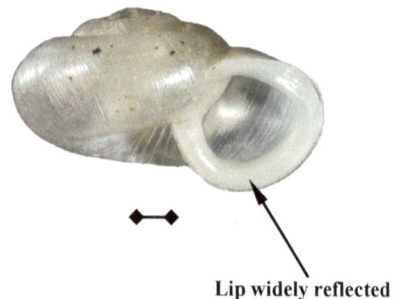

Lip widely reflected

Bronze pinecone

Strobilopsidae

Strobilops aeneus Pilsbry, 1926

Diameter: 2.4-2.8 mm, height 1.5-2 mm

Description: Dome shape; lip reflected; shell with 5.5 tightly coiled whorls; perforate to umbilicate; light to dark brown; elongated lamellae present in the aperture; within the interior are long lamellae of uneven lengths which can be seen through the bottom of live and fresh dead shells or through the cutaway top; transverse striae modified into minute ribs; periphery angular.

Similar Species: *Strobilops labyrinthicus* has a slightly smaller umbilicus, higher shell profile, and a more rounded periphery; *S. affinis* has a much higher shell (see page 241 for comparison of species); *Euconulus* species are beehive-shaped, have smooth surfaces instead of ribs and have simple lips, not slightly reflected as in *Strobilops*.

Habitat: Found in open, glade like areas especially around limestone outcrops but also commonly found in upland woods under the bark of rotting logs often in association with *Zonitoides arboreus*.

Status: G5/S5; Common, in West Virginia this minute species is reported from scattered locations across the state.

Specimen: Larry Watrous photo collection.

lamellae

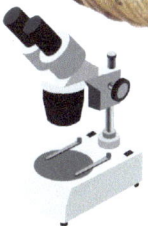

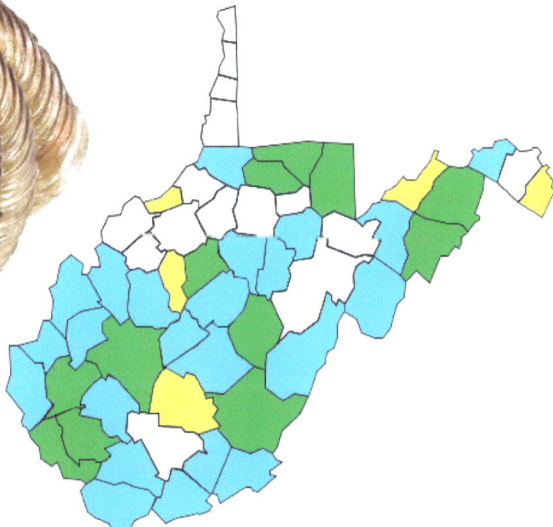

Maze pinecone

Strobilops labyrinthicus (Say, 1817)

Diameter: 2.3-2.5 mm, height 1.7-1.8 mm

Description: Dome shape or beehive in form; lip only slightly reflected; shell with 5.5 tightly coiled whorls; perforate; chestnut-brown; elongated lamellae present in the aperture; within the interior are long lamellae of uneven lengths which can be seen through the bottom of live and fresh dead shells (this page, figure a and page 241, figures c and d); transverse striae are modified into well-developed ribs on the top and side, becoming weaker on the base (this feature is easily view with a hand lens of 10X); periphery of body whorl is rounded or bluntly angular.

Similar Species: *Strobilops aeneus* has a lower shell profile, wider umbilicus, and most importantly, a more angular periphery; *S. affinis* has basal folds that are more or less the same length and line up whereas in *S. labyrinthicus,* these same folds are different lengths and more random in their spatial orientation to each other (see page 241) .

Habitat: Found in mixed hardwood forests and glade-like areas especially around loose soils of limestone substrate with red cedar trees.

Status: G5/S5; Common; this is likely the most common of the two *Strobilops* reliably reported from West Virginia.

Specimen: Larry Watrous photo collection.

lamellae

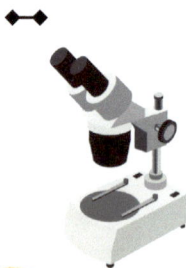

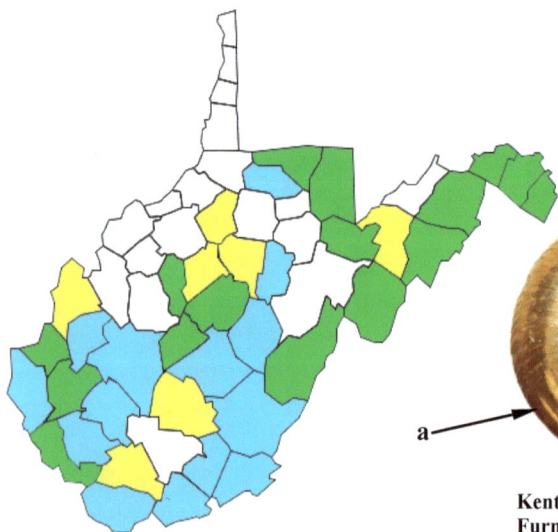

Kentucky, Powell County, Furnace Mountain

238

Eightfold pinecone

Strobilops affinis Pilsbry, 1893

Diameter: 2.7 mm, height 2.5 mm

Description: Dome shape; lip reflected; shell with 6 tightly coiled whorls; perforate to umbilicate; light to dark brown, glossy; elongated teeth or lamellae present in the aperture; elongated lamellae present in the aperture; within the interior are a series of rather short, nearly equal lamellae, that are in line (this page, figures a & b and page 241, figures a & b); these lamellae are easily seen through the bottom of live and fresh dead shells or through the opened top (page 241, figure b); transverse striae modified into minute ribs.

Similar Species: *Strobilops labyrinthicus* has a smaller umbilicus, notably lower shell profile and a more roundish periphery; *S. texasianus* has stronger developed riblets on the base.

Habitat: Found in mixed hardwood forests and in both open and shaded glades.

Status: G5/SNR; even though this mostly northern species is reported from multiple counties in West Virginia by various investigators, Hubricht, who spent a fair amount of time collecting in the state, did not find it, his closest records in Meigs County, Ohio; all specimens examined from the Carnegie Museum labeled as this species were either *S. aeneus* or *S. labyrinthicus*, consequently, its occurrence in West Virginia remains doubtful.

Specimen: New York, Dutchess County, upper Red Hook (FM 234948).

Strobilopsidae

a

Underside showing the aligned lamellae

239

Southern pinecone

Strobilopsidae

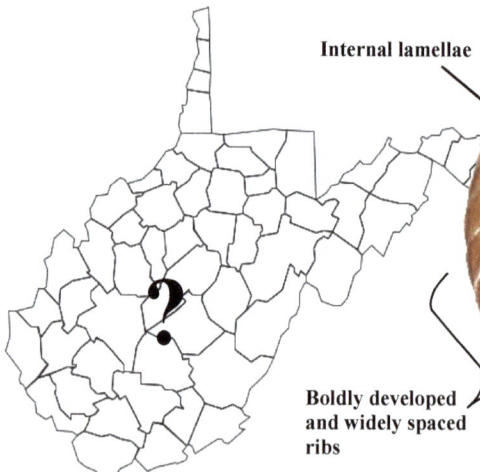

Strobilops texasianus Pilsbry & Ferriss, 1906
Diameter: 2.3-2.4 mm, height 2 mm
Description: Dome shape; lip reflected; shell with 5.5 tightly coiled whorls; perforate to umbilicate; shell pale corneous; elongated lamellae present in the aperture; within the interior are long lamellae of uneven lengths, these lamellae easily seen through the bottom of live and fresh dead shells or through the cutaway top (see page 241); transverse striae modified into strongly developed ribs, continuing boldly onto the base of the shell; periphery very slightly angular or roundish.

Similar Species: *Strobilops labyrinthicus* has a smaller umbilicus, weaker riblets and a more rounded periphery; *S. aeneus* is lower in profile and has a strongly developed angular periphery.

Habitat: A *Strobilops* of wetter ground living among and under leaf litter of mixed hardwood forests.

Status: G5/SNR; although this mostly southern species is reported from multiple counties in West Virginia by various investigators, Hubricht did not include it in his distribution records (1985), his closest accounts being in eastern Maryland and Virginia and to the south, central Tennessee; all specimens examined from the Carnegie Museum labeled as this species were *S. labyrinthicus*; for that reason, its occurrence in West Virginia remains highly questionable.

Specimen: Larry Watrous photo collection.

Internal lamellae

Boldly developed and widely spaced ribs

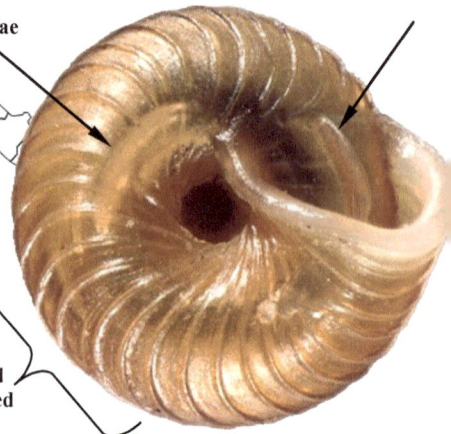

240

Strobilops Shells compared (illustrations from Pilsbry 1948)

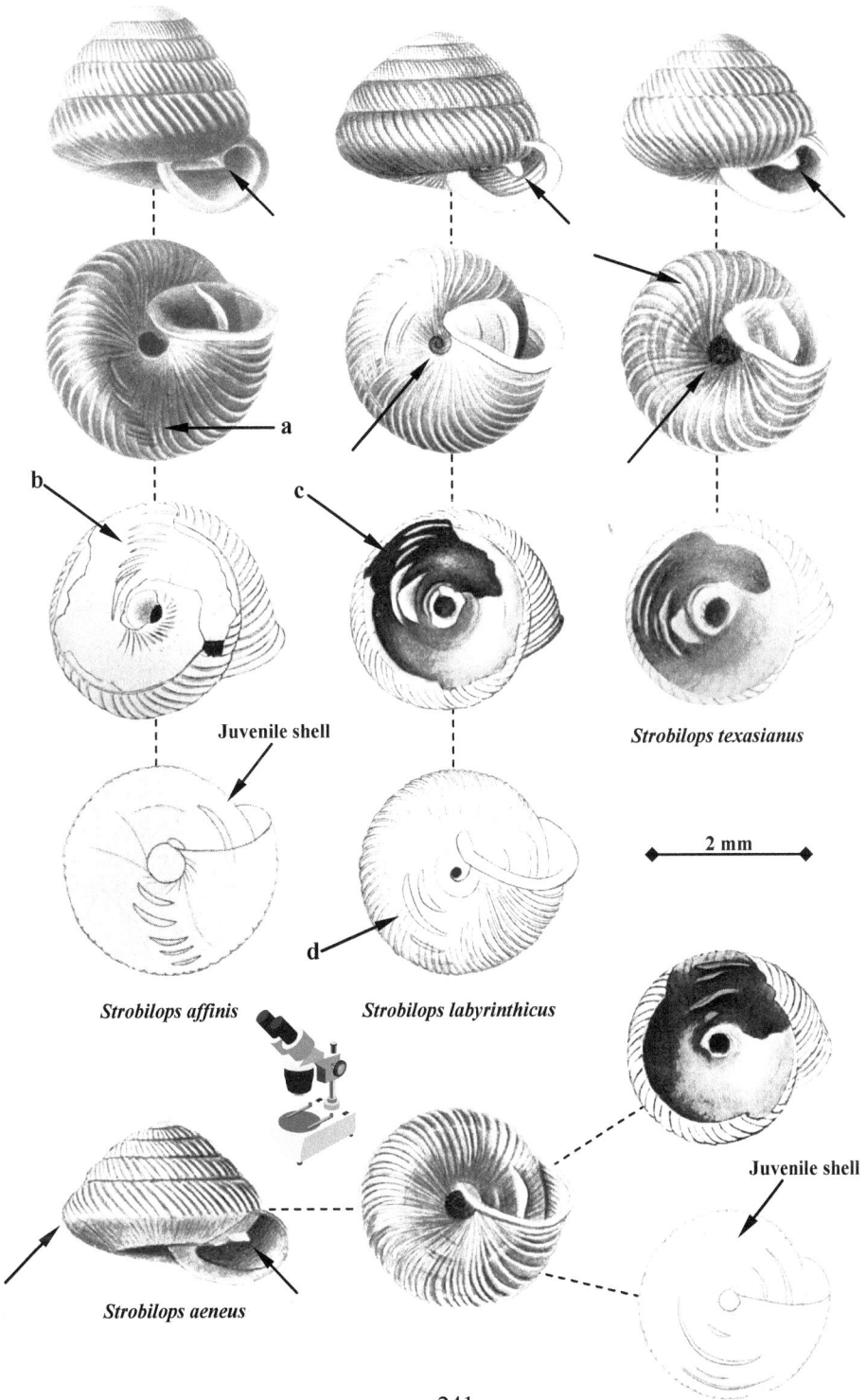

a

b

c

Juvenile shell

Strobilops texasianus

Strobilops affinis

Strobilops labyrinthicus

d

2 mm

Strobilops aeneus

Juvenile shell

241

Iroquois vallonia

Valloniidae

Vallonia excentrica Sterki, 1893

Diameter: 1.8-2.3 mm, height 1.1 mm

Description: Depressed heliciform; lip reflected and heavily thickened within (a); shell with 3-3.5 loosely coiled whorls, the last whorl expanded but not flaring outward (b); umbilicate; pale corneous or white; semi-transparent; shell surface smooth with only faint transverse striae (dissecting scope required); no teeth present at any stage of growth.

Similar Species: *Vallonia pulchella* has a more narrow umbilicus, although this distinguishing feature between the two species appears to vary considerably; most importantly, the outer lip of *V. pulchella* turns outward abruptly at the end of its growth (also see page 247); other *Vallonia* species in West Virginia have ribbed (not smooth) surfaces.

Habitat: Found in open habitat such as glades, grassy edges and roadsides where it is frequently found with *V. pulchella* (Hubricht 1985); like *V. pulchella*, *V. excentrica* is something of a "tramp" snail (Pilsbry 1948).

Status: G5/S3; Uncommon; reported from a dozen counties in West Virginia and like most *Vallonia*; typically occurs in large numbers.

Specimen: All figures from Larry Watrous photo collection.

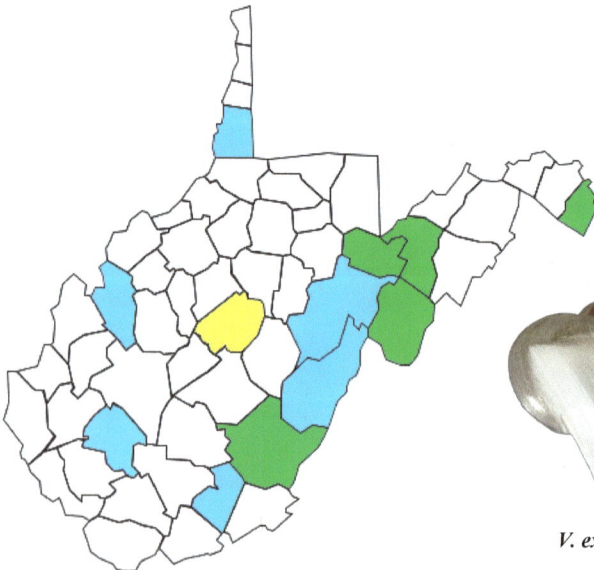

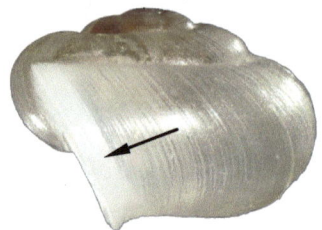

V. excentrica, side view

Lovely vallonia

Valloniidae

Vallonia pulchella (Müller, 1774)

Diameter: 2-2.5 mm, height 1.2 mm

Description: Depressed heliciform; lip re-flected and thickened within; shell with 3-3.5 loosely coiled whorls, the last whorl greatly expanded; the lip flaring outward (a); widely umbilicate; pale corneous or white; semi-transparent with a somewhat oily gloss; shell surface smooth with only faint transverse striae; no teeth present.

Similar Species: *Vallonia excentrica* has a wider umbilicus and a lip that expands only slightly whereas the lip of *V. pulchella* expands abruptly at the end of growth (see page 247); other *Vallonia* species have ribbed surfaces.

Habitat: A species of open habitats such as glades, grassy edges and roadsides but also found around lawns and gardens; this common species is sometimes found crawling about in prodigious numbers after rain events.

Status: G5/S3; Fairly Common; this largely northern land snail is reported from scattered locations across the state.

Specimen: Kentucky, Jefferson County, Echo Farm, 3.8 miles W Simpsonville (FM 230722).

a

V. pulchella

Larry Watrous

V. excentrica

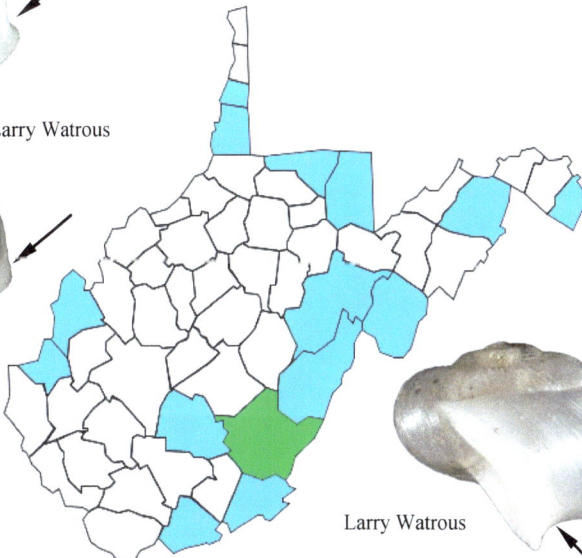

Larry Watrous

243

Costate vallonia

Vallonia costata (Müller, 1774)

Diameter: 2.5 mm, height 1 mm

Description: Depressed heliciform; lip widely reflected, thickened within; aperture lip not continuous (a); shell with 3.5 loosely coiled whorls, the last whorl greatly expanded; widely umbilicate; gray or pale yellowish-corneous; no teeth present; shell surface with 20-35 paper-thin ribs in the last whorl (see page 195 for counting ribs); periphery is well rounded.

Similar Species: *Vallonia perspectiva* is smaller in diameter and has a smaller number of ribs in the last whorl (averaging approximately 35 or more); *V. parvula* (not yet recorded from WV, page 381) is smaller and pale gray or nearly white.

Habitat: A Holarctic species of open habitat such as glades, grassy edges, and roadsides, also found at the edges of abandoned parking lots.

Status: G5/S2; Uncommon; this mostly northern gastropod is reported from eight West Virginia counties.

Specimen: Kentucky, Powell County, parking lot in the city limits of Stanton (author's collection).

Valloniidae

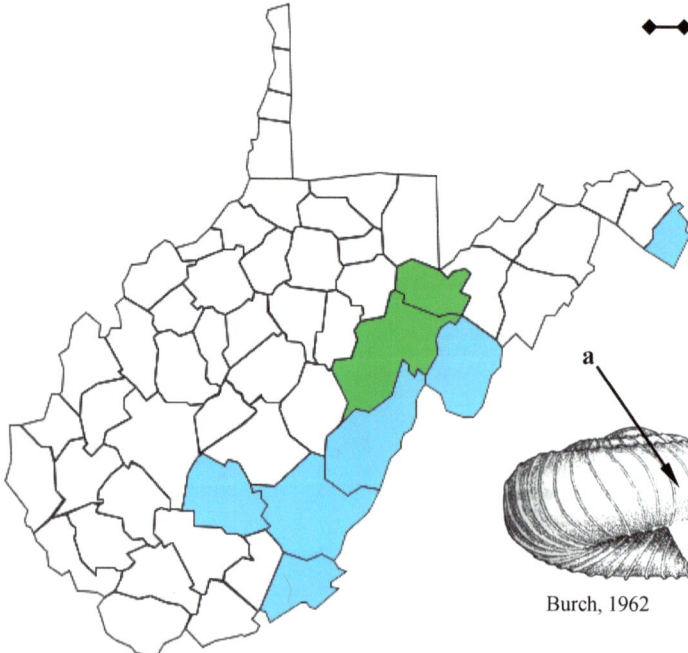

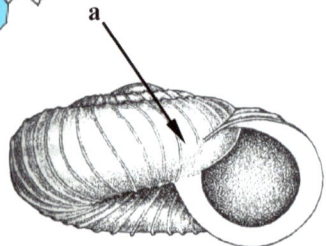

Burch, 1962

Thin-lip vallonia
Vallonia perspectiva Sterki, 1893

Valloniidae

Diameter: 1.7-2 mm, height 0.7 mm

Description: Depressed heliciform; lip only slightly reflected and thin, continuous across the body whorl or on occasion, growing free of it (a); shell with 3.5 loosely coiled whorls, the last whorl expanded; widely umbilicate; pale horn to colorless; no teeth present; shell surface with paper thin ribs (about 35 or more in the last whorl); periphery roundish.

Similar Species: Similar to *Vallonia costata* but with a greater number of ribs in the last whorl and a notably wider umbilicus (these two characters being somewhat variable) and smaller in size; *V. parvula* (page 381) has a strongly thickened lip and smaller umbilicus.

Habitat: A calciphile snail, found on talus slopes, dry glade-like limestone outcrops with mixed cedar trees.

Status: G4G5/S3; Uncommon; in West Virginia this species is currently known only from seven counties, but is likely more widespread than records indicate.

Specimen: West Virginia, Greenbrier County, near mouth of Second Creek (author's collection).

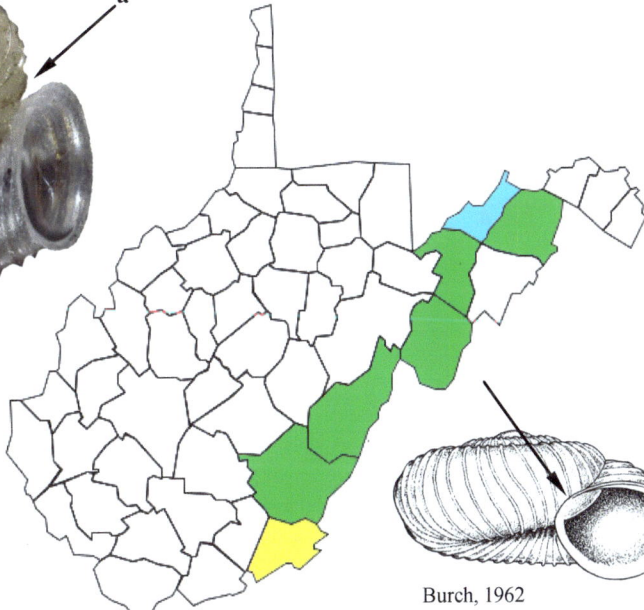

Burch, 1962

245

Vallonia snails usually live in hot dry habitats such as glades and other exposed environments. The extraordinary, paper thin ribs on the shell surface of these little jewels are thought to act like radiators, keeping the animal inside cool. Pictured here is the Thin-lip Vallonia, *Vallonia perspectiva* from Greenbrier County, West Virginia.

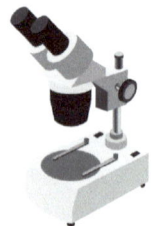

Vallonia pulchella

Vallonia excentrica

Larry Watrous photo collection

247

Shells 6-11 mm, <u>rimate</u>, slightly reflected lip, <u>long, blade-like parietal tooth</u> (*Euchemotrema*)

Euchemotrema are small woodland snails that, in general, are quite common where they are found. Both species in West Virginia have an extended parietal tooth (a) and are always rimate (b), the umbilicus covered, but never completely sealed. This character may be hard to see with the naked eye, but can usually be distinguished with a hand lens of 10X.

Shell 6-11 mm, rimate, long parietal tooth, pg 249-250

Euchemotrema fraternum
Euchemotrema leai

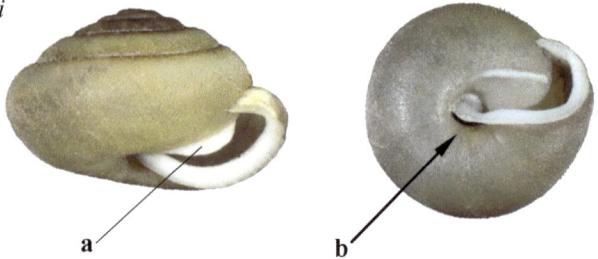

a   b

Shells 15-30 mm, <u>rimate</u>, reflected lip, if present the <u>parietal tooth small and knob-like</u> (*Mesodon*-in part)

<u>Mesodon</u> are the conspicuous land snails of West Virginia, some species reaching 35 mm or larger in diameter. Most *Mesodon* are umbilicate except for the 2 species presented here, which are always rimate. Only *M. thyroidus* is common in the state, and mostly a gastropod of degraded habitat or waste places.

Shell 15-30 mm, rimate, with or without teeth, shell globose, pg 251-252

Mesodon clausus (rare)
Mesodon thyroidus

Parietal tooth small
and knob-like

Upland pillsnail

Polygyridae

Euchemotrema fraternum (Say, 1824)

Diameter: 7.8-11.4 mm, height 5.2-6.9 mm

Description: Heliciform; lip reflected; shell with 5-6 tightly coiled whorls; rimate (a), the umbilicus never completely closed; large parietal tooth that is blade like; shell dull; pale tan to cinnamon-buff; minute hairs present in last whorl but hairs are mostly lost in older shells; transverse striae poorly developed; shell periphery rounded (b).

Similar Species: Similar to *Stenotrema* species but without the basal notch and interdenticular sinus; *E. leai* has a more compressed shell, but most importantly, has a more open umbilicus, not nearly as covered by the peristome (lip) as seen in *E. fraternum*.

Habitat: A species found under forest litter and around logs in upland mixed hardwood forests in both calcareous and acidic soils.

Status: G5/S5; Common; this species is reported from nearly every West Virginia county.

Specimen: West Virginia, Greenbrier County, Jct. 63 and Ft. Spring Pike (author's collection).

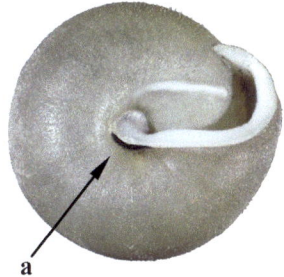

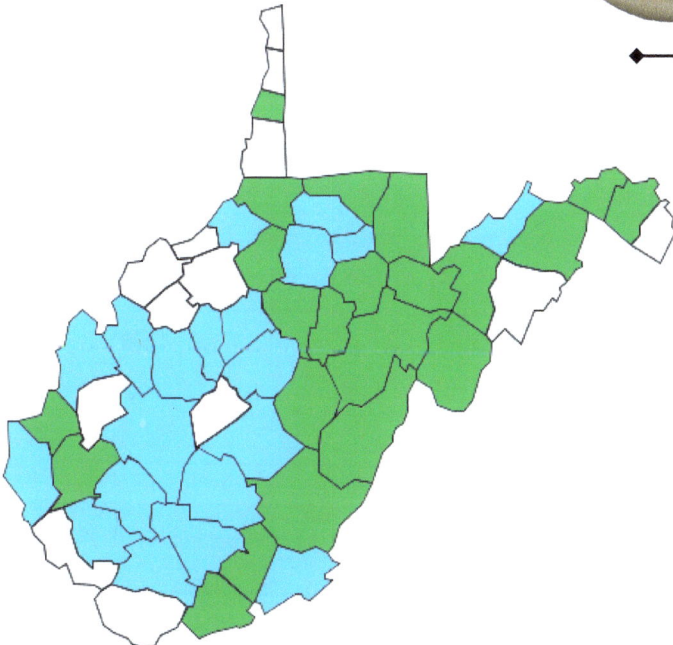

Lowland pillsnail

Polygyridae

Euchemotrema leai (A. Binney, 1841)

Diameter: 6.1-9.4 mm, height 3.9-5.7 mm

Description: Heliciform; lip reflected; shell with 5.5-6.5 closely coiled whorls; rimate (a); large parietal tooth that is blade-like; shell somewhat glossy; tannish-olive to cinnamon-buff; minute hairs present in last whorl, but hairs are often lost in older shells; transverse striae poorly developed; shell periphery shouldered (b).

Similar Species: Like *Stenotrema* species but without the basal notch and interdenticular sinus; *Euchemotrema fraternum* is larger, has a nearly closed umbilicus,, and its shell surface is a more dull glossy.

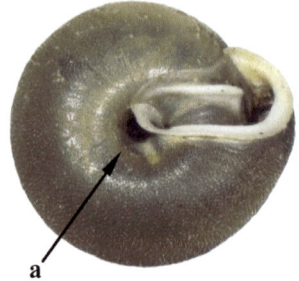

Habitat: A species of low wet areas, marshes; swamps and floodplains, but also found in meadows and along roadsides.

Status: G5/S5; Relatively Common; this species is reported from scattered locations across the state.

Specimen: West Virginia, Tucker County, Canaan Valley around rocky outcrops (author's collection).

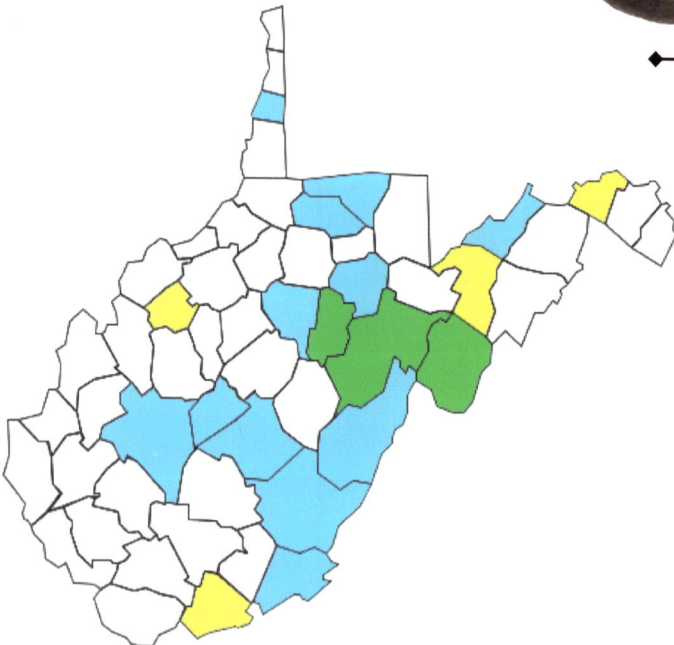

Yellow globelet

Polygyridae

Mesodon clausus (Say, 1821)

Diameter: 15-17 mm, height 10-13 mm

Description: Heliciform; lip reflected; shell with 5-5.5 loosely coiled whorls; rimate, the umbilicus never entirely closed (a); without teeth; shell glossy and thin; pale-yellow to light tan; no hairs; transverse striae moderately developed; spiral striae present; shell periphery well-rounded.

Similar Species: *Mesodon thyroidus* is larger and has a small parietal tooth; by its color, the spiral striae, and the nearly closed umbilicus, *M. clausus* resembles *Inflectarius downieanus* (a more southern species), but in larger form.

Habitat: A calciphile species of low wet areas, marshes, swamps, and floodplains; also found along roadsides and waste places such as old parking lots, damp roadside ditches, grassy slopes, and ditches along railroad tracks; in wet weather found climbing high on herbaceous plants.

Status: G5/S2; Uncommon to Rare; in West Virginia this is a scarce species, becoming more common in central Kentucky and westward.

Specimen: Tennessee, Blount County, Cades Cove, GSMNP (GSMNP collection).

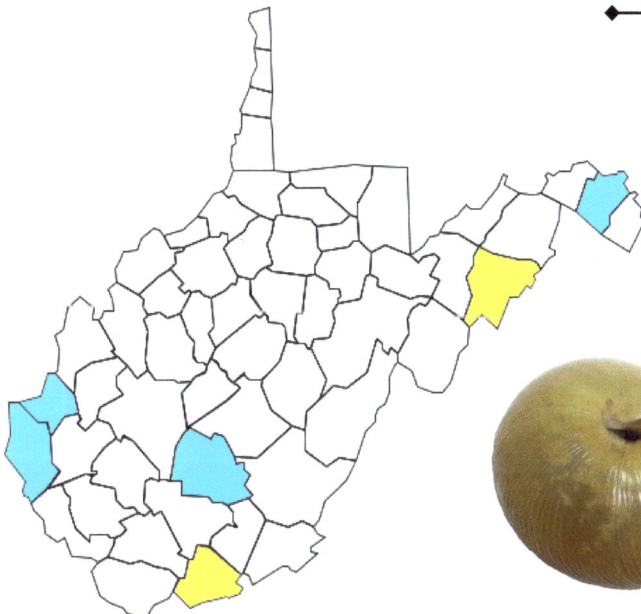

White-lip globe

Polygyridae

Mesodon thyroidus (Say,1816)
Diameter: 15-30 mm, height 11-18 mm
Description: Heliciform; lip reflected; shell with 5-5.5 loosely coiled whorls; umbilicus always rimate (a), never completely closed; small parietal tooth usually present; ivory yellow, pale yellowish green to brown; no hairs; transverse and minute spiral striae ever-present; shell periphery well-rounded.
Similar Species: *Mesodon clausus* is smaller and is without a parietal tooth; the umbilicus of *M. normalis* is always completely closed; *M. zaletus* has a thicker shell and a completely closed umbilicus.
Habitat: This species can be exceptionally common in bamboo slicks (river cane), young floodplain forests, meadows, and waste places such as grassy roadsides; it is not a species of rich woods and old growth forests.
Status: G5/S5; Common; reported from every West Virginia county; interestingly, this species was likely rare before forests were cleared.
Specimen: West Virginia, Logan County, (author's collection).

a

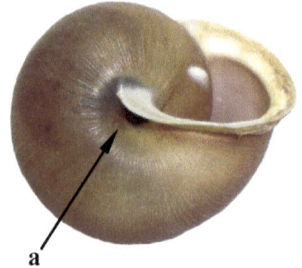

Furnace Mountain, Powell County, Kentucky

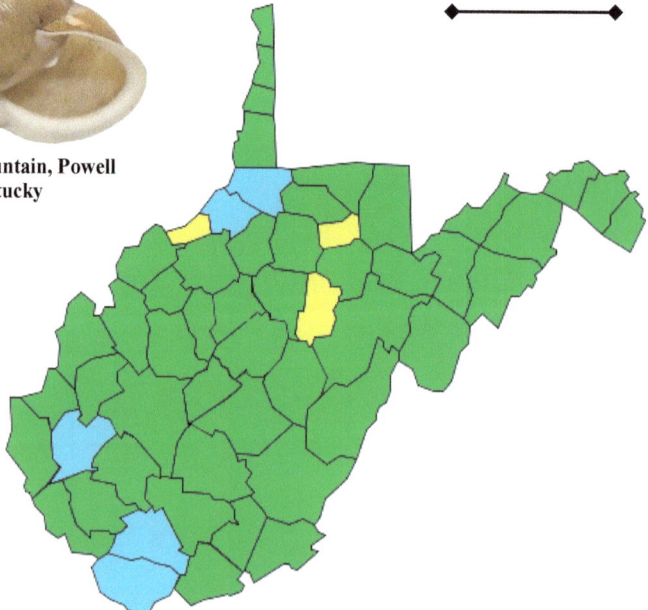

Canaan Valley Bogs, Tucker County, West Virginia

©2014 mdmpix.com

©2014 mdmpix.com

Shells under 7-15 mm, <u>imperforate</u>, reflected lip, teeth always present (*Inflectarius*)

<u>*Inflectarius*</u> species are under 15 mm, imperforate with the lip is less widely reflected than in *Mesodon*. The two *Inflectarius* in West Virginia have tightly coiled shells, are imperforate, and contain three teeth (a), the parietal being the largest. They are best distinguished in frontal view and by examining the aperture for a small basal and palatal tooth. *Inflectarius* are sometimes confused with *Stenotrema*, but a closer examination of the aperture opening reveals their differences. Similar to *Euchemotrema,* but *Inflectarius* have a basal and palatal tooth and are imperforate, not rimate. Juvenile *Inflectarius* shells have periostracal processes or fringes (see below), but these interesting shell features are usually lost in aging shells.

<u>Shell 7-16 mm, always with a parietal, basal and palatal tooth, pg 255-256</u>

*Inflectarius inflectus (uncommon)
*Inflectarius rugeli

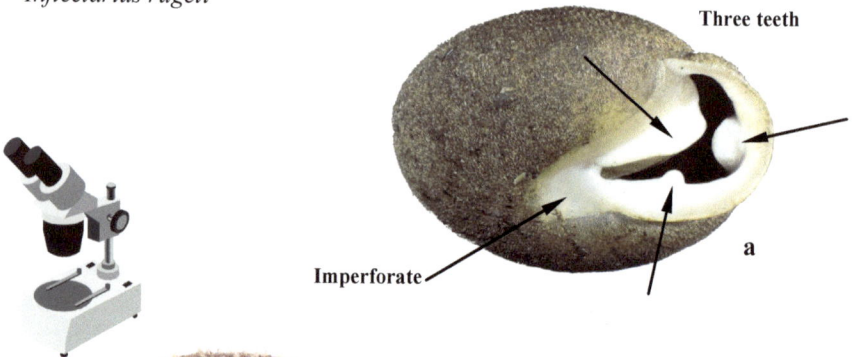

Three teeth

Imperforate

a

A juvenile *Inflectarius rugeli* shell enlarged (around 3 mm in actual size), illustrating the periostracal processes or fringes seen on young shells under 5 mm, but often lost in older shells. Forest litter often sticks to these fringes and is believed to help camouflage the gastropod from potential predators such as birds and salamanders.

Juvenile *Inflectarius rugeli*

254

Shagreen

Inflectarius inflectus (Say, 1821)

Diameter: 7.5-13.8 mm, height 5.4-8.1 mm

Description: Depressed heliciform; lip reflected; shell with 4.5-5.5 whorls; imperforate; large parietal tooth present; basal tooth small; palatal tooth small and only slightly recessed in aperture (a); cream buff colored to light yellowish-horn; shell surface with short periostracal processes and fine transverse striae or wrinkles; periphery rounded.

Similar Species: *Inflectarius rugeli* is slightly larger and the palatal tooth is larger and more deeply positioned in aperture (see figure a, next page); intermediate forms between the two species are not uncommon, occasionally making their separation a challenge.

Habitat: This is a land snail of roadside ditches, railroad rights-of-ways, grassy areas, and shale banks of road cuts; not a species of mature woods.

Status: G5/S2; Uncommon; this otherwise common species west of West Virginia is reported from only 7 counties by Taylor and Hubricht (1985).

Specimen: Kentucky, Hickman County, Obion Creek WMA, Travis Slough (author's collection).

Polygyridae

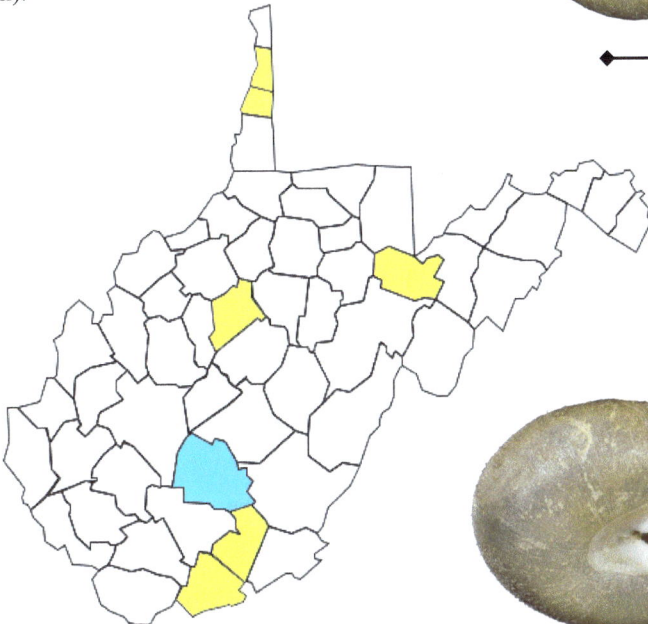

255

Deep-tooth shagreen

Inflectarius rugeli (Shuttleworth, 1852)

Diameter: 7.8-16.4 mm, height 6.4-9.3 mm

Description: Depressed heliciform; lip reflected; shell with 5.5 whorls; imperforate; large parietal tooth present; basal tooth small; <u>palatal tooth large and deeply positioned in aperture</u> (a) and in bottom view somewhat hidden; waxy horn colored; shell surface with short periostracal processes that may at times appear as scales especially in older shells.

Similar Species: *Inflectarius inflectus* is slightly smaller, has a smaller palatal tooth which is less deeply set in aperture; *I. rugeli* is thought to be a species complex, containing multiple species showing little difference in shell morphology except for size differences.

Habitat: A habitat generalist of mixed hardwood forests, found under leaf litter and logs and along forested road edges; a species of mature woods.

Status: G5/S2; Uncommon; although this species is quite common in eastern Kentucky and southward, it is an infrequent resident of West Virginia woodlands.

Specimen: Tennessee, Monroe County, Hawk Knob Bottom, Cherokee NF (author's collection).

Polygyridae

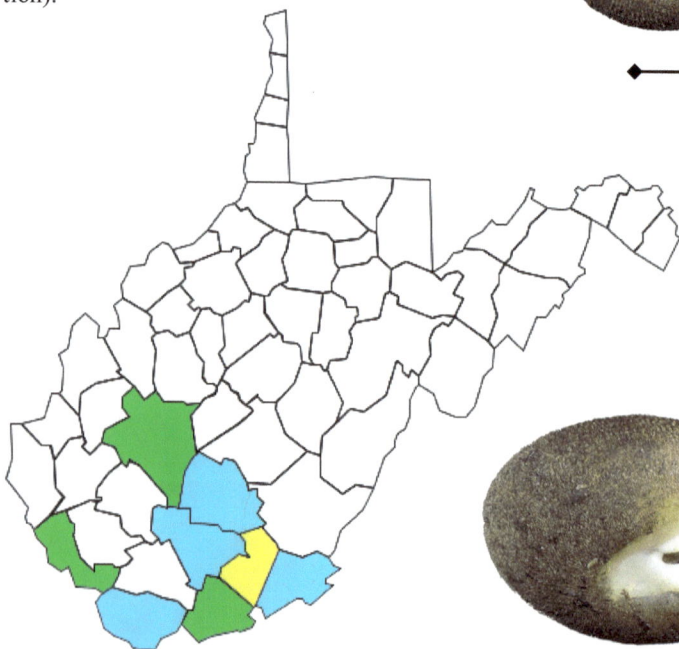

256

Above, top views showing the finer, lower, and more closely spaced transverse striae and glossier shell of *Patera panselenus* and the coarser, wider spaced striae and duller shell of *Patera appressa*. Both specimens from the same location from Breaks Interstate Park, Pike County, Kentucky.

Shells 15-24 mm, imperforate, <u>with long, low basal tooth,</u> adult shells with reflected lip (*Patera*)

<u>Patera</u> species are usually depressed heliciform with the oddball exception of *Patera pennsylvanica* (page 259). This species defies the standard by being heliciform, looking like a small *Mesodon*. All adult *Patera* shells exhibit reflected lips, and in West Virginia, display a large parietal tooth, except for *P. pennsylvanica.* <u>While the long, low basal tooth (a) is a wanting feature on many shells, it is an important diagnostic mark for the group as a whole and should not be disregarded.</u> There are either spirally arranged papillae (b) or spiral striae (c). These diagnostic features may be an obscure attribute on aged shells that have worn surfaces. An area that generally remains protected from abrasions is located just behind the reflected lip of mature shells (d). All *Patera* are imperforate. Immature *Patera* shells will not have reflected lips, and are sometimes confused with young *Mesomphix* shells. The following *Patera* are arranged according to these features and are as follows: **Group 1** shells that are heliciform, **Group 2** shells that are depressed heliciform and have spiral papillae and **Group 3** shells that have spiral striae in the form of incised lines.

<u>Grouping 1: shells that are heliciform, pg 259</u>

*Patera pennsylvanica

<u>Grouping 2: shells with spiral papillae, pg 260</u>

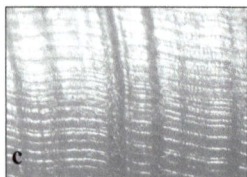

*Patera appressa

<u>Grouping 3: shells with spiral striae, pg 261-262</u>

*Patera laevior (rare)
*Patera panselenus

d

b

Spiral papillae

c

Spiral striae

a

Proud globelet

Polygyridae

Patera pennsylvanica (Greene, 1827)

Diameter: 15-20 mm, height 10.2-15 mm

Description: Heliciform; upper portion of lip points abruptly downward (a) and is only slightly reflected; right side of aperture square-like, not rounded as in other *Patera* species; shell with 5.5-6 whorls; imperforate; no teeth present; light tan to yellowish-olive; no hairs in adult shells; transverse striae well-developed; spiral striae always present and a strong feature; shell periphery rounded; an unusual *Patera* given that most species in this genus are depressed heliciform.

Similar Species: *Mesodon thyroidus* has a more rounded aperture and slightly open umbilicus.

Habitat: Open road cuts and grassy areas along abandoned railroad beds and limestone slopes; fairly common in mature sugar maple woods (pers. comm. John MacGregor).

Status: G3/S2; Uncommon; reported from scattered locations across the state.

Specimen: Kentucky, Henry County, KY River WMA, Raccoon Branch Tract (author's collection).

a

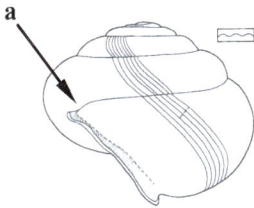

Emberton, 1998

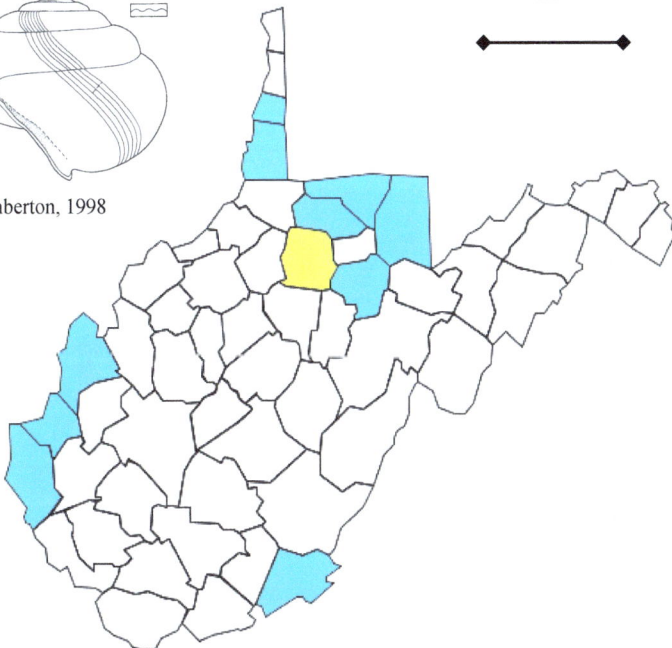

Flat bladetooth

Polygyridae

Patera appressa (Say, 1821)

Diameter: 13-19.5 mm, height 7-9 mm

Description: Depressed heliciform; lip reflected; shell with 4.5-5 whorls; imperforate; large parietal tooth present with a wide base (a); and a long, low basal tooth, which at times may be poorly defined (b); cinnamon-buff to brownish-horn; shell surface somewhat glossy; adult shells without hairs; transverse striae are well developed on top and sides, but weakly defined on the base; spirally arranged papillae typically present; shell periphery either rounded or very slightly angular.

Similar Species: *Patera panselenus* has strongly developed spiral striae (not papillae); *P. laevior* is of the same size and build, but has weakly developed striae (not papillae).

Habitat: Limestone and sandstone rock outcrops and cave entrances.

Status: G4/S4; Common; this species is reported from scattered locations across the state.

Specimen: West Virginia, Mercer County, Brush Creek (all in author's collection) .

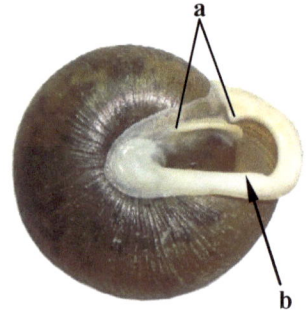

Caldwell Cave, Mercer County, WV

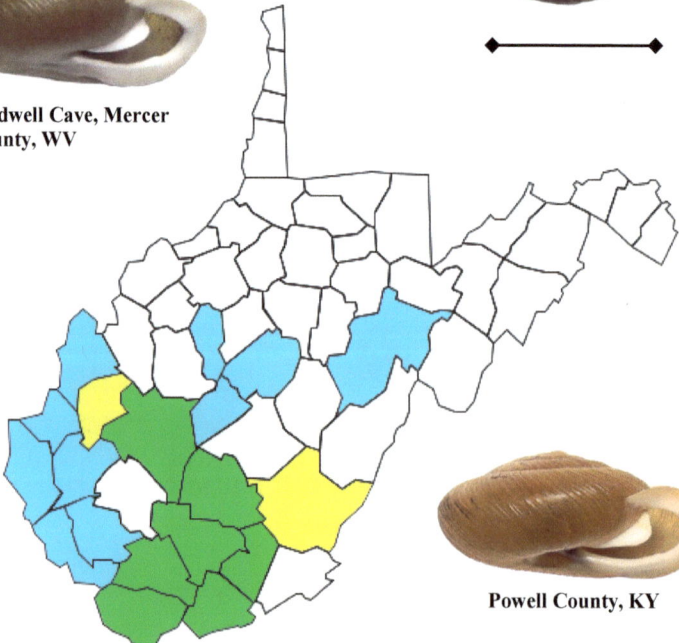

Powell County, KY

Smooth bladetooth

Patera laevior Hubricht, 1968

Diameter: 16-18 mm, height 8-9 mm

Description: Depressed heliciform; lip reflected; shell with 4.5-5 whorls; imperforate; large parietal tooth present; long, low basal tooth, which at times, may be poorly defined (a); cinnamon buff; shell surface somewhat glossy; no hairs in adult shells; transverse striae well developed on top of shell but weakly defined on the base; thin spiral lines on the upper surface and on the base, but are sometimes only faintly visible; shell periphery rounded.

Similar Species: *Patera appressa* has moderately to well developed spiral papillae; *Patera panselenus* is flatter, glossier and has stronger developed spiral striae.

Habitat: Mixed hardwood forests, rocky areas, both sandstone and limestone clifflines and steep wooded hillside; common along railroad beds and in old quarries (MacGregor pers. comm.).

Status: G4/SH; Rare; collected in Logan County near Blair by Emberton (1991) who was able to distinguish its internal anatomy.

Specimen: Kentucky, Franklin County; intersection of Routes 1900 and 1262, base of limestone cliffs (all figures from author's collection).

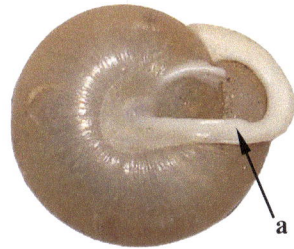

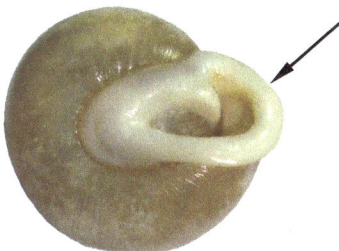

Some populations of *P. laevior* have an excessively thickened peristome, for unknown reasons, Mammoth Cave NP, Edmonson County, KY

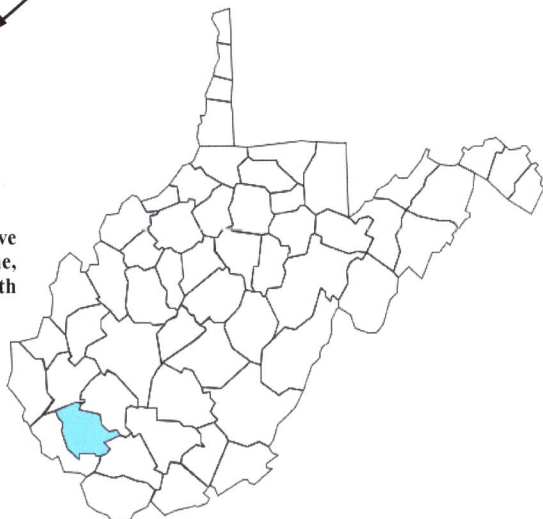

West Virginia bladetooth

Polygyridae

Patera panselenus (Hubricht, 1976)

Diameter: 18-20 mm, height 8.3 mm

Description: Depressed heliciform; lip reflected; shell with 5 whorls; imperforate; parietal tooth large with a wide base; long, low basal tooth, which at times, may be poorly defined 9 (a); shell pale brownish, glossy; no hairs in adult shells; through a hand lens of 10X the transverse striae appear closely spaced, but weakly developed; spiral striae are a strong and reliable feature; periphery roundish.

Similar Species: *Patera appressa* has spirally arranged papillae, bolder and more widely spaced transverse striae (see page 257); the two species occasionally occur together; *Patera perigrapta,* which was widely reported in West Virginia by past collectors, are now considered to be individuals of the more recently described *P. panselenus* (Hubricht 1976); in general, *P. perigrapta* has a more inflated shell, a parietal tooth with a narrower base (b) and stronger developed traverse and spiral striae.

Habitat: Shale banks adjacent to small streams; under rocks and logs in mixed hardwoods forests, damp limestone and sandstone rock faces.

Status: G3/S3; Relatively Common; in West Virginia this species appears to be at least as common as *P. appressa.*

Specimen: West Virginia, Kanawha County, WV near the town of Charleston (all figures in author's collection).

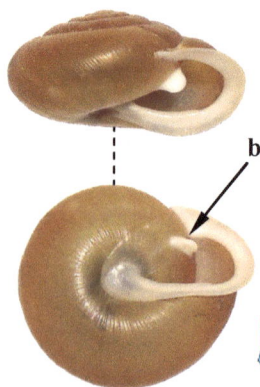

Patera perigrapta, Swain County, North Carolina

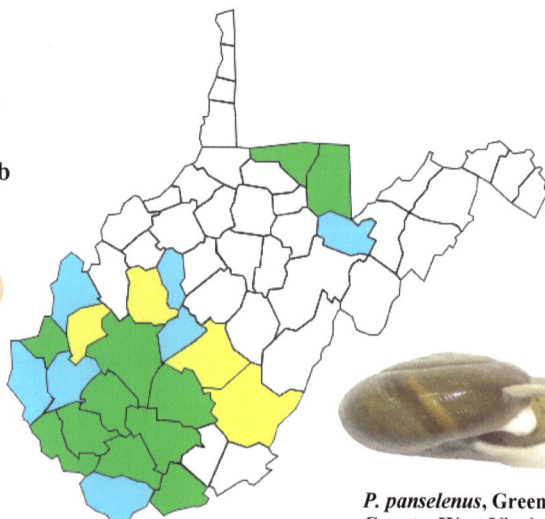

P. panselenus, Greenbrier County, West Virginia

Shell 9-22 mm, 3 teeth, <u>umbilicate</u> (*Triodopsis*)

Triodopsis in West Virginia have done especially well, containing 13 species, 4 which are endemic. Most are depressed heliciform, having widely reflected lips at maturity and all species are umbilicate. Shells typically contain three teeth, the parietal tooth being the largest. In some species, the basal and palatal teeth may be wanting features as in *T. platysayoides* and very rarely in *T. tennesseensis*. <u>It is important to determine if the parietal tooth points above or below the palatal tooth (figures a and b).</u> Always use the bottom view to determine this attribute. The transverse striations are usually a relatively strong character. The following *Triodopsis* are grouped as follows: **Group 1** parietal tooth points above the palatal tooth, **Group 2** parietal tooth points at or below the palatal tooth and **Group 3** without prominent basal and palatal teeth.

Group 1: umbilicate, parietal tooth points above palatal tooth, pg 264-278

*Triodopsis fallax
*Triodopsis hopetonensis
*Triodopsis picea (rare)
*Triodopsis vulgata
*Triodopsis fraudulenta
*Triodopsis rugosa (rare)
*Triodopsis anteridon
*Triodopsis juxtidens
Triodopsis juxtidens robinae, new subspecies (rare)

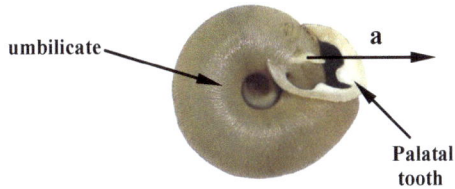

umbilicate

a

Palatal tooth

Group 2: umbilicate, parietal tooth points at or below palatal tooth, pg 279–281

*Triodopsis species (undetermined) (rare)
*Triodopsis tridentata
*Triodopsis tennesseensis

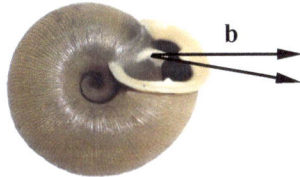

b

Grouping 3: umbilicate, without prominent basal and palatal teeth, pg 283

*Triodopsis platysayoides (rare)

Shell 19-25 mm, 3 teeth, <u>imperforate</u> (*Xolotrema*)

Xolotrema are very similar to *Triodopsis* in terms of size, build and number of teeth, but are imperforate, having a completely sealed umbilicus, pg 289.

*Xolotrema denotatum

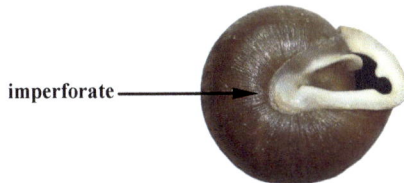

imperforate

Mimic threetooth

Polygyridae

Triodopsis fallax (Say, 1825)

Diameter: 9.9-13.5 mm, height 7.5-8.2 mm

Description: Depressed heliciform to heliciform, shell more compact than other *Triodopsis* species; lip reflected; shell with 5.5 whorls, the last whorl expanding slightly; umbilicate; in frontal view, the large, curving parietal tooth (a) crowds the aperture and in bottom view, points well above the palatal tooth; basal tooth sits on a wide buttress (b) that reaches deeply into the shell (c & d); palatal tooth recessed; olive-buff; transverse striae are not well developed; periphery rounded.

Similar Species: *Triodopsis tridentata* is larger, its parietal tooth points at or below the palatal tooth.

Habitat: A species of open ground such as clearings, along roadsides and railroads and on waste ground in urban areas.

Status: G5/S2; Uncommon to rare; in West Virginia this species is confirmed from three eastern panhandle counties.

Specimen: First 4 figures are from West Virginia, Mineral County, Patterson Creek Road N. of Ft. Ashby and Figure (d) from North Carolina, Cumberland County, Fort Bragg (author's collection).

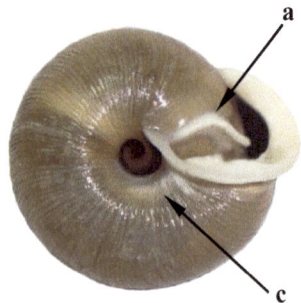

a

c

b

d

Magnolia threetooth

Triodopsis hopetonensis (Shuttleworth, 1852)

Diameter: 9.2-13 mm, height 5.5 mm

Description: Depressed heliciform, shell rather compact for a *Triodopsis*; lip reflected; shell with 4.5-5.5 whorls, the last whorl only slightly expanding; umbilicate; the curving parietal tooth (a) is thin and points well above the palatal tooth; palatal tooth not situated on a buttress and is recessed in aperture; olive-horn; transverse striae are well developed on the top and sides of the shell continuing well into the umbilical region; adult shells without hairs; periphery rounded.

Similar Species: *Triodopsis tridentata* is larger, its parietal tooth points at or below the palatal tooth; *T. picea* is a species of higher elevation and has a straight or nearly straight parietal tooth.

Habitat: A species of waste places, roadsides, also found in and around scrap piles of construction lumbers (one mode of transportation of the species to new locations, pers. obs.); natural habitats include low wet woodlands.

Status: G5/S1; Rare; in West Virginia this mostly southern species of the coastal plains is reported from only two counties.

Specimen: Kentucky, Martin County, a motel parking lot in the town of Inez (author's collection).

a

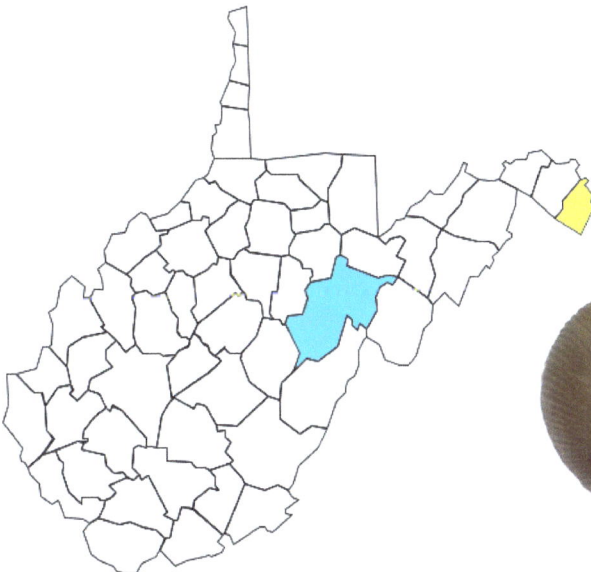

265

Spruce Knob threetooth

Polygyridae

Triodopsis picea Hubricht, 1958

Diameter: 10.7-14.8 mm, height 7.9-9.7 mm

Description: Depressed heliciform; lip reflected; shell with 5-5.5 whorls; umbilicate, but small; the nearly straight parietal tooth (a) points well above the palatal tooth; basal tooth and palatal tooth substantially smaller; the small basal tooth seated on a nearly straight callus ridge (b); shell dull snuff brown; transverse striae are well developed on the top and sides of the shell continuing well into the umbilicus region; entire surface, except nuclear whorls, covered with dense papillae; adults without hairs; periphery is rounded.

Similar Species: *Triodopsis picea* teeth are similar to *T. fraudulenta* but not as large; it resembles *T. juxtidens* in having a smaller umbilicus; *T. picea* is mostly unique in having a generous covering of papillae on the entire upper surface of the shell, except nuclear whorl.

Habitat: Found in higher elevation, rocky hardwood and spruce forests under leaf litter around rocks or close to logs.

Status: G3/S3; Uncommon; most records for this scarce *Triodopsis* are from nine West Virginia counties.

Specimen: West Virginia, Pocahontas County, Red Spruce Knob, 4379 feet (author's collection).

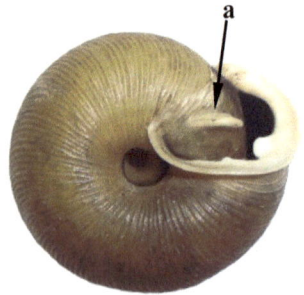

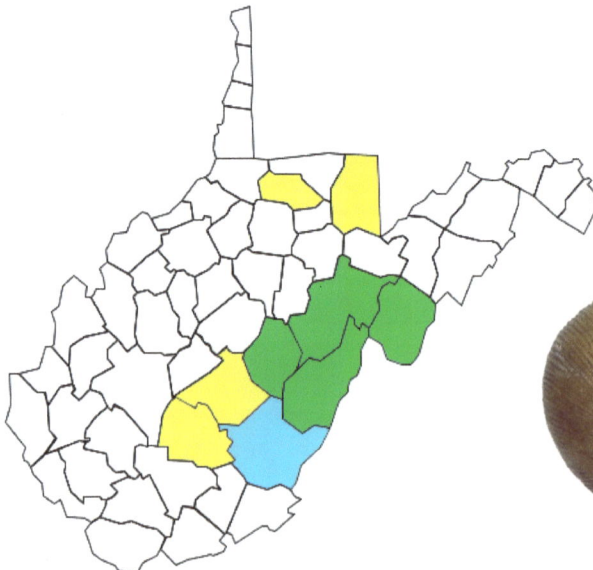

Dished threetooth

Polygyridae

Triodopsis vulgata Pilsbry, 1940

Diameter: 13.5-19.5 mm, height 7.3-10.4 mm

Description: Depressed heliciform; lip reflected and abruptly turning downward (a); shell with 5-6 whorls; umbilicate; parietal tooth large and curving (b), points well above the palatal tooth; basal tooth small; palatal tooth larger and with a wider base, sitting deeper in the aperture; the three teeth crowd the aperture in frontal view; cream-buff to cinnamon buff; transverse striae are well developed on the top, sides and bottom of the shell; scattered papillae are present but generally a weak feature, being strongest around the umbilicus; adults without hairs; periphery rounded.

Similar Species: The lip of most other *Triodopsis* species lacks the steep downward curve seen in side view (a).

Habitat: Usually found on soil and rubble among limestone in mixed hardwood forests.

Status: G5/S2; Uncommon; confirmed from six counties in West Virginia.

Specimen: Kentucky, Powell County, Furnace Mountain (author's collection).

b

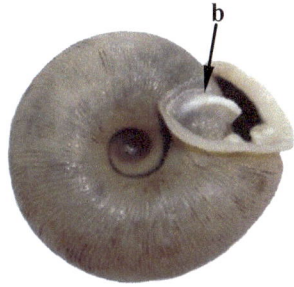

a

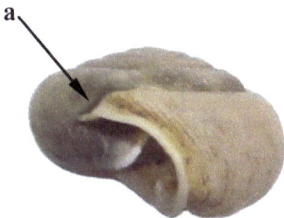

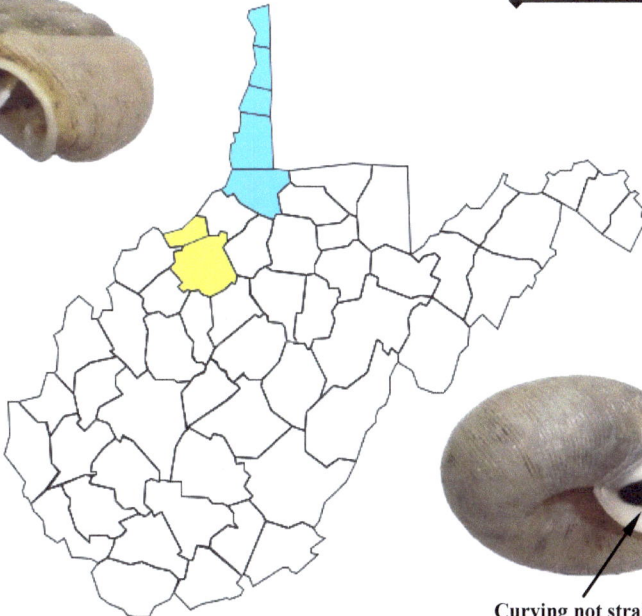

Curving not straight

Baffled threetooth

Triodopsis fraudulenta (Pilsbry, 1849)

Diameter: 14-16 mm, height 7.7-9 mm

Description: Depressed heliciform; lip reflected; shell with 5-5.5 whorls; umbilicate; parietal tooth (a) large and nearly straight, pointing well above the palatal tooth; basal and palatal teeth smaller, the palatal deeply set; the small basal tooth resting on a wide, straight ledge (b); aperture distinctly dished, almost closed by the crowded teeth; cinnamon-buff; transverse striae are fairly well developed on the top and sides of the shell continuing well into the umbilical region; scattered papillae are generally a strong feature especially near sutures and umbilicus; adults without hairs; periphery rounded.

Similar Species: *Triodopsis vulgata* has the same build and size, but with less crowded teeth, and its parietal tooth has a greater curvature, *T. fraudulenta* being nearly straight.

Habitat: Found in mixed hardwood forests under leaf litter and close to logs on hillsides but also along roadsides and in old pastures.

Status: G4/S4; Common; this species is more common and widespread in the state than *T. vulgata;* specimens from the "Trough" along the South Branch of the Potomac River are larger than usual (page 269).

a

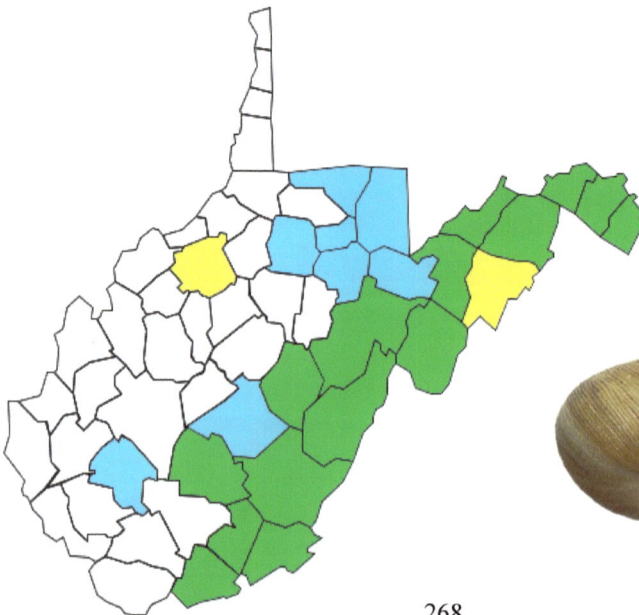

b

Some Interesting *Triodopsis fraudulenta* Shells Compared

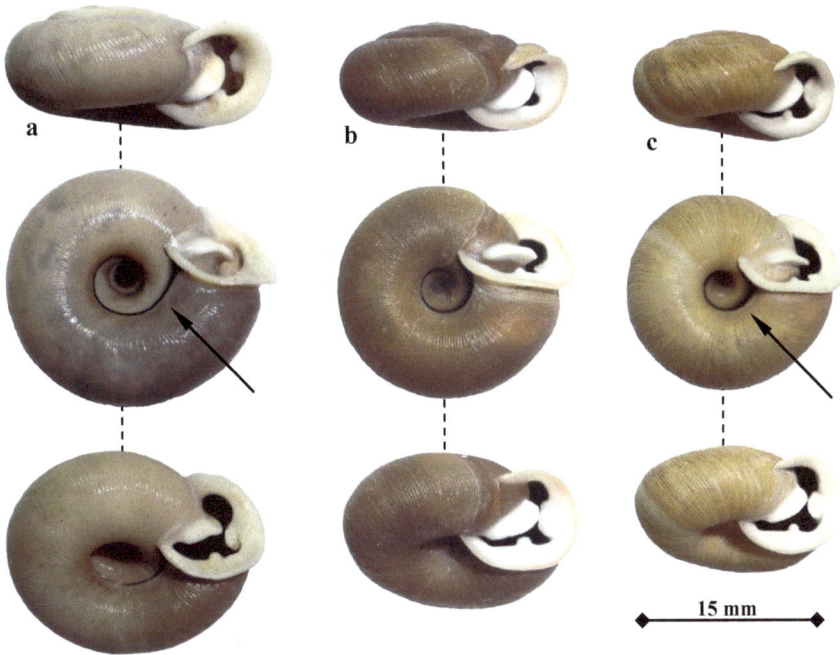

Figures (a & b) *Triodopsis cf. fraudulenta,* from the "Trough" area along the South Branch of the Potomac River, Hampshire County, West Virginia. These specimens range from 17 mm to 21 mm, larger than the reported standard of 14 -16 mm; additional specimens needed for a final deliberation. Figure (c) represents a more typical size and shell of *T. fraudulenta* seen across the state; Grant County, West Virginia (all specimens in author's collection).

Triodopsis tridentata Shells Compared (proportionate)

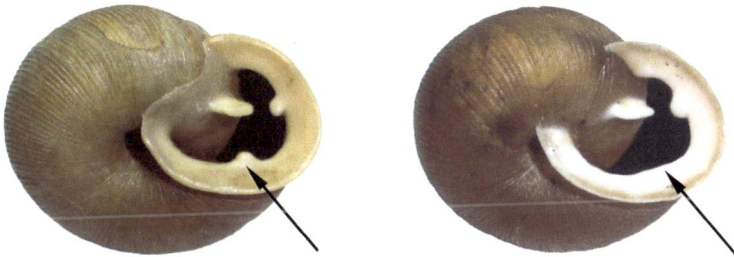

These two shells of *Triodopsis tridentata* collected by Donna Mitchell are from the same hillside, near Summersville Dam, below the campground, Nicholas County, West Virginia. Note that the specimen on the right is missing its basal tooth, while the one on the left is with the archetypal tooth. Finding uncharacteristic shells is not uncommon and can bewilder the investigator (speaking from my own experience, of course). This is why collecting a series of shells is so central to a correct identification.

Land Shell Freaks!

Mesodon zaletus

Mesodon thyroidus

Land shell "freaks" can be the result of falls from high places or attacks from predators, which can cause everything from simple hair-line fractures to severe structural damage to the shell. If the snail survives, it will repair the breach by using calcium carbonate. Small fractures are sealed with little effort, but a severely damaged shell will require a great deal of time and skill to fully restore. If the damage is around the aperture, the renovated peristome will take on some rather interesting and bizarre shapes, as seen above. Both shells from below a limestone outcrop on Furnace Mountain, Powell County, Kentucky.

Similar *Triodopsis* Shells Compared (proportionate)

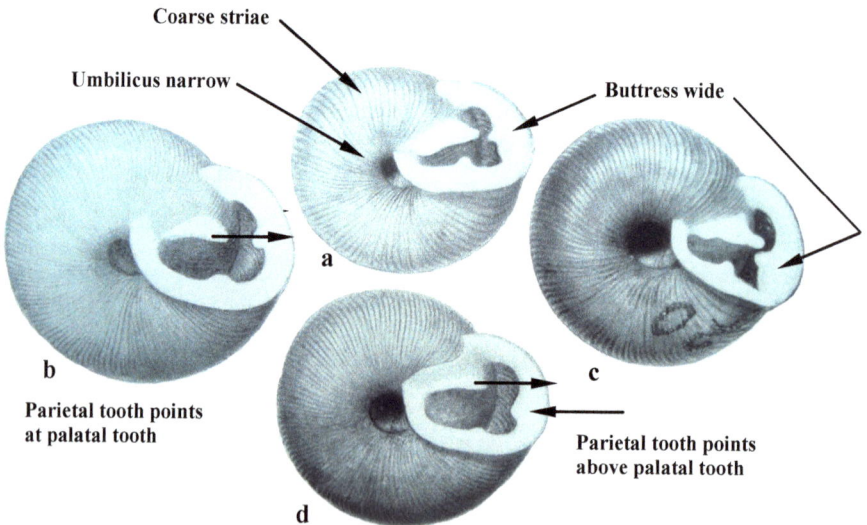

Above images of four West Virginia *Triodopsis* species of similar size and build; a) *Triodopsis rugosa,* Paratype; b) *Triodopsis tridentata,* typical; c) *Triodopsis anteridon,* Holotype and d) *Triodopsis juxtidens,* Paratype. Images from Pilsbry's classic works of 1940. Arrows indicate special features that aid in separating the four shells.

Triodopsis Shells Compared, Illustrations are proportionate
(by K. C. Emberton, 1998)

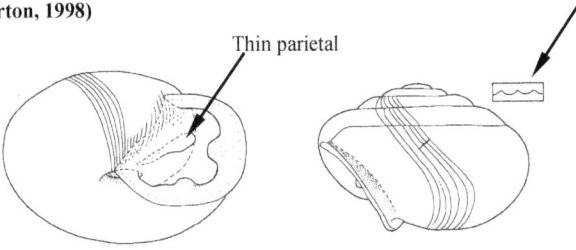

Thin parietal

Triodopsis hopetonensis

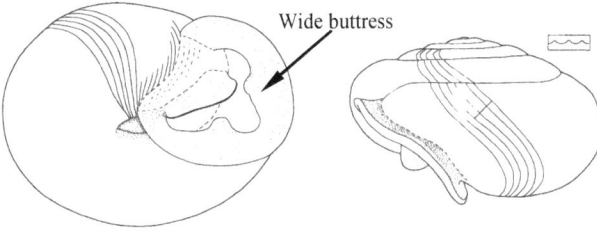

Wide buttress

Triodopsis anteridon

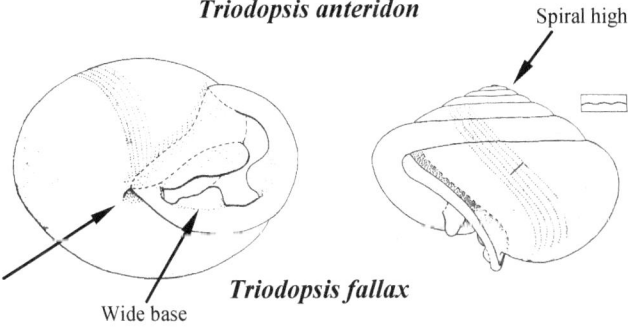

Spiral high

Wide base

Triodopsis fallax

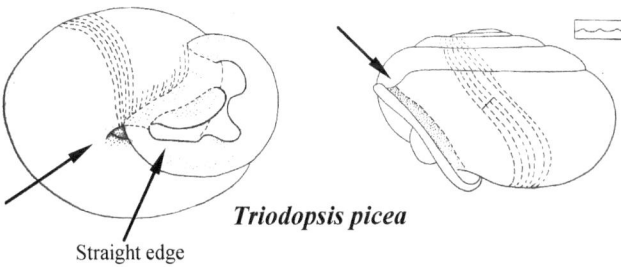

Straight edge

Triodopsis picea

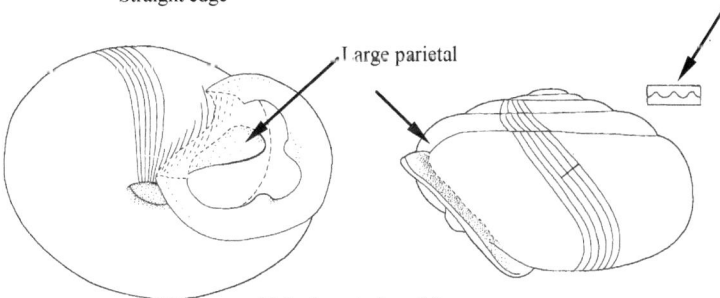

Large parietal

Triodopsis juxtidens

10 mm

271

Small parietal

Sharp angle

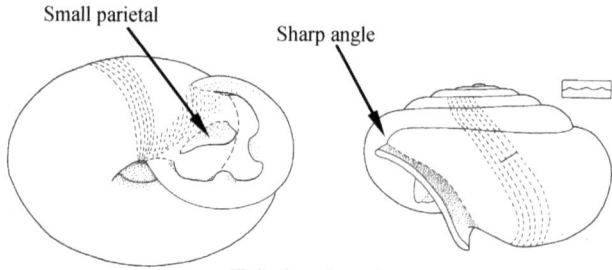

Triodopsis vulgata

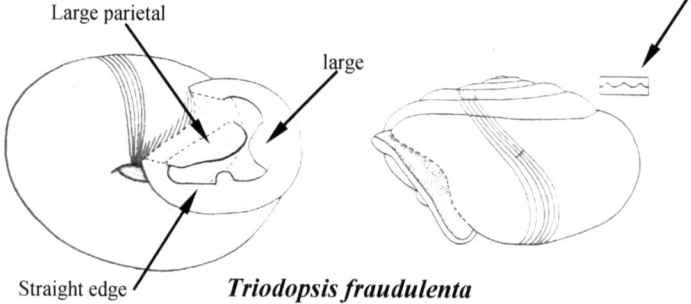

Large parietal

large

Straight edge

Triodopsis fraudulenta

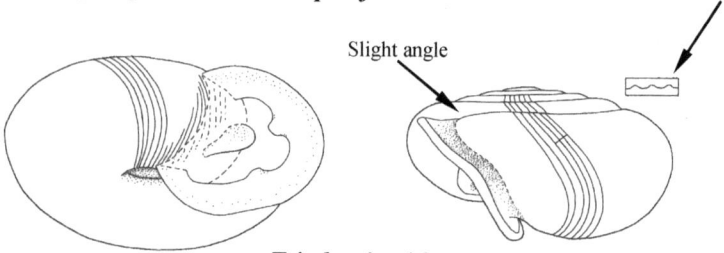

Slight angle

Triodopsis tridentata

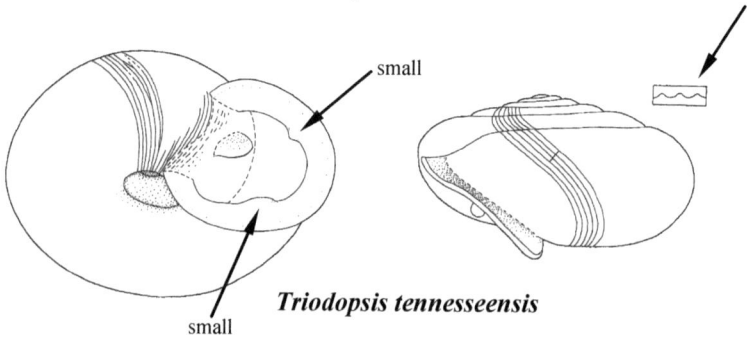

small

small

Triodopsis tennesseensis

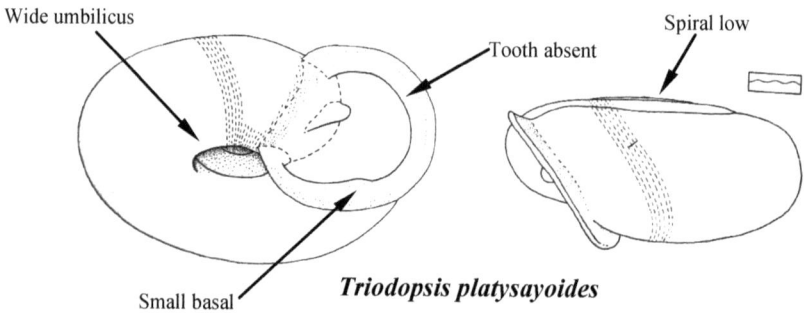

Wide umbilicus

Tooth absent

Spiral low

Small basal

Triodopsis platysayoides

10 mm

Buttressed threetooth

Triodopsis rugosa Brooks and MacMillan, 1940

Diameter: 9.2-10.9 mm, height 5.6 mm

Description: Depressed heliciform; lip reflected and thickened within; shell with 5-5.5 whorls; umbilicate; large parietal tooth, points above the palatal tooth; basal tooth and palatal tooth substantially smaller, of nearly equal size; palatal tooth situated on a wide buttress; shell brown; the best diagnostic feature of the shell is the well developed and widely spaced transverse striae observed on the top and sides, continuing as strongly to the base (see page 275); scattered papillae; adults without hairs; periphery is rounded or shouldered.

Similar Species: *Triodopsis tridentata* is larger, the parietal tooth is smaller and points at or below the palatal tooth; *T. anteridon* has a larger shell with a slightly wider umbilicus and finer, more closely set transverse striae.

Habitat: Found in damp ravines and on hillsides of mixed hardwood forests under leaf litter and close to logs.

Status: G1/S1; Rare; Endemic to West Virginia; although this rare *Triodopsis* is reported from several WV counties, all specimens I have examined from outside Logan County refer to *T. anteridon,* therefore *T. rugosa* should always be very closely scrutinized.

Specimen: West Virginia, Logan County, Blair Mountain (FM 264690).

T. rugosa, Logan Co, WV (FM 264690)

Carter threetooth

Polygyridae

Triodopsis anteridon Pilsbry, 1940

Diameter: 11-14 mm, height 6.6-7.2 mm

Description: Depressed heliciform; lip reflected; shell with 5-5.5 whorls; umbilicate; large parietal tooth, points above the palatal tooth; basal tooth and palatal tooth substantially smaller; palatal tooth situated on a wide buttress (a), a widening of the outer lip; shell brown; transverse striae are well developed, closely spaced on the top and sides of the shell continuing well into the umbilicus region; scattered papillae are generally a strong feature especially near sutures and the umbilicus; periphery is rounded.

Similar Species: *Triodopsis tridentata* is larger, with a smaller parietal tooth that points at or below the palatal tooth.

Habitat: Found in mixed hardwood forests under leaf litter and close to logs; occasionally under logs and rocks.

Status: G3/S3; Uncommon; an infrequent gastropod of West Virginia woodlands.

Specimen: Tennessee, Carter County, Cherokee National Forest (author's collection).

T. anteridon, Grayson County, Kentucky

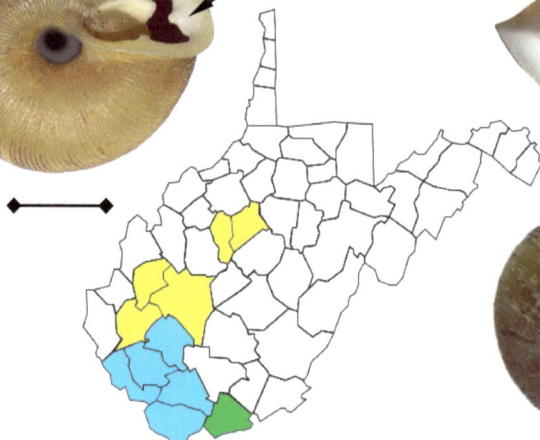

Triodopsis rugosa and *Triodopsis anteridon* Compared

Triodopsis rugosa, note that the transverse striae are more widely-spaced than seen in *T. anteridon*, Logan County, West Virginia

Triodopsis anteridon, note the closely-spaced transverse striae, Putnam County, West Virginia

Small mammals are aggressive feeders on a variety of terrestrial snails. Above, a golden mouse examines a grand globe, *Mesodon normalis* for consumption; eastern Kentucky. Below, a hairy-tailed mole gripping a Magnolia threetooth, *Triodopsis tridentata,* and bearing its imposing teeth; Carvers Gap, Roan Mountain, Mitchell County, North Carolina.

Atlantic threetooth

Triodopsis juxtidens (Pilsbry, 1894)

Diameter: 11.9-19.1 mm, height 6.6-10.2 mm
Description: Depressed heliciform; lip reflected; shell with 5-5.5 whorls; umbilicate; the large parietal tooth points directly at or in most cases above the palatal tooth; basal tooth and palatal tooth smaller; palatal tooth not situated on a buttress; reddish-brown; transverse striae are well developed on the top and sides of the shell continuing into the umbilicus region; scattered papillae are on top and base of shell, strongest around the umbilicus region; adults without hairs; periphery rounded.

Wide space

Similar Species: *T. tennesseensis* has a notably smaller parietal tooth, is less compact, and the basal and palatal teeth are typically smaller; *T. juxtidens robinae* has a wider lip, larger umbilicus, and is restricted to the upper Bluestone River watershed in Mercer and Summer Counties, West Virginia.

Habitat: A species found under leaf litter, logs and rocks of upper elevation in a variety of mixed hardwood forests; also a snail of waste places, pastures, roadsides and urban areas.
Status: G5/S5; Common; found throughout most of West Virginia.
Specimen: West Virginia, Pendleton County, near Franklin (author's collection).

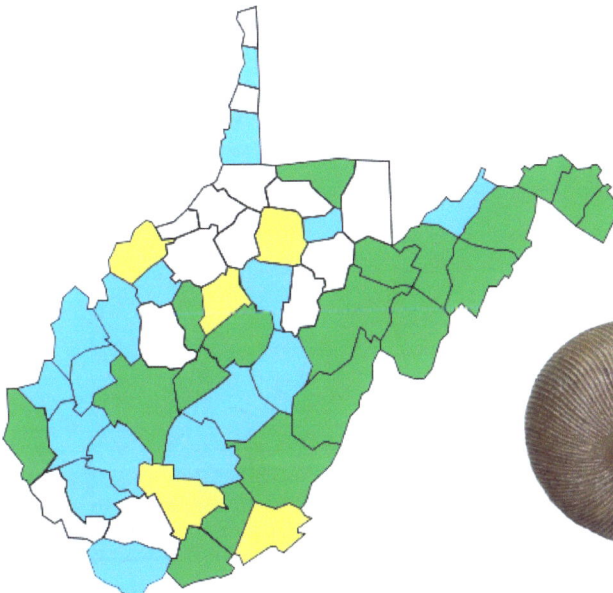

Wide space

Brush Creek threetooth

Polygyridae

Triodopsis juxtidens robinae Hotopp, 2014

Diameter: 13-17.6 mm, height 7-9 mm

Description: Depressed heliciform; lip reflected; shell with 5-5.5 whorls; umbilicate; the large parietal tooth points directly at or in most cases above the palatal tooth; basal tooth and palatal tooth smaller; palatal tooth not situated on a buttress; reddish-brown; transverse striae are well-developed on the top and sides of the shell continuing into the umbilicus region; scattered papillae found across the shell surface, strongest around the umbilicus region; adults without hairs; periphery rounded.

Similar Species: *Triodopsis juxtidens robinae* differs from *T. juxtidens* by having a wider lip, broader umbilicus and a narrower gap between the columellar insertion and basal tooth (a); the Piney Creek threetooth (page 279) has a smaller parietal tooth, glossier shell, finer, more flattened transverse striae, a narrower umbilicus, and its parietal tooth points at or below the palatal tooth.

Habitat: A species of rich woods found living among leaf litter in ravines of lower hillsides.

Status: G5T1/S5T1;Rare; Endemic to West Virginia; restricted to the upper Bluestone River and Brush Creek watershed in Mercer and Summer Counties, West Virginia.

Specimen: West Virginia, Mercer County, along Brush Creek (CMNH 103371, all images by Charlie Sturm & Tim Pierce).

a

Wide lip

Wide umbilicus

Narrow space

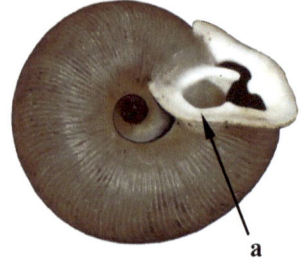

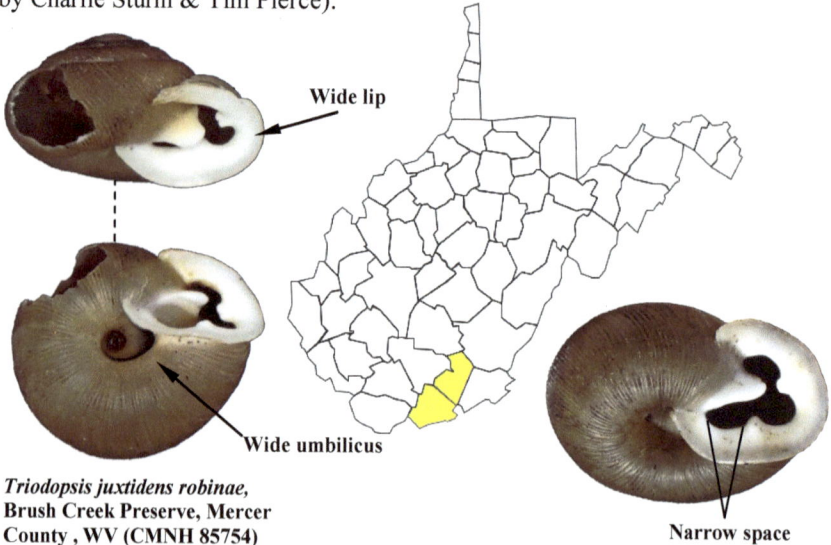

Triodopsis juxtidens robinae,
Brush Creek Preserve, Mercer
County , WV (CMNH 85754)

Piney Creek threetooth

Triodopsis species (undetermined)

Diameter: 14-17 mm, height 9-10 mm

Description: Depressed heliciform; lip widely reflected, somewhat concave, many specimens having a distortion on upper lip (a); shell with 5 -5.5 whorls; umbilicate; parietal tooth of moderate size which points at or below the palatal tooth; basal tooth and palatal tooth substantially smaller, nearly equal in size; shell brown and rather glossy; transverse striae are moderately developed on the top and sides of the shell continuing, but weaker into the umbilicus; papillae are strongest near sutures lines and umbilicus, but sometimes quite prevalent across the entire shell surface; periphery is rounded.

Similar Species: Most like *T. tridentata,* but with a glossier shell, more compact form, and in the lower lip, a smaller, roundish gap (2-3 mm wide) between the columellar insertion and basal tooth (b); in *T. tridentata* of the same size, this space is 3.5–5 mm wide, giving the aperture and shell a slightly wider build.

Habitat: Found in mixed hardwood forests under leaf litter below sandstone outcrops along Piney and Brush Creeks.

Status: G1/S1; Rare; Endemic to West Virginia; live specimens are needed for dissection and DNA work to make a final determination; first discovered by Jeff Hajenga from around his cabin in Mercer County, WV.

Specimen: West Virginia, Mercer County, Brush Creek (author's collection).

Polygyridae

b

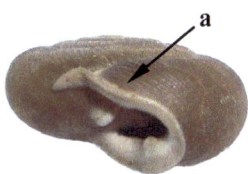

a

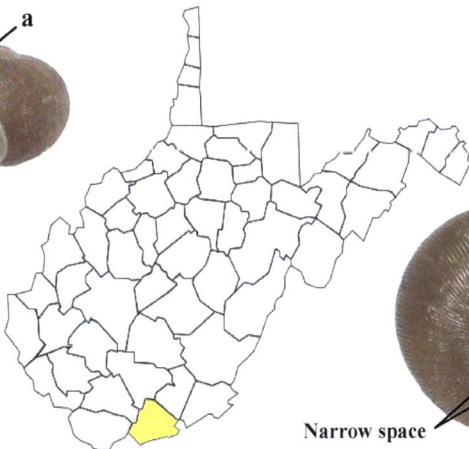

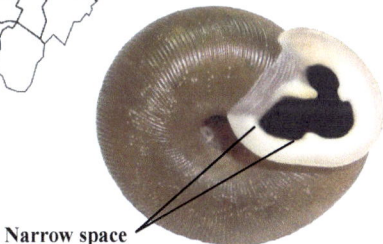

Narrow space

279

Northern threetooth

Polygyridae

Triodopsis tridentata (Say, 1816)

Diameter: 12-20 mm, height 6.4-7 mm

Description: Depressed heliciform; lip reflected, not abruptly turning downward; shell with 5-6 whorls; umbilicate; parietal tooth points at or below the palatal tooth; basal and palatal teeth small, not crowding the aperture in frontal view; light cream to pale cinnamon-buff; transverse striae are well developed on the top, sides and bottom of the shell; like most *Triodopsis* species there are scattered papilla on top and on the base; without hairs in adult shells; periphery rounded.

Similar Species: Most similar to the Piney Creek threetooth but in bottom view, the basal tooth (a) is usually positioned to the right of the inside body whorl (b), while the basal tooth in the Piney Creek threetooth sets at or in most cases to the left of the inside body whorl (c). *T. vulgata* has larger more crowded teeth and an aperture that turns abruptly downward;

Habitat: Found in rich upland woods under forest litter, rocks and logs, but also a snail of roadsides and urban areas.

Status: G5/S5; Common; the most widespread *Triodopsis* found in West Virginia woodlands.

Specimen: West Virginia, Greenbrier County near Lost Cave (author's collection).

b

T. tridentata

a

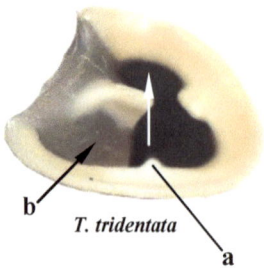

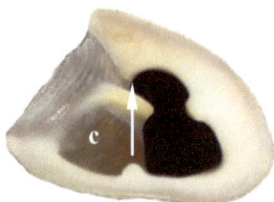

Piney Creek threetooth, *Triodopsis* species (undetermined)

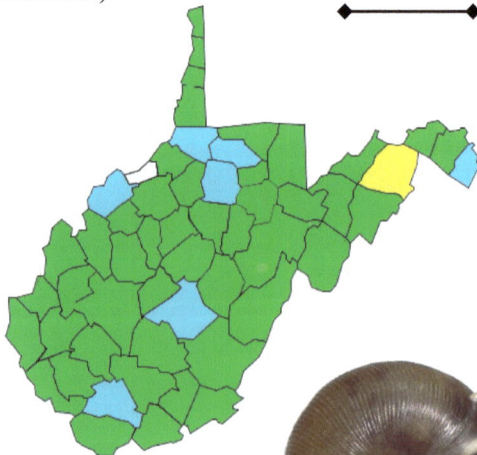

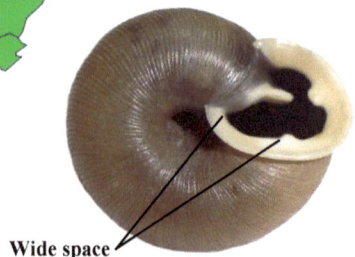

c

Wide space

Budded threetooth

Polygyridae

Triodopsis tennesseensis (Walker & Pilsbry, 1902)

Diameter: 19-24 mm, height 10-10.4 mm

Description: Depressed heliciform; lip reflected and wide; shell with 5-5.5 whorls; umbilicate; parietal tooth moderate in build, points directly at or below the palatal tooth; basal and palatal tooth small, roughly the same size and do not crowd the aperture in frontal view; occasionally without well defined basal and palatal teeth (a & b); shell snuff brown; transverse striae are well developed on the top, sides and bottom of the shell; scattered papillae mostly along the transverse striae and around the umbilicus; adults without hairs; periphery rounded.

Similar Species: *Triodopsis tridentata* is typically smaller, more compact in build and has a smaller umbilicus and larger basal and palatal teeth; *T. juxtidens* has larger teeth (especially the parietal) which crowd the aperture in frontal view, and in bottom view, the parietal tooth usually points notably above the palatal.

Habitat: Found under leaf litter in mixed hardwood forests on hillsides and in ravines.

Status: G4/S3; Uncommon; this species becomes more common in southern counties.

Specimen: Kentucky, Pike County, Breaks Interstate Park (author's collection).

T. tennesseensis, Pike County, Kentucky

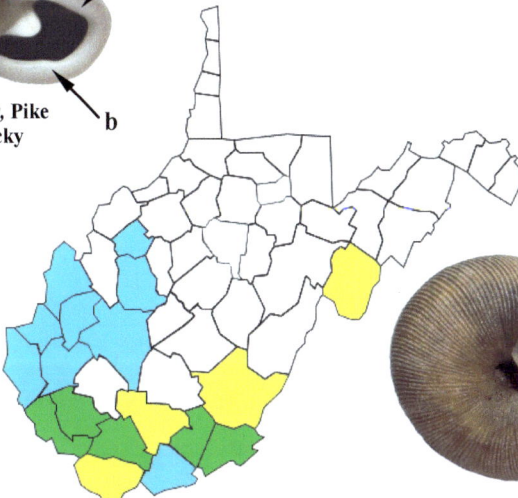

281

Cheat threetooth, *Triodopsis platysayoides*

Flat-spired threetooth

Polygyridae

Triodopsis platysayoides (Brooks, 1933)

Diameter: 18-25 mm, height 8 mm

Description: Depressed heliciform or even discoidal (a); lip reflected; shell with 5-5.5 whorls; umbilicate; parietal tooth of moderate size; without a palatal tooth; basal tooth generally absent or scarcely detectible; shell brown; transverse striae are not as well developed as in other *Triodopsis*; adults without hairs; periphery is rounded.

Similar Species: Unlike any other *Triodopsis* in West Virginia primarily by being flatter and lacking notable parietal and basal teeth.

Habitat: This is a highly restricted and very specialized land snail, found almost entirely in cool sandstone talus and cliffline habitat, occasionally associated with limestone caves; rarely found more than a meter from a rock feature (Dourson 2008).

Status: G1/S1, Rare; **Federally Threatened**; Endemic to West Virginia; known only from the Cheat River Canyon in Monongalia and Preston Counties.

Specimen: West Virginia, Monongalia County, Coopers Rock State Forest (WVDNR collection).

a

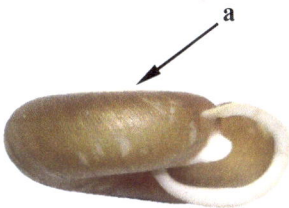

Known Global Range

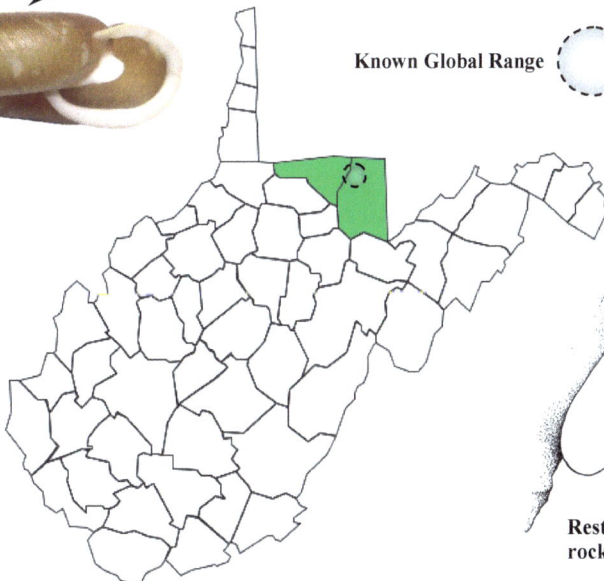

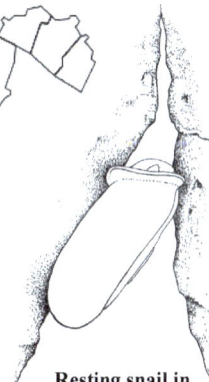

Resting snail in rock crevice

Hiding Among Boulders is an Extraordinary WV Animal

Cool boulder talus considered important habitat for the species

In 1940, Henry A. Pilsbry, a world renowned malacologist, described *Triodopsis platysayoides* as remarkable. He was right in a number of ways: the snail's splendid adaptation to deep talus and its global scarcity puts it in a class with few other species. It's either an amazing survivor of a past era or rather recent to these times. Little did Pilsbry know that this enigmatic backwoods snail would provoke such emotion among our kind.

The Diet of the Cheat threetooth

Rhododendron blossoms

The Cheat threetooth has been observed eating no less than twenty-seven different foods (Dourson, 2008), many of which are likely consumed by most other land snail species of similar size.

A large variety of fungi.

The Diet of the Cheat threetooth

Dead cave crickets.

Fresh woodrat scat, yummy!

Decomposing rhododendron blossoms.

The Diet of the Cheat threetooth

Crustose Lichen growing on sandstone.

Mold filaments growing from woodrat scat, a favorite.

The Diet of the Cheat threetooth

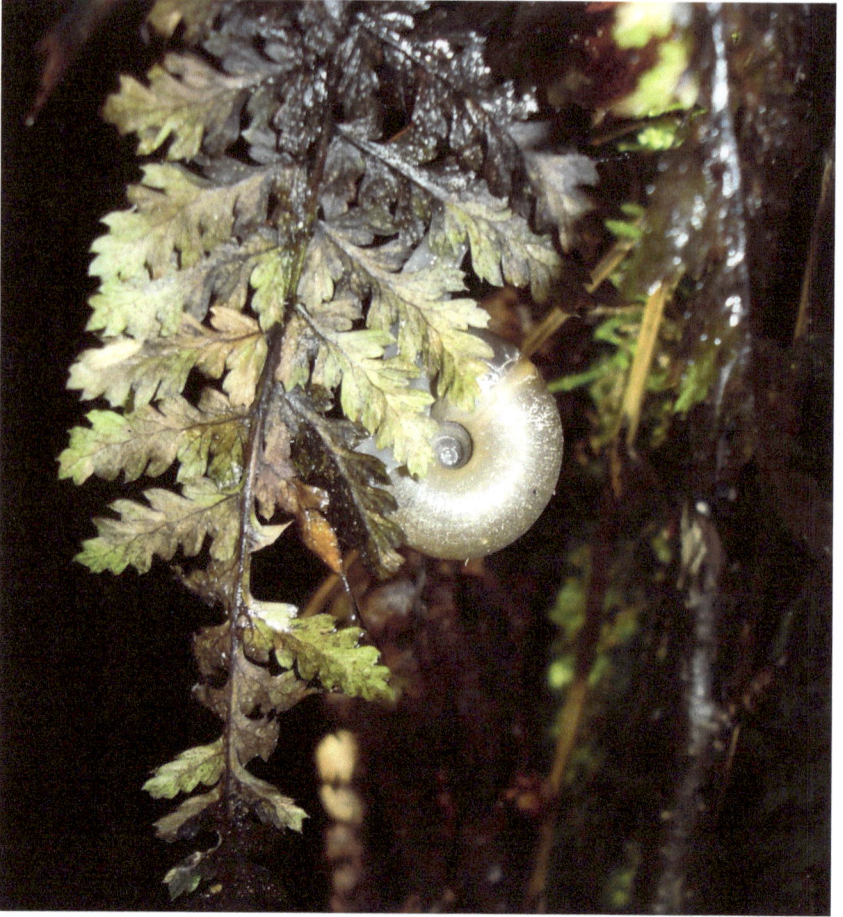

Decaying fronds of hay-scented and wood ferns.

Gleaning old shells for the calcium carbonate

Velvet wedge

Polygyridae

Xolotrema denotatum (Fèrussac, 1821)

Diameter: 19-25.6 mm, height 10-13.7 mm

Description: Depressed heliciform; lip reflected; shell with 5.5 whorls; imperforate; parietal tooth large with a wide base, points well above the palatal tooth; basal tooth small, may be absent or so poorly defined it is barely detectable; palatal tooth of medium build; tawny olive to snuff brown; transverse striae weakly developed, shell surface with close set periostracal processes or thickened hairs which pick up forest debris (see next page) typically seen on live and fresh dead shells only; periphery rounded in most specimens, but in some populations the shell periphery may be weakly angular.

Similar Species: Although *Triodopsis* species also contain three teeth, they differ by having an open, not completely sealed umbilicus.

Habitat: Often associated with rotting hardwood trees in advanced stages of decay in a variety of upland mixed hardwood sites; species appears indicative of good quality habitat.

Status: G5/S5; Fairly Common; this species which is reported from across the state is rarely ever found in large numbers.

Specimen: Tennessee, Sevier County, Roaring Fork, GSMNP (GSMNP collection).

Small tooth

Closed

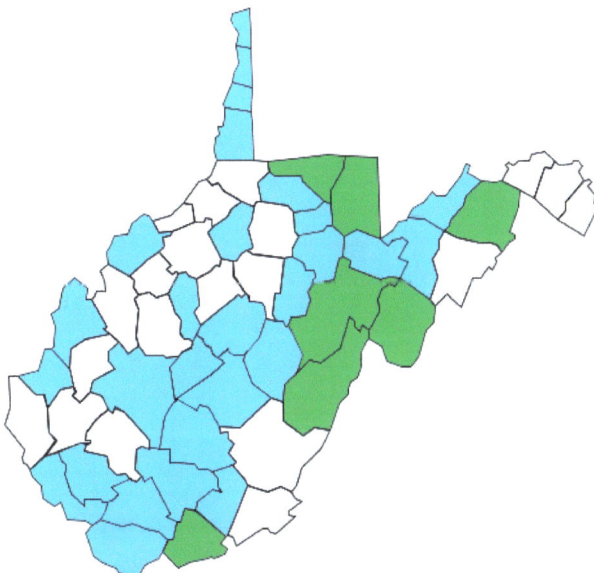

A live Velvet wedge, *Xolotrema denotatum* crawling on rock face at the Tate Lohr WMA, near Caldwell Cave, Mercer County, West Virginia.

Dead Trees are Exceptional Land Snail Habitat

Philomycus carolinianus

Philomycus venustus

Philomycus flexuolaris

Anguispira alternata

Zonitoides elliotti

Gastrodonta fonticula

Discus patulus

Zonitoides arboreus

Strobilops aeneus

Xolotrema denotatum

It's been said that there's more life in dead trees than live ones. Without question, dead trees in advanced stages of decay are outstanding habitats for several species of common land snails, especially slugs. These snails can be found on both standing or downed trees or under the tree's exfoliating bark. As rotting trees age, so does land snail cuisine including fungi, slime-molds, and lichens. The above species are commonly associated with dead trees.

Shells large, mostly heliciform, 15-45 mm, most over 20 mm, imperforate, with or without a parietal tooth, no basal or palatal teeth, reflected lip (*Mesodon, Neohelix* and *Webbhelix*)

Mesodon, Neohelix and *Webbhelix* species are among the largest and most conspicuous land snails in West Virginia. Mostly heliciform and always having reflected lips in mature shells, several species develop a small parietal tooth but none contain basal or palatal teeth. Transverse and spiral striation are a constant and well-developed features, and the umbilicus is generally imperforate. Top views of these genera are not typically diagnostic for separating species. The following species are arranged according to some of these features and are as follows; see page 293 for more detailed key.

Shells large, globose, imperforate, pg 293-304

Mesodon mitchellianus
Mesodon aff. andrewsae (rare)
Mesodon zaletus
Mesodon normalis (rare)
Neohelix albolabris
Neohelix dentifera
Webbhelix multilineata (rare)

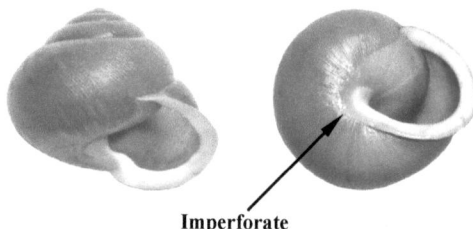

Imperforate

Shells large, depressed heliciform, 20-35 mm, umbilicate, with or without a parietal tooth, basal tooth maybe present, reflected lip (*Allogona* and *Appalachina*)

Allogona and *Appalachina* are also large land snails, but differ from the land snails above by being umbilicate and having more depressed shells. They always have reflected lips in maturity, and both species below have a basal tooth; see page 293 for more detailed key.

Shells large, depressed, umbilicate, pg 305-308

Allogona profunda
Appalachina sayana

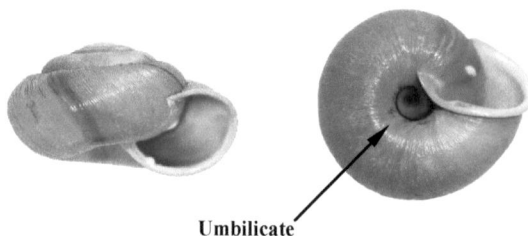

Umbilicate

292

Key to Large Shells with Reflected-Lips (proportionate)

◆———— 30 mm ————◆

Shell <u>imperforate</u>, globose, no color bands (*Mesodon*)

1

15-20 mm

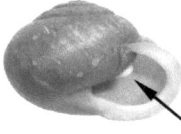
Parietal tooth

25 mm

30-40 mm

M. mitchellianus   *M. zaletus*   *M. aff. andrewsae*   *M. normalis*

Shell <u>imperforate</u>, somewhat depressed, no color bands (*Neohelix*)

2

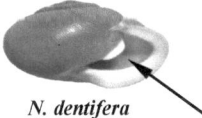
Parietal tooth

Wide lip

N. dentifera   *N. albolabris*

Shell <u>imperforate</u>, globose, color bands (*Webbhelix*)

3

imperforate

W. multilineata

Color bands

Shell <u>umbilicate</u>, depressed (*Allogona* and *Appalachina*)

4

Color bands

Bands

No bands

A. profunda   *A. sayana*

Sealed globelet

Polygyridae

Mesodon mitchellianus (I. Lea, 1839)

Diameter: 15-17.2 mm, height 11.7– 13 mm

Description: Heliciform; lip reflected; shell with 5 whorls; imperforate, umbilicus always completely sealed; no teeth present; pale yellowish-tan; adults without hairs; transverse and spiral striae always present; shell periphery well rounded; river island specimens tend to be thinner and translucent, figures (b and c).

Similar Species: *Mesodon clausus* has a rimate umbilicus that is slightly open whereas the umbilicus in *M. mitchellianus* is always completely closed (a key feature); *M. thyroidus* is larger, usually has a parietal tooth and has a rimate umbilicus.

Habitat: Found in low places such as floodplains, meadows, and roadsides occurring under leaf litter and other forest debris.

Status: G4/S3; Uncommon; reported from thirteen counties in West Virginia.

Specimen: Figures (a) from West Virginia, Raleigh County, Mill Creek, and figures (b and c) from West Virginia, Wood County, Muskingum Island (author's collection).

a

b

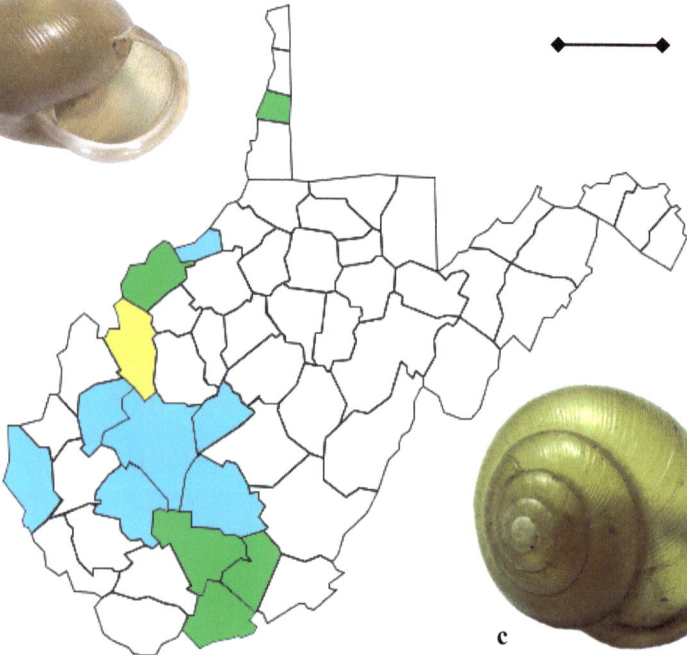

c

Balsam globe (?)

Polygyridae

Mesodon aff. andrewsae (W.G. Binney, 1879)

Diameter: 25 mm, height 17 mm

Description: Heliciform; lip reflected; shell with 5.5-6 whorls; imperforate; shell thin for its size and somewhat glossy; without a parietal tooth; coppery brown; adults without hairs; transverse striae not well developed; fine spiral striae closely spaced and easily observed with a hand lens of 10X; shell periphery well-rounded.

Similar Species: *Mesodon thyroidus* has a rimate umbilicus that is slightly open; *M. normalis* is larger with a thicker shell and generally found below 1400 m; *M. zaletus* has a thicker shell and a parietal tooth.

Habitat: In West Virginia, a species restricted to higher elevation, northern hardwood and spruce forests.

Status: G2/S1; Rare; in West Virginia, the species pictured here has been reported as *M. andrewsae,* a land snail of more southern mountain ranges, but there is uncertainty as to its precise identity. It may, in fact, represent an undescribed species, This hypothesis, however, awaits further investigation and DNA work. In general it has a slightly more compressed, glossier shell but otherwise similar in other shell characteristics seen in *M. andrewsae.*

Specimen: West Virginia, Webster County, along Williams River (author's collection).

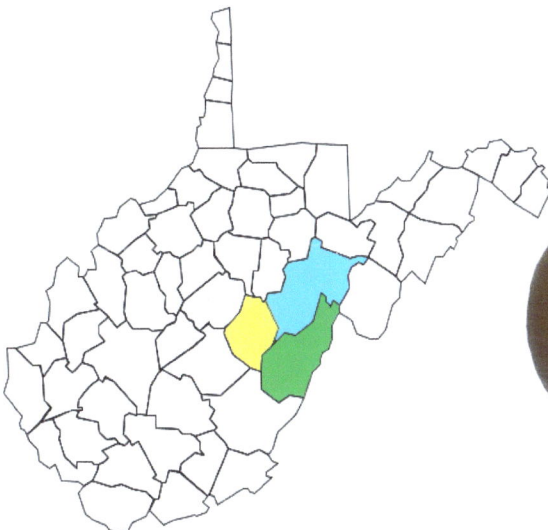

Shell thin and fragile

Balsam globe

Polygyridae

Mesodon andrewsae (W.G. Binney, 1879)

Diameter: 20-30 mm, height 15-18.2 mm

Description: Heliciform; lip reflected; shell with 5.5-6 whorls; imperforate; shell thin for its size, some shells easily dented with finger tips (a); without a parietal tooth; coppery to chocolate-brown, figure (b) illustrating a beautiful chocolate color morph with live animal inside; adult shells without hairs; transverse striae only weakly developed; spiral striae a strong feature; shell periphery well-rounded; this species has been observed feeding on dead snails of its own kind and deceased millipedes, hypothetically, for the calcium content of the exoskeleton, both feeding events observed on Roan Mountain, North Carolina (pers. obs.).

Similar Species: *Mesodon thyroidus* has a rimate umbilicus that is slightly open; *M. normalis* is larger with a thicker shell and generally found below 1400 m; *M. zaletus* has a thicker shell and a parietal tooth.

Habitat: A species of high elevation northern hardwood and spruce/fir forests (from 1600 to 2000 meters).

Status: G2/S1; Rare; this species is currently reported from scattered high elevation mountains over 1600 meters, including Roan Mountain (North Caronia and Tennessee), Mount Rogers, Virginia north to the West Virginia highlands (Hubricht, 1985).

Specimen: All specimens from North Carolina, Mitchell County, in spruce/fir forest, near Carvers Gap, Roan Mountain (author's collection).

a

b

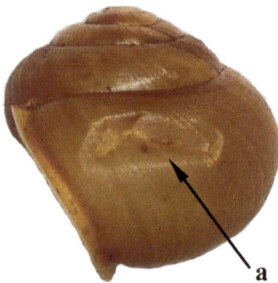

This page provided to compare West Virginia specimens of *Mesodon aff. andrewsae* to confirmed populations of *Mesodon andrewsae* from North Carolina

296

Toothed globe

Mesodon zaletus (A. Binney, 1837)

Diameter: 19-31 mm, height 16.9-24.5 mm

Description: Heliciform; lip reflected; shell with 5.5-6 whorls; imperforate; small parietal tooth typically present, but may be lacking in some individuals in the same populations; shell solid (thick walled); cream color to cinnamon-buff; no hairs in adults; transverse striae are well developed; spiral striae always present and well defined; shell periphery rounded.

Similar Species: *Mesodon normalis* is larger with a slightly higher profile, has a thinner shell, and never has a parietal tooth; *M. thyroidus* has a thinner shell and is rimate to perforate; *Neohelix albolabris* is generally larger, having a thicker shell and is usually without a parietal tooth.

Habitat: Most common on rich, mesic, wooded slopes with mature to old growth forest cover.

Status: G5/S5; Relatively common; in West Virginia, a species scattered throughout the state and typical of quality woodlands.

Specimen: Tennessee, Sevier County, Roaring Fork, GSMNP (GSMNP collection).

Polygyridae

M. zaletus, McCreary County, Kentucky

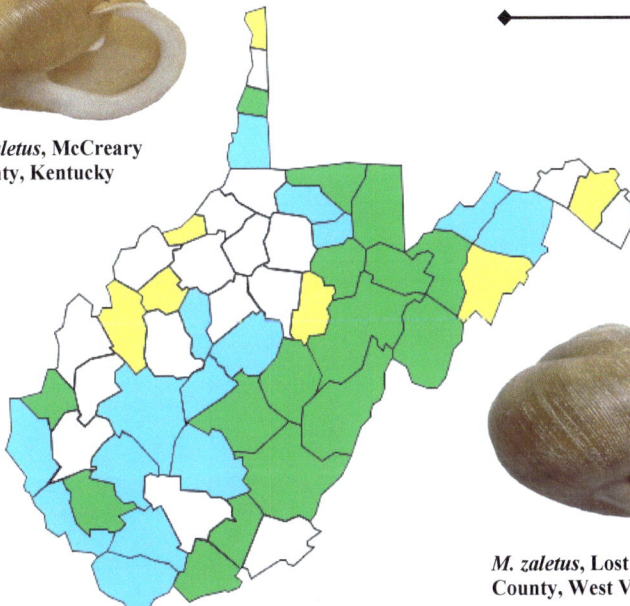

M. zaletus, Lost Cave, Greenbrier County, West Virginia

Grand globe

Polygyridae

Mesodon normalis (Pilsbry, 1900)
Diameter: 30-39 mm, height 26.1-26.5 mm
Description: Heliciform, shells varying considerably in height; lip reflected; shell with 5.5-6 whorls; imperforate; no teeth present; horn colored to tannish olive; without hairs in adult shells; transverse and spiral striae always present; shell periphery well rounded; shell height will vary between different populations; despite its large size, *M. normalis* is reported to be an annual species (Hubricht, 1985).
Similar Species: *Mesodon thyroidus* has a rimate umbilicus that is slightly open; *M. zaletus* has a thicker shell and has a parietal tooth; *N. albolabris* has a thicker, more compressed shell and wider lip.
Habitat: A species of acidic soils; found in upland habitats (up to 1400 m), usually in mixed hardwoods, but also found in pine woods around logs.
Status: G5/S1; Rare, in West Virginia, this species was collected near Premier in McDowell County by Leslie Hubricht and by Craig Stihler on Peters Mountain in Monroe County.
Specimen: Tennessee, Cocke County, Cosby Campground, GSMNP (GSMNP collection).

M. normalis, Macon County, North Carolina

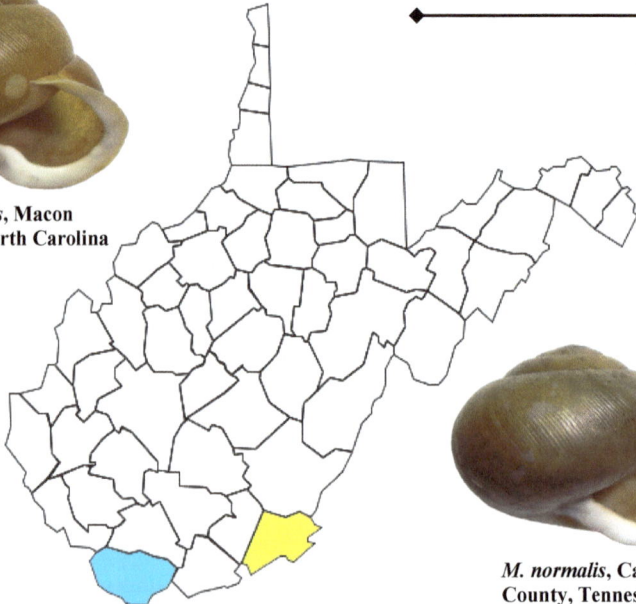

M. normalis, Carter County, Tennessee

Chrissy Mann

Above image of grand globes, *Mesodon normalis*, exchanging sperm during copulation and below the same species feeding on fresh black bear dung. Both images from the Great Smoky Mountains Natural Park of Tennessee and North Carolina.

Above a newly hatched grand globe, *Mesodon normalis,* feeding on the unopened flower spikes of a Dwarf rattlesnake plantain orchid, *Goodyera repens,* Nolichucky River Gorge, Unicoi County, Tennessee.

Whitelip

Neohelix albolabris (Say, 1817)

Diameter: 17.6-45.3 mm, height 16-18.7 mm

Description: Depressed heliciform, although some shells of this large snail will sometimes be more globose in form, figure (a); lip reflected, thickened and wide, varying in its overall shape; shell with 5-6 whorls; shell solid with a dull surface; imperforate; typically without teeth or very occasionally, with a weak parietal tooth (b); cream buff to pale tan; transverse striae well developed, impressed wavy spiral striae well developed; adults without hairs; periphery well rounded; this beast of a snail is one of the largest in eastern North America.

Similar Species: *Mesodon normalis* has a higher shell profile and thinner shell; *M. zaletus* has a higher shell profile and a parietal tooth.

Habitat: A species found in a wide range of upland mixed hardwood sites; at the base of limestone cliffs, but also found on dry acidic ridge tops, in waste places and urban areas.

Status: G5/S5; Common; one of the most common and widespread large land snails in West Virginia, reported from nearly every county.

Specimen: West Virginia, Monroe County, along Wayside Creamery, Indian Draft Creek (author's collection).

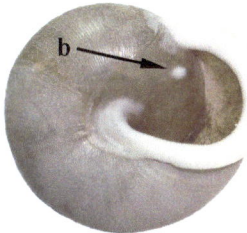

Hampshire County, West Virginia

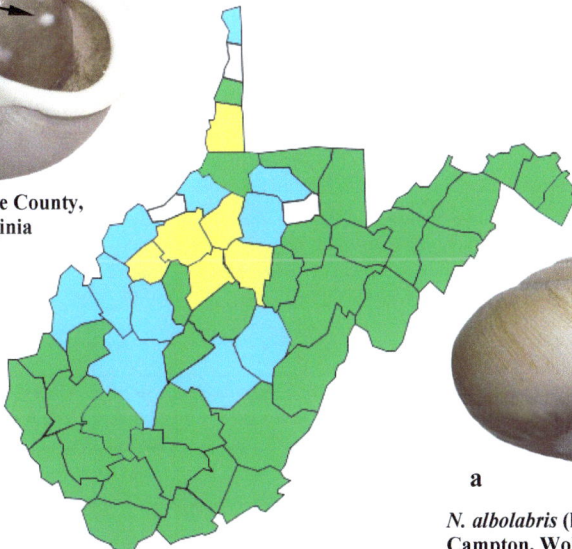

N. albolabris (life size), near town of Campton, Wolfe County, Kentucky

301

Big-tooth whitelip

Polygyridae

Neohelix dentifera (A. Binney, 1837)

Diameter: 20-30.5 mm, height 11.4-16.5 mm

Description: Depressed heliciform; lip reflected; shell with 5-5.5 whorls; imperforate; shell surface dull; parietal tooth low and wide; shell color variable from pale olive to chestnut brown; transverse striae are well developed, impressed spiral striae present; no hairs in adult shells; periphery rounded.

Similar Species: *Neohelix albolabris* is larger, with a higher shell profile and no parietal tooth; *Mesodon normalis* has a thinner, higher profile shell and no parietal tooth; *M. zaletus* has a higher shell profile and contains a smaller parietal tooth.

Habitat: Generally restricted to higher elevation mixed northern hardwood forests occurring under leaf litter, around rocks and logs on acidic soils.

Status: G5/S5; Common; this mostly northern land snail is reported from many counties in West Virginia and likely occurs in all counties except a few along the Ohio River.

Specimen: Kentucky, Letcher County, Pine Mountain (author's collection).

N. dentifera, Red Spruce Knob Trail, Pocahontas County, West Virginia

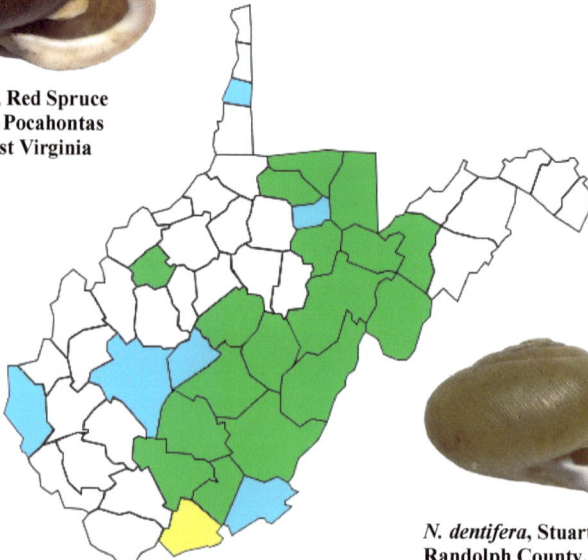

N. dentifera, Stuart Recreation Area, Randolph County, West Virginia

Striped whitelip

Polygyridae

Webbhelix multilineata (Say, 1821)

Diameter: 14.5-32 mm, height 17-20 mm

Description: Heliciform; lip thin and reflected; shell with 5.5-6 whorls; imperforate; ivory yellow to olive buff, with multiple dark reddish brown bands that vary in width, shell glossy and thin; transverse striae are well developed, impressed spiral striae present; adults shells without hairs; periphery rounded; one of the most colorful native land snails in West Virginia.

Similar Species: *Allogona profunda* is flatter, umbilicate and with a small basal tooth.

Habitat: A species of low moist habitats including floodplains, marshes, and margins of lakes and ponds; also occurs on forested islands in the Ohio River where it is can be surprisingly common.

Status: G5/S1; Rare; in West Virginia this interesting gastropod is reported from Ohio River islands, (Taylor and Counts 1976) and Cranesville Swamp on the Maryland border.

Specimen: Kentucky, Carlisle County, island in the Mississippi River (author's collection).

It's believed that color streaks in shells act as disruptive patterns, making snails more difficult to spot by predators.

303

Crypsis in Thin Shells?

What appears to be the dappled shell of a Balsam globe, *Mesodon andrewsae* (above and below images), is in fact the live mottled animal, seen through a very thin shell. This may be simply a consequence of life in a calcium poor environment, or perhaps, an evolutionary advantage to the snail surrounded by the usually high number of predatory shrews found on Roan Mountain. In the picture above, the patterns of the shell (a) are surprisingly similar to the patterns of the stone (b) on which the snail crawls. In the dim light of nightfall, the snail shell may be nearly impossible for a shrew to distinguish. Roan Mountain, Mitchell County, NC.

Broad-banded forestsnail

Polygyridae

Allogona profunda (Say, 1821)

Diameter: 19-34 mm, height 15-17.1 mm

Description: Depressed heliciform; lip reflected; shell with 5.5 whorls; umbilicate; a small basal tooth usually present; shell light tan with a few reddish-brown bands that vary in their width, but shells are also found without color bands; shell glossy and thick; transverse striae well developed, spiral striae are better developed on top, but nearly absent on the base; adults shells without hairs; periphery rounded; live animal tricolor (figure a, page 306).

Similar Species: *Appalachina sayana* is similar in size and build, but is without color bands and has a parietal tooth, in limestone sites both species often occur together.

Habitat: A species of both limestone and acidic mountainsides in mixed hardwood forests; on Bickle Knob live animals have been found climbing on sunflowers and nettles in July at 1189 m (Stihler pers. comm.).

Status: G5/S5; Fairly Common; this beautiful land snail is reported from across West Virginia, but in general, found only sparsely where it occurs.

Specimen: All specimens including next page from Kentucky, Powell County, Furnace Mountain (author's collection).

Non banded form

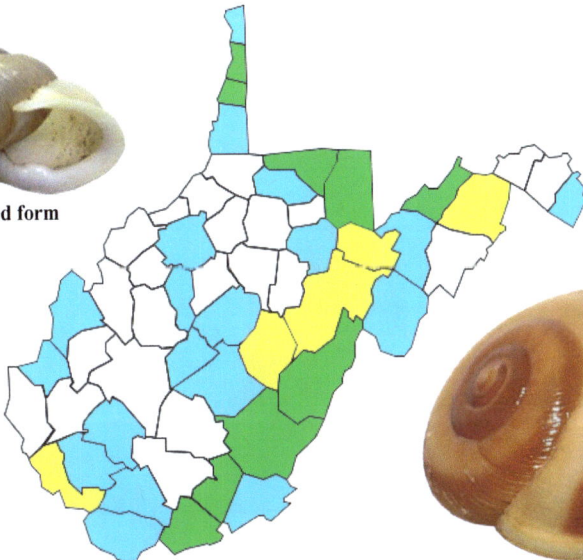

305

Spike-lip crater

Appalachina sayana (Pilsbry, 1906)

Diameter: 19.4-27 mm

Description: Depressed heliciform to heliciform; lip reflected, thin and wire like; shell with 5.5 whorls; umbilicate; shell thin for its large size; parietal tooth variable in size; basal tooth small (a); pale yellow to pale olive-tan, sometimes with darker streaks; no hairs in adult shells; transverse and minute spiral striae always present; periphery well-rounded.

Similar Species: *Mesodon normalis* is larger with a closed umbilicus and thicker shell; *Appalachina sayana kentucki* (reported in Kentucky counties that border West Virginia) is smaller, has a smaller umbilicus and a larger basal tooth and is typically without a parietal tooth (see page 384 for comparison).

Habitat: A common species of rich upland mixed hardwood forests under leaf litter and other forest debris on both limestone and acidic sites, the species is generally indicative of intact and quality forests.

Status: G5/S5; Common; found throughout much of West Virginia.

Specimen: Typical form illustrated to the upper right (3 views), from Kentucky, Powell Co, Furnace Mountain (author's collection).

Yahoo Hollow, Randolph County, West Virginia

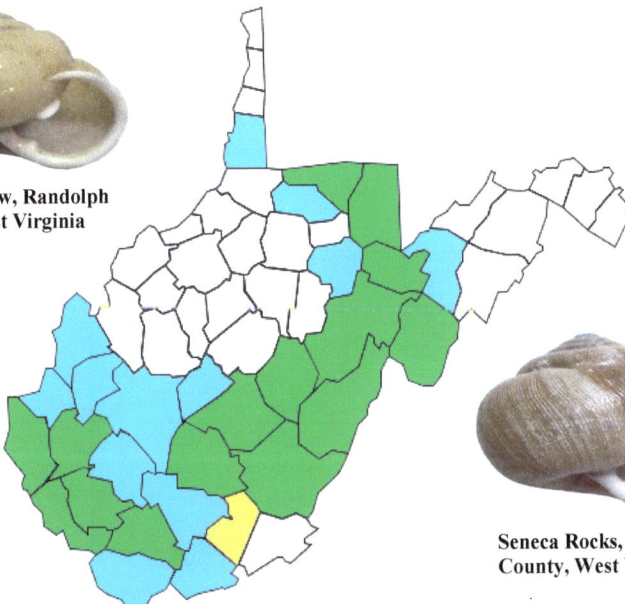

Seneca Rocks, Pendleton County, West Virginia

307

Symbiosis Between Land Snails and Slime Molds?

Above image of a Spike-lip crater, *Appalachina sayana,* feeding on immature slime mold sporangium in Red River Gorge, Kentucky. There is some speculation that symbiotic relationships exist between land snails and particular species of slime molds (a hypothesis proposed during a discussion between the author, Adam Rollins and Ron Caldwell in Belize, 2010). For example, slime molds in the order Physarales precipitate amorphous calcium carbonate on some portion of their mature fruiting bodies (Townsend *et al.* 2005). *Badhamia utricularis* (figure a) not only coats the outer surface of its fruiting body with calcium, but has a system of calcareous tubes running throughout its core (figure b). This may be coincidence or a clever strategy for attracting land snails who require the mineral for shell building. What if Physarales slime molds sequester and store calcium for the singular purpose of attracting land snails? To ensure the spores are eaten by the visiting snail, the calcium deposits are intermingled with the spores (b). Pushing the hypotheses a bit further, what if slime molds were one of several conduits that help facilitate land snail migration into more acidic environments? These hypotheses await investigation. Figures a & b images from Eumycetozoan Research Project, University of Arkansas.

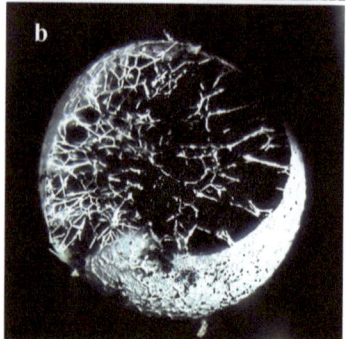

Coopers Rock State Forest, Monongalia County, West Virginia

In general, native slugs are the least studied land snails in many regions of the world, and West Virginia is no exception. These under-appreciated gastropods are often all categorized as pests, but nothing could be farther from the truth. Native slugs are rarely a nuisance and typically become scarce or disappear altogether when the natural vegetation has been eliminated. Even in ecosystems that are more or less pristine, native slugs can be hard to find. In contrast, introduced slugs are quick to overpopulate an area, particularly in degraded habitat and without natural controls in place. These inadvertent guests are at home in the most degraded habitats such as small isolated woodlots, yards, roadsides, piles of trash and other waste places.

Considered excellent indicators of a healthy forest, native slugs by and large are only common where there are significant decomposing structure (logs), a habitat also used by other groups of organisms such as small mammals, salamanders, micro-invertebrates, fungi, and slime molds. All slugs, native or otherwise, can vary considerably in color and mantle patterns, even within the same population. Many gastropods in the family *Philomycidae* can only be reliably separated by dissection or DNA sequencing (Fairbanks pers. comm. 2010).

Genera Included:
(in order of appearance in text)

Philomycus
Megapallifera
Pallifera
Deroceras

Key to the PHILOMYCIDAE Family

Philomycidae are native terrestrial slugs found primarily in the eastern half of the United States, and even today, there remains a number of undescribed species (Fairbanks 1998). Many of the more conspicuous PHILOMYCIDAE are large, reaching crawling lengths of more than 100 mm. There are three North American genera in this family (see below): *Philomycus, Megapallifera,* and *Pallifera*. **Philomycus** species are large (75-100 mm), obese slugs. In terms of their reproductive anatomy, the species of *Philomycus* are the most interesting, characterized by the presence of a dart sac and dart (referred to as love darts). Some *Philomycus* species, in isolation, become self fertilizing (Nicklas and Hoffman 1981). In *Philomycus* species, the mantle covers the entire body including the head (a). **Megapallifera** species are smaller in size (approximately 80 mm long or less), are also obese but lack the dart sac and dart (Fairbanks 1998) and the head is not entirely covered by the mantle (b). **Pallifera** species are the smallest among the group (50 mm or less), are slender and, like *Megapallifera,* lack the dart sac and dart. The head remains uncovered by the mantle. Great color variation exists in Philomycidae slugs making their ID very difficult. Young slugs of all species will generally have the same mantle patterns as the adults. One last detail for separating slugs includes their defensive slime. To view this feature, first gently irritate the animal (in one spot) with a small stick until a small mass of slime accumulates (d). In natural lighting, determine the color of the slime. This is an important diagnostic mark and varies by species. The crawling slime of slugs is typically clear.

Defense mucus on a stick

Mantle

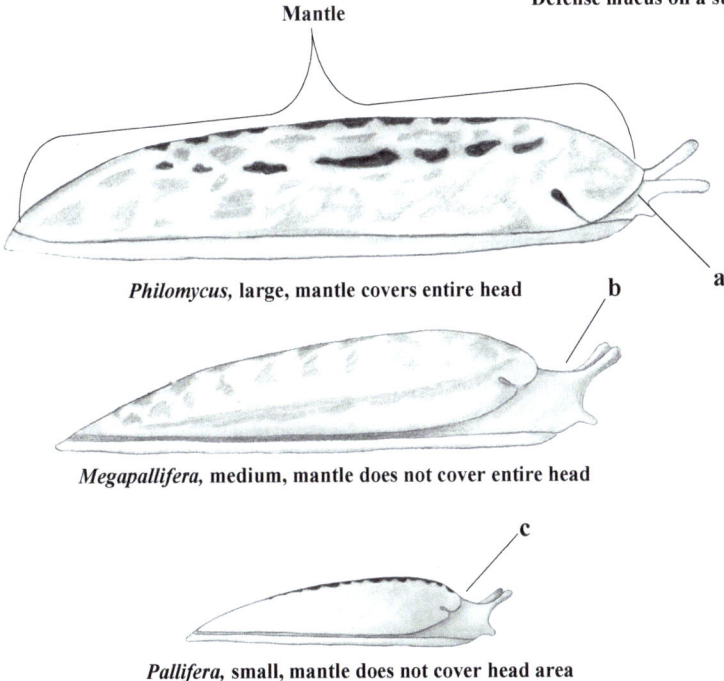

Philomycus, large, mantle covers entire head   b   a

Megapallifera, medium, mantle does not cover entire head

c

Pallifera, small, mantle does not cover head area

311

Native Slugs of West Virginia Compared (Proportionate)

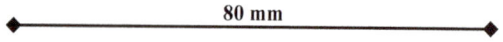

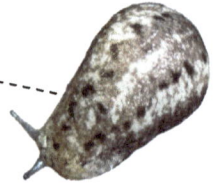

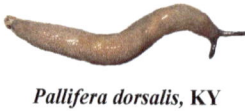

80 mm

Philomycus carolinianus, KY

Defense posture

Philomycus togatus, NC

Philomycus venustus, NC

Defense posture

Philomycus virginicus, WV

Megapallifera mutabilis, KY

Pallifera dorsalis, KY

Pallifera ohioensis, WV

Pallifera fosteri, NC

Pallifera secreta, NC

Philomycus flexuolaris, KY

Deroceras laeve

Carolina mantleslug   Philomycidae
Philomycus carolinianus (Bosc, 1802)

Length: Adults 50-100 mm while crawling

Description: Mantle uniformly mottled with brown; usually two dorsal longitudinal rows of black spots (figure a, page 314), the best diagnostic feature for the species; animal carries a dart sac and dart (figure b, page 314); defense mucus whitish (figure d, page 314); when touched with a dry finger, the mucus immediately seeps out of the entire mantle having a bitter taste much like the skin secretions of a spring peeper (MacGregor pers. comm. 2010); sole of foot white (figure c, page 314).

Similar Species: *P. flexuolaris* is smaller, without the dark rows of black spots; the two species are often found together.

Habitat: Found in floodplain forests over much of its range, it becomes an upland species in the mountains, but does not occur much over 600 m; found hiding under exfoliating bark of large rotting hardwood logs in advanced stages of decay, especially logs forming log bridges over small ravines, also found inside hollow trees such as large American beech.

Status: G5/SNR; Uncommon; reported from scattered locations across West Virginia, but likely more common than current records indicate.

Specimen: All slug images from Powell County, Red River Gorge, Kentucky (author's photo collection); figure (b) SEM of *P. carolinianus* love dart by Jodi White-McLean.

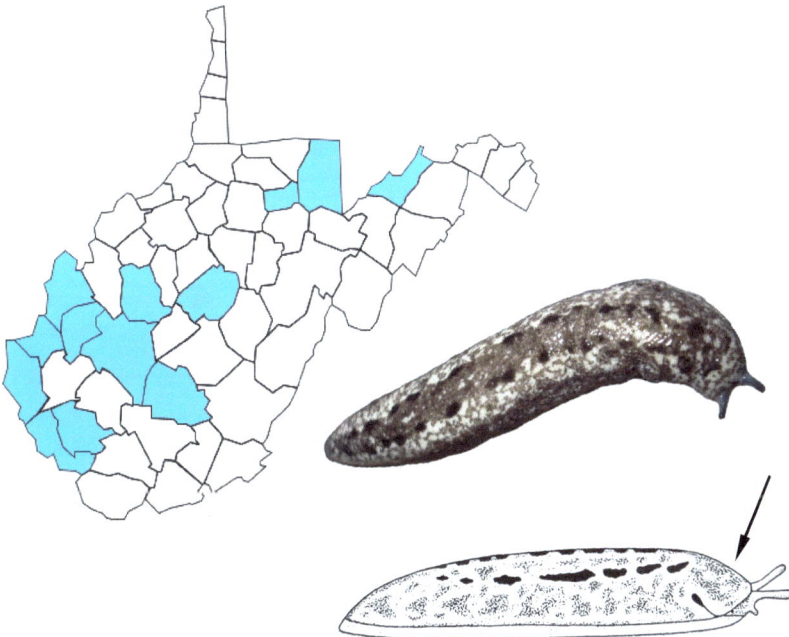

Burch, 1962

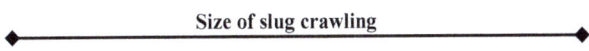

Size of slug crawling

Philomycus carolinianus

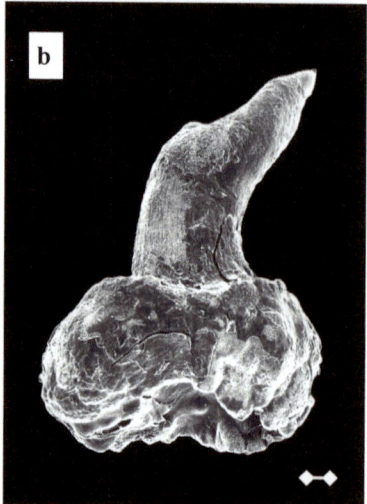

Above image of two Carolina mantleslugs, *Philomycus carolinianus,* illustrating two color morphs from the same log; below the same slugs in defensive posture and stick showing the color of their whitish defense mucus. Both images from the Nolichucky River Gorge, Unicoi County, Tennessee.

Two Carolina mantleslugs, *Philomycus carolinianus,* interested in sperm exchange for the purpose of fertilizing the internal eggs each individual slug carries. Red River Gorge, Powell County, Kentucky.

Winding mantleslug   Philomycidae

Philomycus flexuolaris Rafinesque, 1820

Length: Adults 50-80 mm while crawling

Description: Three longitudinal mottled bands (dorsal and lateral), are usually distinct, but sometimes markings are so pale as to be scarcely discernible, and in some specimens, they may be nearly lost in a general dusky shade (figure b page 318); although the base color of the mantle varies considerably (as can be seen in the images on next three pages) the darker color patterns remain more or less the same; this species possesses a dart sac and dart; defense mucus pale-yellow (figure b, page 318); foot margin pale, sometimes with a hint of yellow; sole of foot cream to flesh color (figure c, page 318); for images of copulating *P. flexuolaris* see pages (319 & 320).

Similar Species: *Philomycus carolinianus* is generally larger, having two longitudinal rows of black or brown spots running down the center of mantle or back; *P. carolinianus* and *P. flexuolaris* can often be found living on the same large American beech log.

Habitat: A species of upland mixed hardwood forests up to 1500 m; as with most native slugs, during the day it goes into hiding under the exfoliating bark of large rotting logs in advanced stages of decay, especially large American beech trees.

Status: G5/SNR; Uncommon; like most native slugs, *P. flexuolaris* has been under reported, due in part to a lack of interest by collectors who often ignore these interesting but obscure creatures.

Specimen: Figures (a, b and c) from Powell County, Red River Gorge, Kentucky; figure (d) from Pike County, Kentucky (author's photo collection).

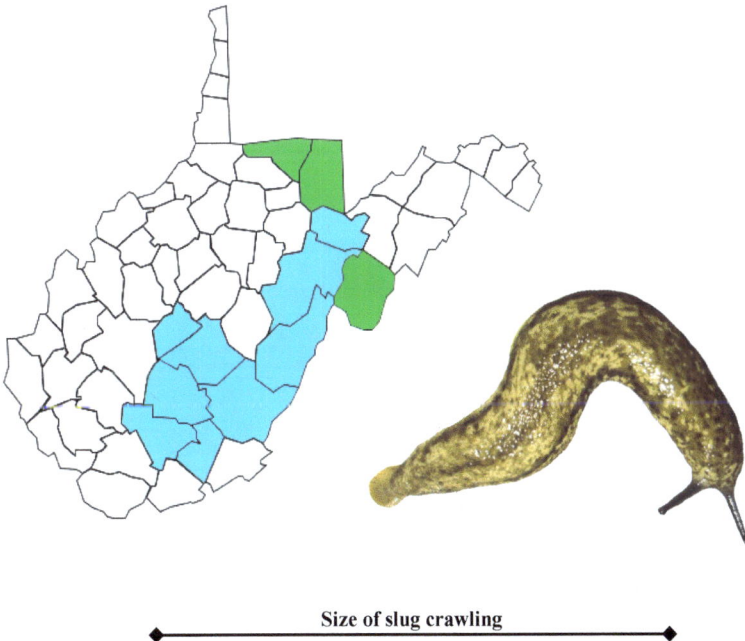

Size of slug crawling

Philomycus flexuolaris

Copulating *Philomycus flexuolaris*

Two Winding mantleslugs, *P. flexuolaris,* seeking a sperm exchange in Red River Gorge, Powell County, Kentucky

Copulating *Philomycus flexuolaris*

The sperm exchange

Variable mantleslug   Philomycidae

Philomycus togatus (Gould, 1841)

Length: Adults 60-100 mm while crawling

Description: Top of animal heavily mottled , consisting of a broad dorsal band (sometimes broken into two bands) and narrower lateral bands (bands in some specimens may be hard to detect at times) on each side and scattered, irregular, small spots between the bands (Hubricht 1951); as can be seen on the next page (figures a-d, page 322) there is considerable variation in the species; defense mucus orange (figure b, page 322); foot margin orange-red to orange (figure c, page 322), probably one of the best external features of the species.

Similar Species: Differs from *P. carolinianus* in being darker brown without the two rows of black spots (Hubricht 1951) and having orange not whitish defense mucus; *P. flexuolaris* is smaller, has a yellowish defense mucus and its foot margin is pale not orange-red.

Habitat: An upland (to 1200 m) species found on wooded hillsides and in ravines, under loose bark of logs; in wet weather found on trunks of smooth-barked trees at night.

Status: G5/SNR; Uncommon; reported from scattered locations across West Virginia, but no doubt occurs in additional counties.

Specimen: Figures (a & b) are from Dickenson County, Breaks Interstate Park, Virginia (author) and figure (d) from Surry County, North Carolina (Wayne Van Devender).

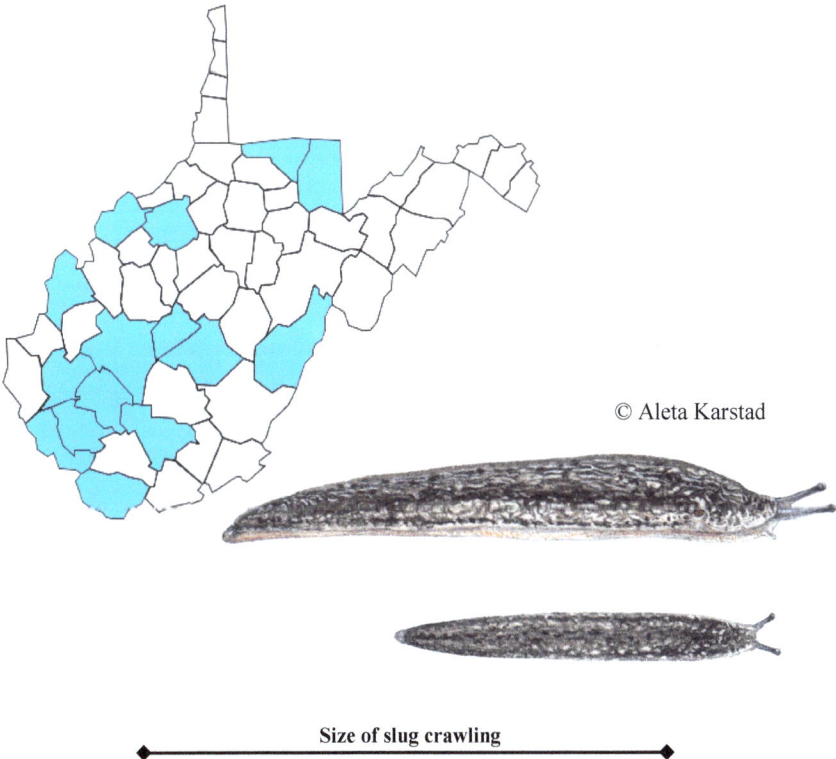

© Aleta Karstad

Size of slug crawling

Philomycus togatus

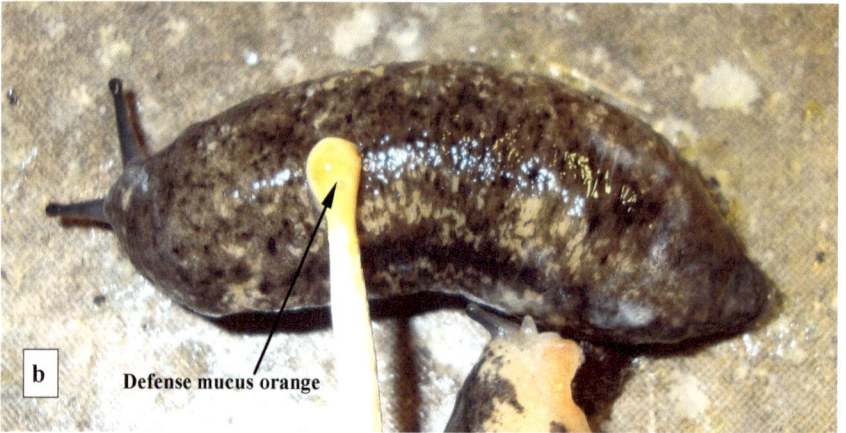

a

c

b
Defense mucus orange

d
Juvenile

Brown-spotted mantleslug   Philomycidae

Philomycus venustus Hubricht, 1953

Length: Adults 75-120 mm while crawling

Description: The color pattern varies from individuals having two contrasting dark chestnut brown to black dorsal bands, a narrow lateral band on each side, connected to the dorsal bands by a series of oblique bands or spots (figures a and c, page 324), forming chevrons; to individuals in which this pattern is broken up into a series of spots (figure d, page 324); defense mucus white; sole of foot creamy white (figure b, page 324).

Similar Species: *Philomycus virginicus* is darker brown in color and has lighter longitudinal and oblique bands; *P. carolinianus* does not have the oblique bands or spots (forming chevrons) seen in *P. venustus*.

Habitat: An upland species (up to 1800 m on Roan Mountain in North Carolina) found on wooded hillsides and mountains; during the day it can be found hiding under the exfoliating bark of large rotting logs in advanced stages of decay; also found in crevices and cavities of beech trees.

Status: G4/S4; Uncommon, reported from just 4 counties in West Virginia.

Specimen: Figure (a) from Forney Creek, GSMNP, North Carolina; figures (b and c) from Pike County, Kentucky and figure (d) from Mitchell County, Roan Mountain, North Carolina.

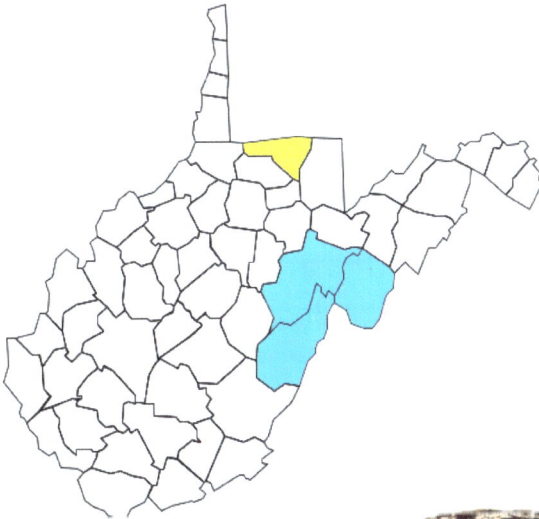

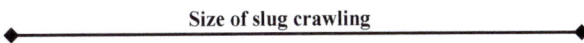

Size of slug crawling

Philomycus venustus

Virginia mantleslug

Philomycus virginicus (Hubricht, 1953)

Length: Adults 75-100 mm while crawling

Description: Base color of mantle is a grayish-white; color pattern consisting of a darker (brownish) broad dorsal band, sometimes bordered by a row of darker spots and a narrow lateral band on each side, connected to the dorsal band by a series of faint oblique (diagonal) stripes (figure d, page 326), the whole pattern obscured by a general fine flecking (figures a-c, page 326); young slugs with the same pattern, but brownish-gray, becoming chestnut-brown with age (Hubricht 1953); defense mucus whitish; foot margin off-white; sole of foot gray to whitish.

Similar Species: *Philomycus venustus* has three darker longitudinal bands often times broken into spots which are also connected by darker oblique bands or spots, all of which are notably darker on a usually lighter colored animal, than is seen in *P. virginicus*.

Habitat: An upland species found in mixed hardwood under the exfoliating bark of large rotting hardwood logs in advanced stages of decay.

Status: G3/S2; Rare; in West Virginia this slug is reported from six counties, but likely occurs in additional localities.

Specimen: Figure (a & b) from Coopers Rock State Forest, Monongalia County, West Virginia (author's photo collection) and figure (c) is from Lutherock, Avery County, North Carolina (Wayne Van Devender).

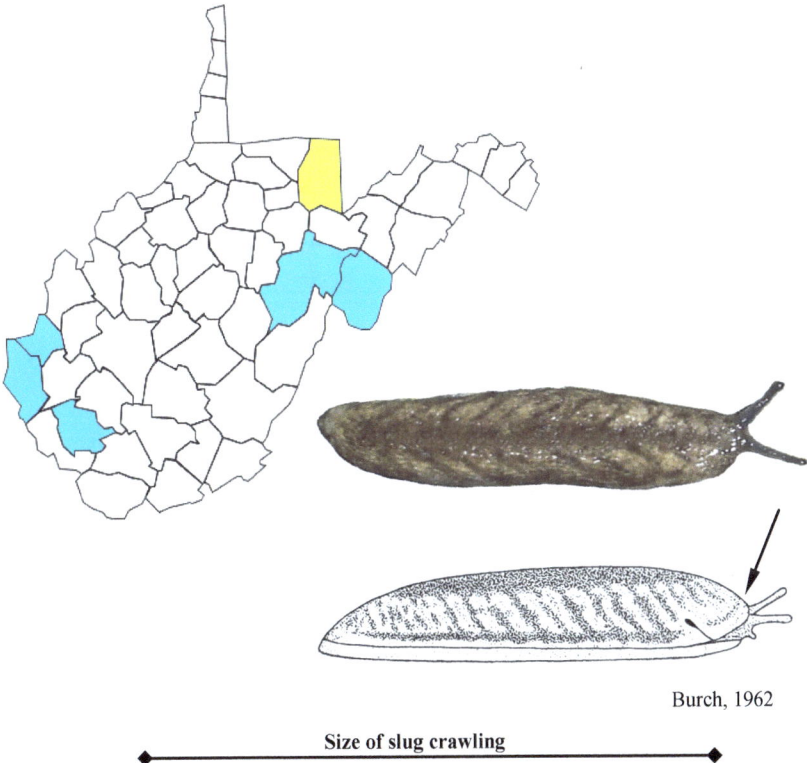

Burch, 1962

Size of slug crawling

325

Philomycus virginicus

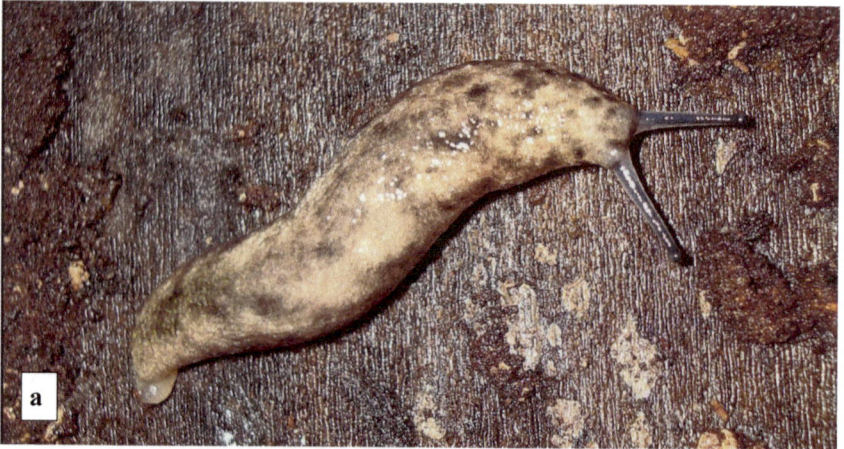

Changeable mantleslug   Philomycidae

Megapallifera mutabilis (Hubricht, 1951)

Length: Adults 60 mm while crawling

Description: Mantle color is fawn or tan, not covering entire head; but this feature may be difficult to discern, requiring an observer to rely on the use of a hand lens of at least 10X power; mantle usually has two (on occasions three) interrupted dorsal dark gray-brown longitudinal pigment bands (more often spots (figures a-c page 328) and a continuous wavy lateral band on each side (figures d page 328), small scattered pigment spots between the bands; spots sometimes producing vague chevrons on the mantle; defense mucus tannish-white or white (figure b, page 328); margins of the foot olive, gray or sometimes whitish; sole of foot cream.

Similar Species: *Philomycus* species are slightly larger in length and notably more robust when crawling and the mantle covers the entire body.

Habitat: Found in upland woods; during warm wet weather at night, it can be found crawling on the trunks of smooth bark trees like the American beech; also a species of urban settings.

Status: G5/Snr; Common; reported from scattered locations across the state, but expected to be more common than current records indicate.

Specimen: Figure (a) crawling on limestone cliff face, above Greenbrier River, Greenbrier County, West Virginia; figure (b) defense mucus on a stick, Rowan County, Kentucky, figure (c) crawling on sandstone cliff, Monongalia County, Coopers Rock State Forest, West Virginia (author's photo collection).

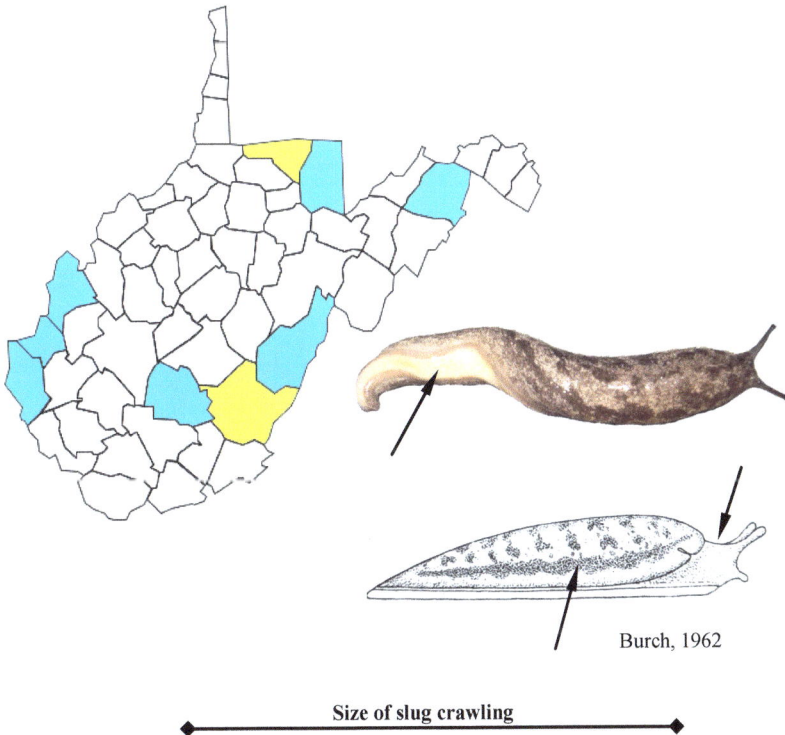

Burch, 1962

Size of slug crawling

327

Megapallifera mutabilis

a

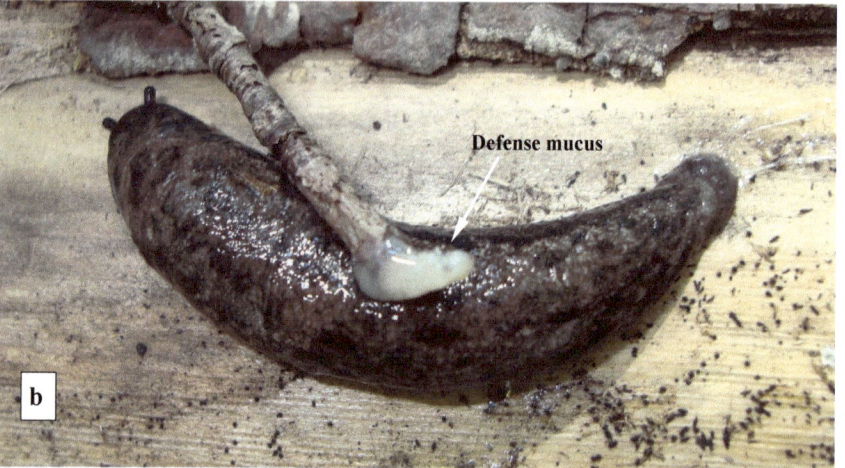

Defense mucus

b

d

c

Severed mantleslug　　　　　　　Philomycidae

Pallifera secreta (Cockerell, 1900)

Length: Adults 20-30 mm while crawling

Description: Mantle very dark gray or gray blue with numerous and well scattered small, round, black spots, often more profuse at the mantle edges near the foot margins and on the anterior end of the mantle (figures a-c, page 330); sole whitish with an ochreous (yellowish) tint (Cockerell 1900); defense mucus whitish.

Similar Species: *Pallifera dorsalis* is smaller, with or without an interrupted black line down the center of the mantle; *P. fosteri* is smaller, flesh colored, with interrupted clusters of spots down the center (dorsal) of the mantle and the anterior margins of its foot is brownish red.

Habitat: This slug lives deep down in drifts of damp leaves next to logs, leaf-filled depressions, and rock talus in upland mixed hardwood forests (up to 1500 m in the southern Appalachians); the species reported to rarely come to the surface (Pilsbry 1948).

Status: G4/SNR; Rare, like other species of *Pallifera,* these slugs are secretive and remain well hidden, as a result, few records exist for this small slug in West Virginia.

Specimen: Figures (a and b) from North Carolina, Transylvania County, Rosman (John Slapcinsky photo collection) and figure (c) from North Carolina, Mitchell County around Carvers Gap, Roan Mountain (author photo collection).

Burch, 1962

Size of slug crawling

Pallifera secreta

Pale mantleslug   Philomycidae

Pallifera dorsalis (A. Binney, 1842)

Length: Adults 20-30 mm while crawling

Description: The live animal is ashy blue, gray or brownish with or without an interrupted black line down the center of the mantle (in my experience, more often without this feature); defense mucus translucent amber to whitish; margins and sole of the foot are white (figure c, page 332); adult slug shown in a typical defensive posture (figure b, page 332).

Similar Species: *Pallifera fosteri* has distinct black spots, which are more prominent at the mantle edges, its mantle slightly humped in front making the neck and head longer than seen in *P. dorsalis*; *P. ohioensis* has distinct red edges running along the sole of its foot (figure d, page 332).

Habitat: A slug of humid forests, generally found in deep, moist leaf litter in and around rock talus or large piles of rotting wood; like other *Pallifera,* very secretive, rarely seen crawling about on the surface.

Status: G5/SNR; Uncommon, only reported from scattered locations throughout the state but the species is likely more widespread than current records indicate.

Specimen: Figure (a, b & c) from Buncombe County, North Carolina); figure (d) comparing *P. dorsalis,* from Powell Co, Kentucky and *P. ohioensis* from Blackwater Falls State Park, West Virginia.

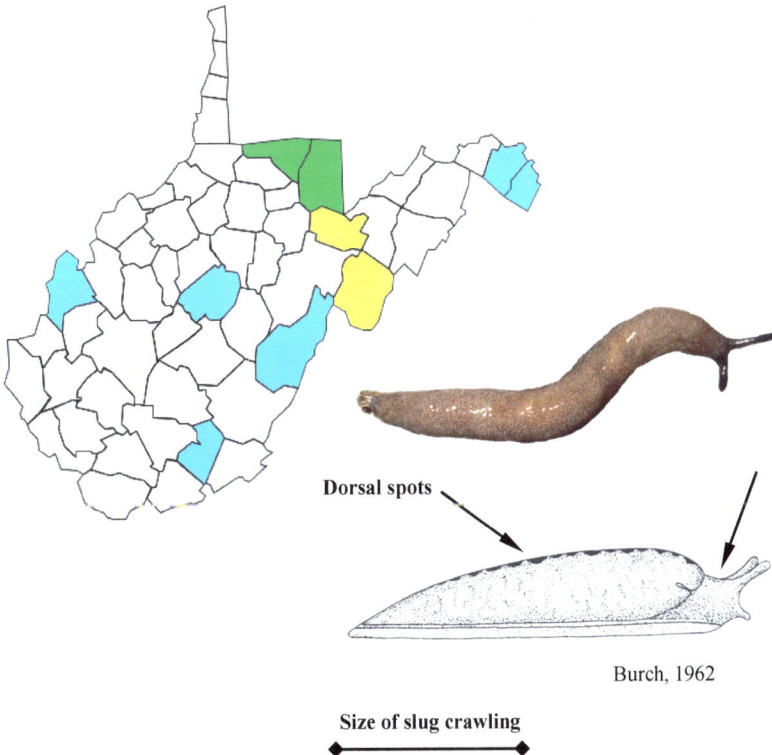

Dorsal spots

Burch, 1962

Size of slug crawling

331

Pallifera dorsalis

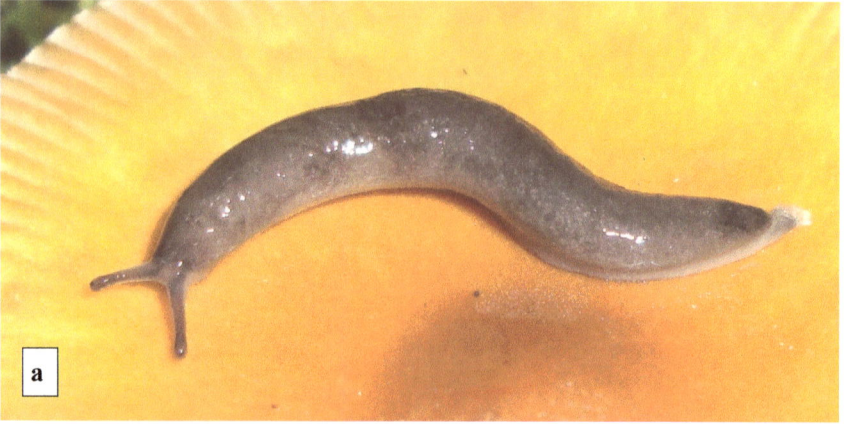

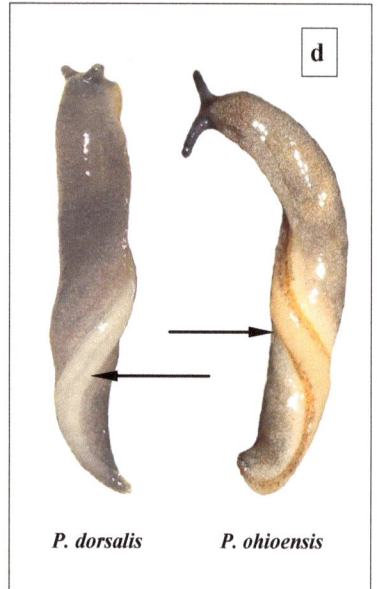

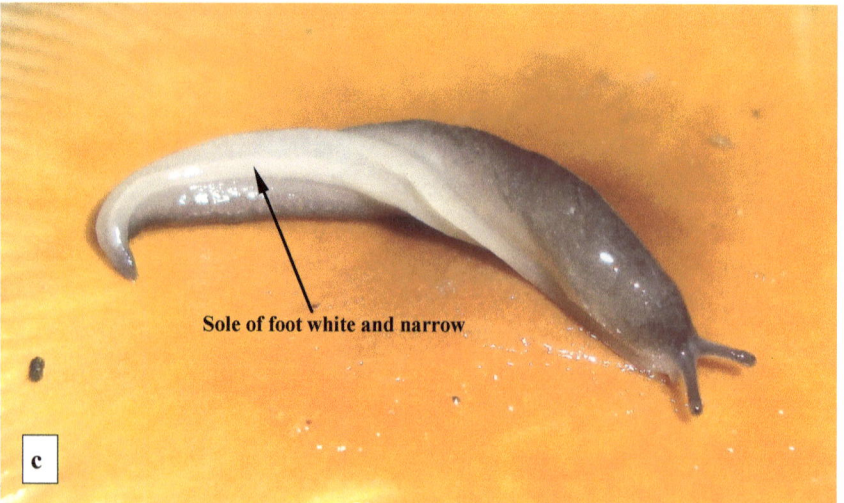

a

b

Defensive posture

d

P. dorsalis   P. ohioensis

Sole of foot white and narrow

c

Redfoot mantleslug

Philomycidae

Pallifera ohioensis (Sterki, 1908)

Length: Adults 15-30 mm while crawling

Description: The live animal has a speckled grayish back without an inter-rupted black line down the center of the mantle; sole of foot with two blood-red bands running lengthwise along the edge (figure a, page 334); defense mu-cus whitish.

Similar Species: *Pallifera dorsalis* is around the same size, but without the red bands, and on occasion, an interrupted set of black spots running down the center of its back.

Habitat: A slug of humid forests, usually found under bark, deep leaf litter under decaying logs or in rock talus; in West Virginia, it is only known in up-per elevation red spruce forests around Blackwater Falls State Park.

Status: G3/SNR; Rare; this species is not recognized by some authors and it may only represent an interesting race of *P. dorsalis*; this species is also re-ported from Ohio, Indiana, and North Carolina (Dourson 2013).

Specimen: All figures from Blackwater Falls State Park, Tucker County, West Virginia, slugs on rock in spruce/hemlock forest (author's photo collection).

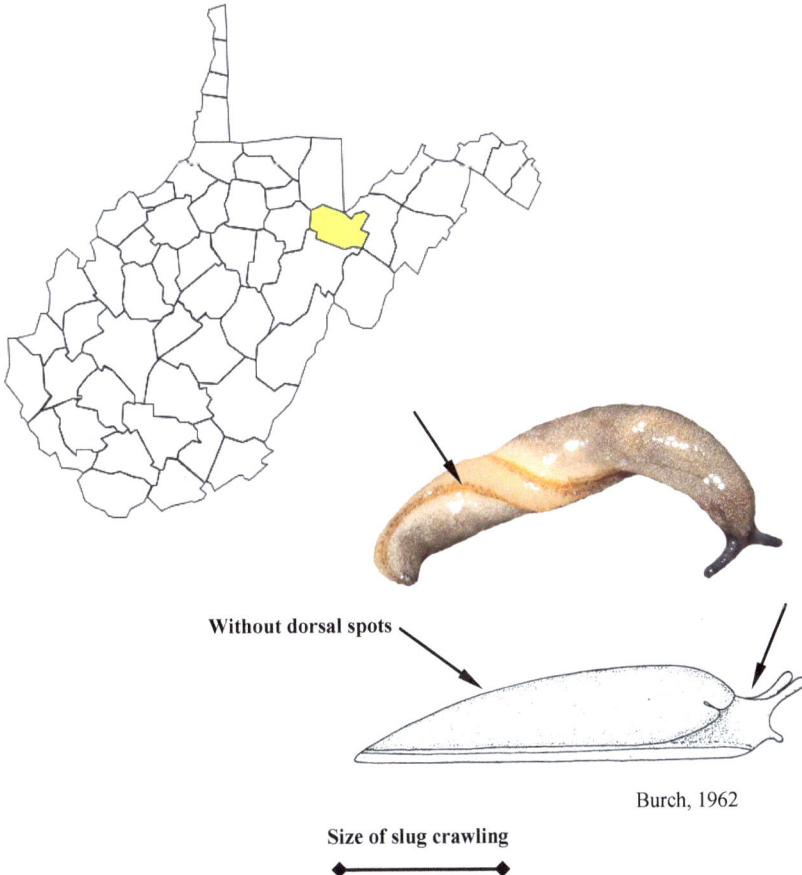

Without dorsal spots

Burch, 1962

Size of slug crawling

333

Pallifera ohioensis

Bottom view

a

Foster mantleslug

Philomycidae

Pallifera fosteri F. C. Baker, 1939
Length: Adults 15-25 mm while crawling
Description: The species may be recognized by the blackish spots on a whitish or flesh colored mantle, these spots or blotches often form interrupted, irregular longitudinal lines, especially near the base of the mantle, or irregularly spaced clusters of small dot like spots scattered over the dorsal surface; in some specimens the black spots form coalescing blotches elongated in form (Pilsbry 1948); Grimm (1961) reported that the mantle is light tan, spotted and reticulate with dark brownish gray, reticulations heaviest in the middle of the back (figure d, page 336), but no solid, dorsal line is formed; at the sides, the reticulations often form two broken lines; tentacles are slate gray and the anterior margin of the foot is brownish red (figure b, page 336, three white arrows); other characters include a slightly humped mantle (see below); defense mucus transparent amber; in juvenile slugs (below image) the tentacles are proportionally large for their body size, but mantle patterns are like that of adults (this is true for most slug species).
Similar Species: *Pallifera dorsalis* is smaller, is a different color, and most importantly, is without the spots on the mantle edges.
Habitat: A species found in a variety of habitats from floodplains to mountain tops under forest litter and around rotting logs in advanced stages of decay.
Status: G5/S; Uncommon; in West Virginia this secretive slug is reported from eight counties, it is likely more common than records currently indicate.
Specimen: Figures (a, b and c) from North Carolina, Swain County, Nantahala Gorge (all images by Wayne Van Devender).

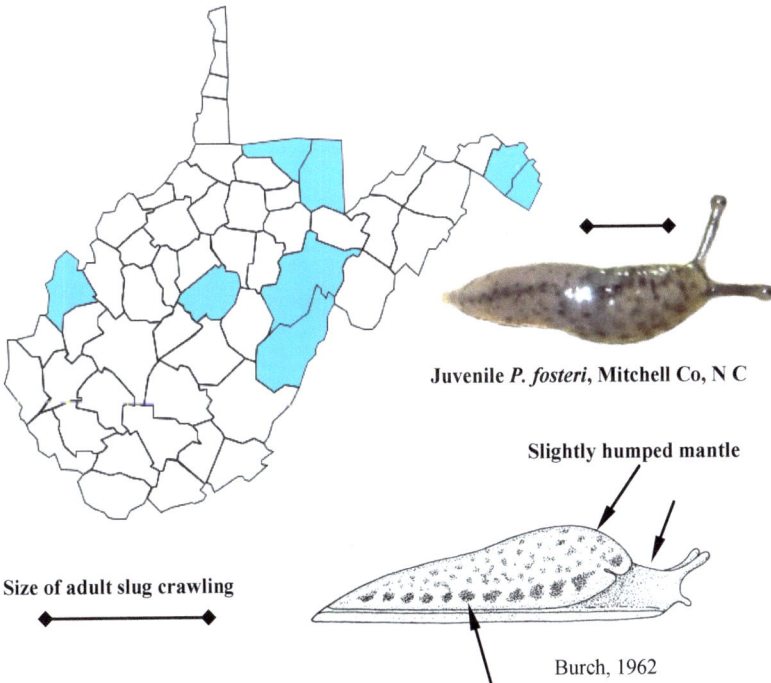

Juvenile *P. fosteri*, Mitchell Co, N C

Slightly humped mantle

Size of adult slug crawling

Burch, 1962

335

Pallifera fosteri

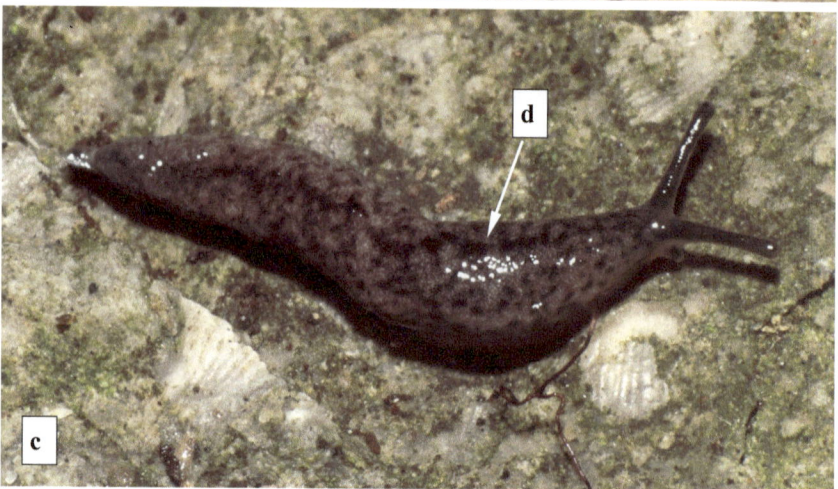

336

Meadow slug   Agriolimacidae

Deroceras laeve (Müller, 1774)

Length: Adults 25 mm while crawling

Description: Color of animal varies from yellowish to pale gray, dark gray to nearly black, sometimes flecked with gray; the mantle (figure b, page 338) is situated on the anterior (front) end of the dorsal surface (back); pneumostome (breathing hole) behind middle of mantle (as opposed to those of *Arion* species, which are in front of the middle of mantle); defense mucus colorless and watery; margins of the foot tan; this slug is a favorite food of the brown snake, *Storeria dekayi* (MacGregor pers. comm. 2010).

Similar Species: Exotic *Arion* species are similar (see page 356); native slugs of the eastern US have mantles that cover or nearly cover the entire animal back.

Habitat: A Holarctic species often found under cardboard and rubbish along roadsides and in vacant lots, also found in meadows and natural glades; even though *D. laeve* is considered native to eastern North America, it has become somewhat of a garden pest.

Status: G5/SNR; Uncommon; reported from around four scattered counties in West Virginia, but like most native slugs it has likely been under reported.

Specimen: Figures (a, c and d) from Madison County, Kentucky (all Wayne Van Devender).

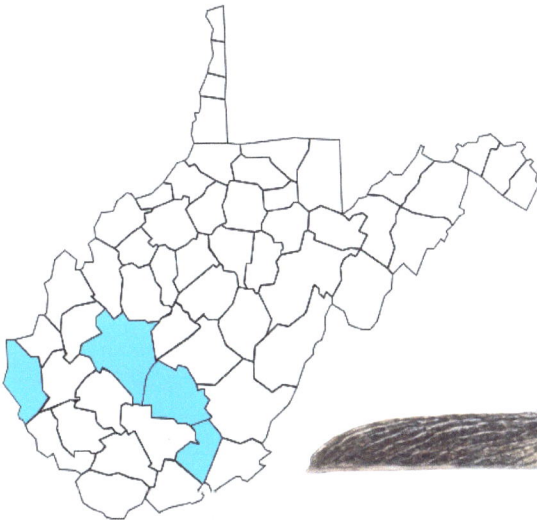

© Aleta Karstad

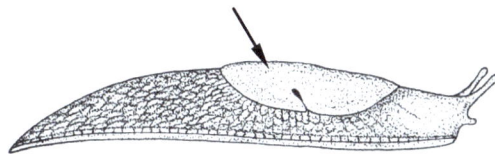

Burch, 1962

Size of slug crawling

Deroceras laeve

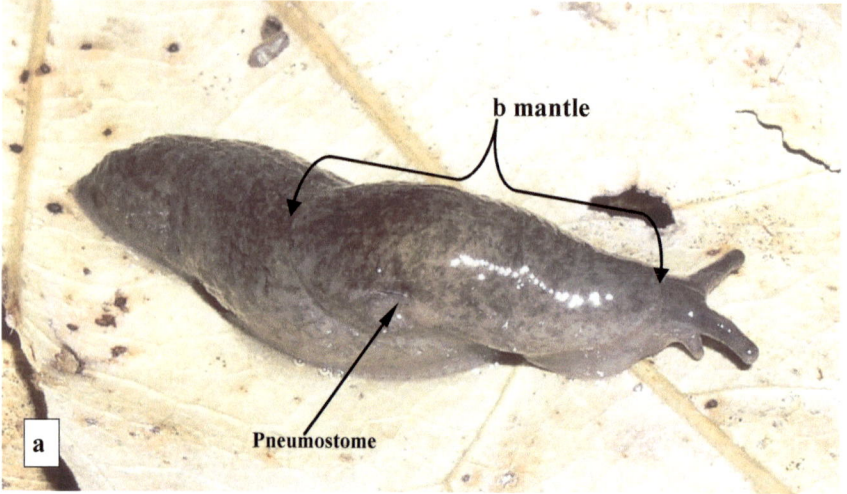

b mantle

Pneumostome

a

c

Mantle edge

d

The colorful redfoot mantleslug of West Virginia's highlands, Blackwater Falls State Park, Tucker County, West Virginia.

Exotic land snails including slugs that have been reported from West Virginia are included in this section. Most have been accidently introduced by way of nursery plants, construction lumber, potting soils and fruit trees. In addition, a full color plate of exotic slugs that may eventually find their way to degraded lands of West Virginia can be found on page 356.

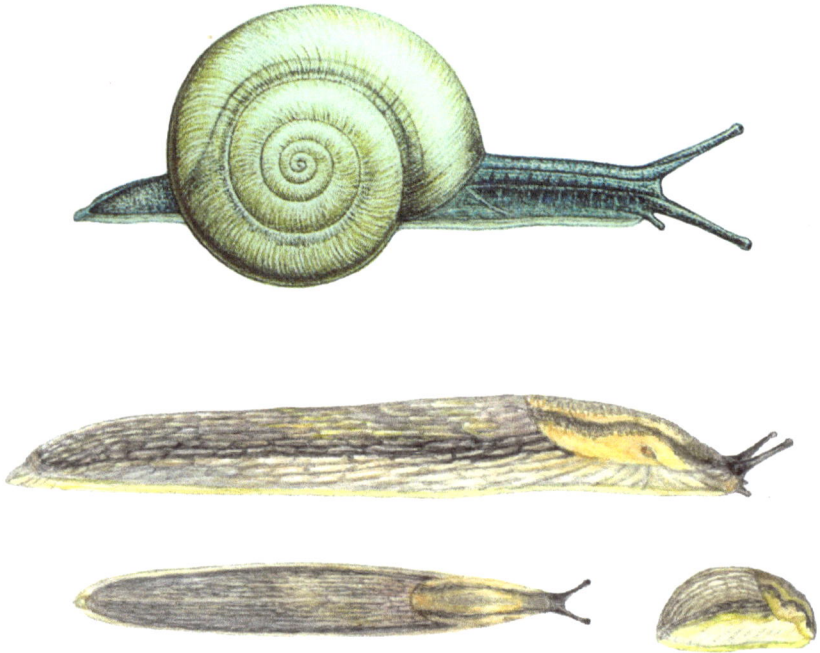

Grovesnail or Woodsnail

Helicidae

Cepaea nemoralis Linneaus, 1758

Diameter: 22-24 mm

Description: Heliciform; lip slightly reflected and dark; shell with 5 whorls; imperforate; without teeth in all stages of growth; shell yellow, olive or red with 1-5 solid reddish-brown or black bands, although some specimens are without color bands; transverse striae moderately developed.

Similar Species: *Cepaea hortensis* is slightly smaller and has a white lip.

Habitat: Found in mostly degraded habitats of cities and urban areas, usually living among or under stones, old wood piles, window wells, but seems especially fond of old limestone fences; the species is less problematic in gardens than exotic slugs due to their preference for dead material (Whitson 2005), *C. nemoralis* can reach large numbers.

Status: A European species, reported from scattered locations across the state.

Specimen: Ohio, Hamilton County, near Mt. Airy Forest in Cincinnati (author's collection).

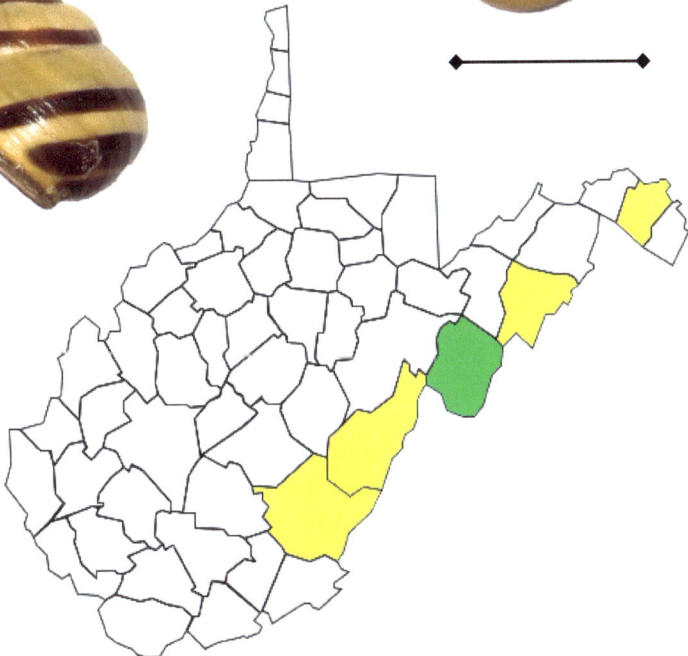

White-lip gardensnail

Helicidae

Cepaea hortensis Linnaeus, 1774

Diameter: 16-21 mm

Description: Heliciform; lip slightly reflected and white; shell with 5 whorls; imperforate; without teeth in all stages of growth; shell yellow, with 1-5 solid reddish-brown or black bands, although some specimens are without color bands.

Similar Species: *Cepaea nemoralis* is slightly larger and has a dark lip.

Habitat: Found in mostly degraded habitats of cities and urban areas, usually living among or under stones, old wood piles, window wells,

Status: There is some disagreement on whether this species is native or introduced to the US; Both Burch (1962) and Hubricht (1985) considered it native to the offshore islands of the New England Coast, Maine, New Hampshire and Massachusetts; it was in North America before the Vikings (Pearce 2010); although not yet reported in West Virginia, the species has been introduced to cities across the US and will likely find its way to urban areas of the state.

Specimen: (FLMNH collection).

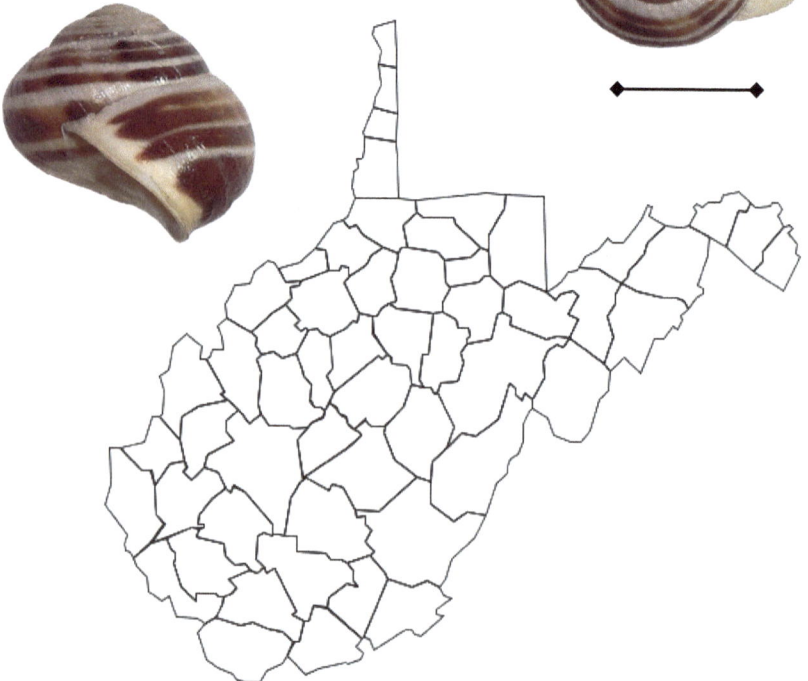

Milk snail

Helicidae

Otala lactea Müller, 1774

Diameter: 27-36 mm

Description: Heliciform; lip slightly reflected, elongate and dark brown; the interior of shell also dark; shell with 5 whorls; imperforate; without teeth in all stages of growth; shell white, with broken reddish-brown spiral color bands, transverse striae are well developed.

Similar Species: *Cepaea nemoralis* is smaller and has a dark lip that is less reflected. *C. hortensis* is smaller and has a white lip.

Habitat: Found in mostly degraded habitats of cities and urban areas, usually living among or under stones, old wood piles and window wells; the species is less problematic in gardens than the exotic slugs, but can sometimes reach large numbers.

Status: Not yet reported from West Virginia; however, Branson (1973) found this species living in the city of Louisville, Kentucky.

Specimen: (FLMNH collection).

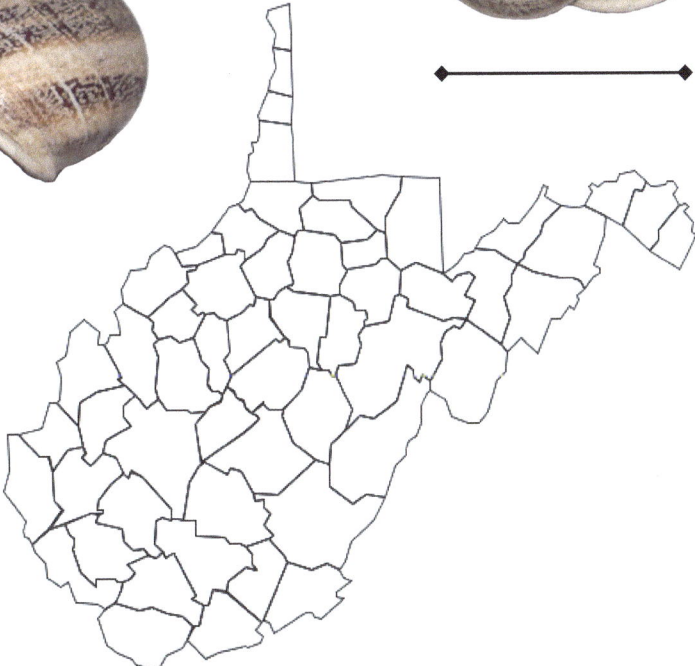

343

Brown gardensnail

Helicidae

Cornu aspersum (Müller, 1774)

Diameter: 25-35 mm

Description: Heliciform; lip exceptionally large and slightly reflected; shell with 5 whorls; imperforate; without teeth in all stages of growth; shell usually pale brown, occasionally yellow, with dark spiral bands, variable and often flecked with white, although some specimens are without color bands.

Similar Species: *Cepaea hortensis* and *C. nemoralis* are smaller, have bolder color bands and have notably smaller apertures.

Habitat: Found in mostly degraded habitats of cities and urban areas, old wood piles, window wells; the species can be problematic in gardens.

Status: A European species, reported from scattered locations including Maine, Massachusetts, Michigan, South Carolina, Texas, Utah and Virginia, but not yet recorded from West Virginia.

Specimen: California, in the city of Los Angeles (author's collection).

Maritime gardensnail

Cernuella cisalpina

Diameter: 8-10 mm, height 5-6 mm

Description: heliciform; lip simple with a slight reflection in mature shells; shell with 5.5-6 whorls; perforate to umbilicate; shell whitish with light color bands (a); no teeth at any stage of growth; transverse striae poorly developed, no hairs; periphery well rounded.

Similar Species: Similar to *Anguispira kochi* but in miniature.

Habitat: In West Virginia this exotic species from Europe was recently discovered (2014) by Jeff Hajenga along a roadway cut.

Status: The species was reported to be especially common at that location (pers. comm. Jeff Hajenga); an aggressive invader.

Specimen: West Virginia, Fayette County, Exit 66 off I-77 (author's collection).

Hygromiidae

a

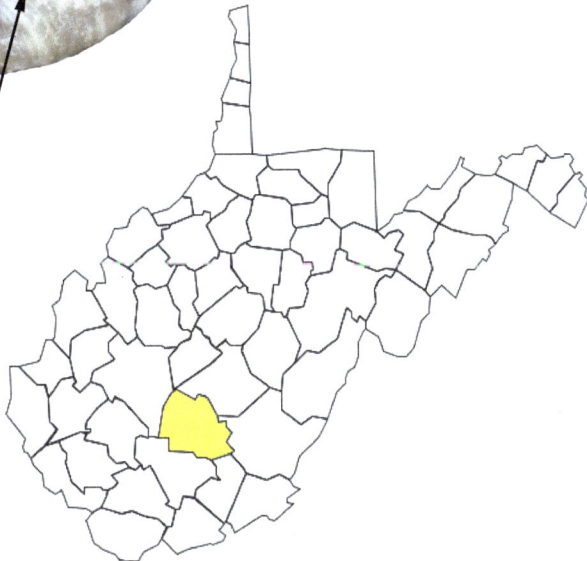

Garlic glass-snail

Oxychilidae

Oxychilus alliarius (Miller, 1822)

Diameter: 6-7 mm

Description: Depressed heliciform; lip simple; shell with 4-4.5 closely coiled whorls; umbilicate; shell glossy, thin, pale yellowish-brown or greenish, often white below; semi-transparent; no teeth at any stage of growth; transverse striae poorly developed; no spiral striae present; no hairs; live animal dark bluish-gray, smelling strongly of garlic if disturbed; periphery well rounded.

Similar Species: Young shells of *Mesomphix inornatus* are similar, but have minute spirally arranged papillae and a less open umbilicus; other *Mesomphix* species are more globose in overall form; *O. cellarius* has 1-2 more loosely coiled whorls.

Habitat: As with most *Oxychilus* species, it is found in degraded habitats and waste places of cities and urban areas; tolerant of poor acidic places (Kerney and Cameron 1979).

Status: Although the species has not yet reported in West Virginia, it has been introduced to many cities and towns across the eastern US.

Specimen: England, High Beach (FM 45218).

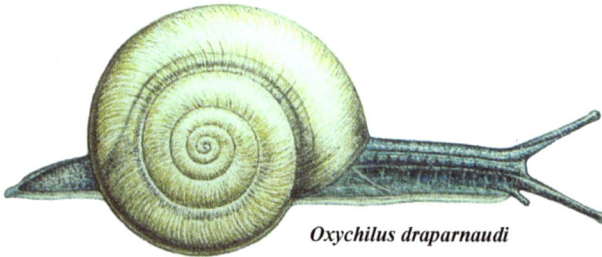

Oxychilus draparnaudi

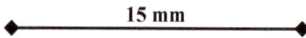

15 mm

Oxychilus alliarius

Illustrations by
Gordon Riley (1979)

Oxychilus cellarius

Cellar glass-snail

Oxychilus cellarius (Müller, 1774)

Oxychilidae

Diameter: 9-14 mm

Description: Depressed heliciform; lip simple; shell with 5.5-6 whorls; umbilicate; shell glossy, thin and semi-transparent; no teeth at any stage of growth; transverse striae poorly developed, no spiral striae present; no hairs; live animal pale bluish-gray, but dark forms also occur; periphery well rounded.

Similar Species: Young shells of *Mesomphix inornatus* are similar, but have minute spirally arranged papillae and a less open umbilicus; other *Mesomphix* species are more globose in overall form; in *O. draparnaudi,* the last whorl is notably wider.

Habitat: Found in mostly degraded habitats of urban areas; under rubbish; frequently found in caves in its native continent of Europe.

Status: Even though reported from only four counties in West Virginia, the species is expected to be more common and widespread than current records indicate; it often hides in plant material used in landscaping, the main means of its transportation to new location.

Specimen: England (FM 90855).

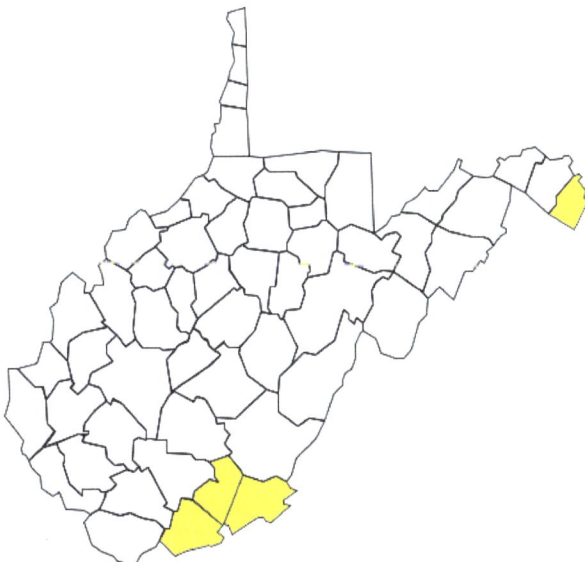

347

Dark–bodied glass-snail

Oxychilus draparnaudi (Beck, 1837)

Oxychilidae

Diameter: 12-16.5 mm

Description: Depressed heliciform; lip simple; shell with 5-5.5 whorls; umbilicate; shell glossy, thin and semi-transparent; no teeth at any stage of growth; transverse striae poorly developed, no spiral striae present; no hairs; live animal cobalt blue, with gray mantle; periphery well rounded; all *Oxychilus* species are carnivorous.

Similar Species: Young shells of *Mesomphix inornatus* are similar, but are with minute spirally arranged papillae and a less open umbilicus; *O. alliarius* is much smaller and has a higher shell profile.

Habitat: Found in mostly degraded habitats and waste places of cities and urban areas; this species can reach considerable numbers.

Status: Although reported from only one county in West Virginia, the species is expected to be more widespread; like *O. cellarius* and *O. alliarius,* it often hides in plant material used in landscaping projects.

Specimen: Virginia, Norfolk County, vacant lot in Norfolk (FM 251405).

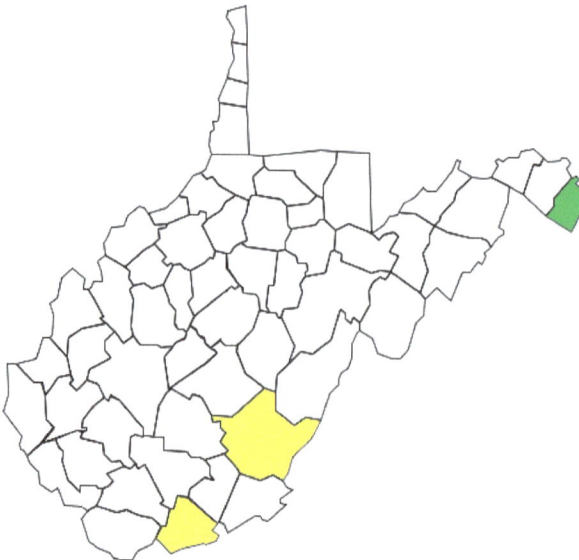

Sharp awlsnail

Subulinidae

Opeas pyrgula Schmacker & Boettger, 1891

Height: 5-8 mm, width 2 mm

Description: Shell cylinder-shaped; lip simple; shell with 6-7 whorls; nearly umbilicate; shell surface glossy and translucent in live animals and fresh shells; color pale corneous to almost white; with live animal inside, shell greenish-yellow (figure a); densely sharply sculptured with irregular and rather strong transverse striae which are very deeply curved; the retraction of the outer lip at the suture is deep (figure a); outer lip thin; last whorl rounded.

Similar Species: Other similar looking native species of West Virginia are much smaller.

Habitat: A tropical and sub-tropical species; in temperate climates, a snail of green houses and indoor gardens, not capable of surviving outside in West Virginia's harsh winters.

Status: First collected in Fayette County, West Virginia by Ralph Taylor.

Specimen: Belize, Maya Mountains, Toledo District (author's collection).

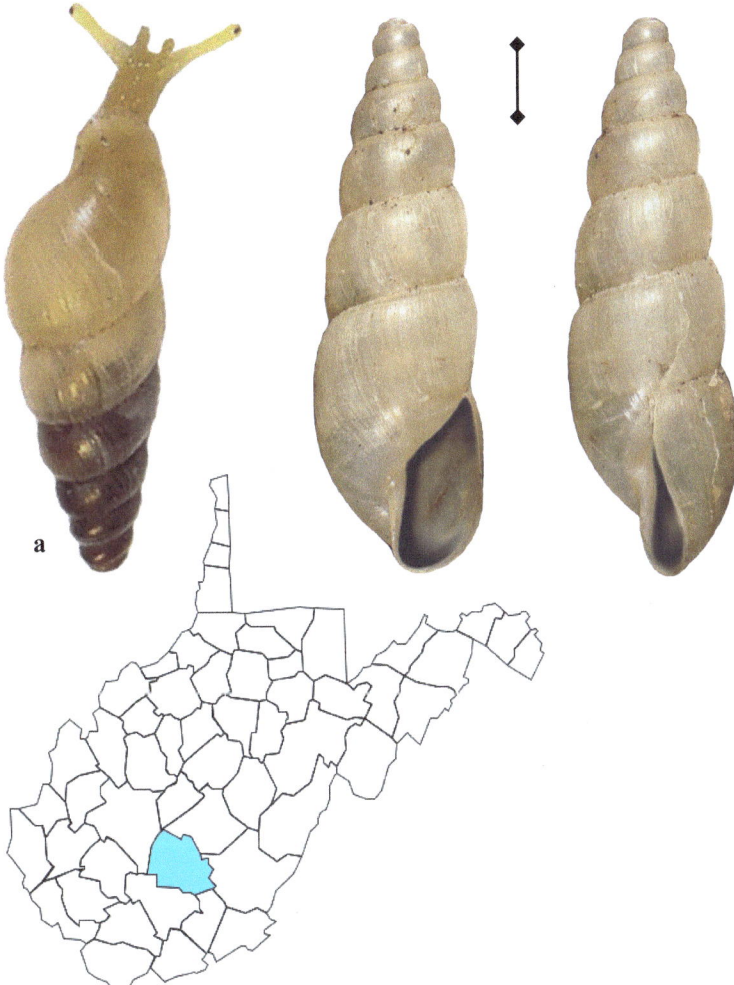

a

Gray fieldslug   Agriolimacidae

Deroceras reticulatum (Müller, 1774)

Length: Adults 35-50 mm while crawling,

Description: Color ranges from nearly white or pale tan to mottled brown or nearly black; mantle or saddle farther back on the slug's body than seen in *Arion* species; posterior end strongly keeled (a); tubercles large and distinct; a whitish border surrounds the pneumostome (b); defense mucus milky and adhesive-like; margins and sole of foot are white, pale and or gray.

Similar Species: Distinguished from *D. laeve* and *D. panormitanum* by its larger size, thicker skin, pale grayish color, slower movements, and its ability to produce sticky, milky-colored mucus when bothered (Grimm *et al*. 2009); it is most reliably separated by differences in their reproductive organs.

Habitat: *Deroceras reticulatum* is likely the most serious molluscan plant pest in North America, a result of its insatiable appetite for live plant material. The species is ubiquitous and generally abundant in fields and gardens, but also in practically every other anthropogenic habitat, including roadsides, forest campsites, and wasteland. It is rarely found in completely undisturbed habitats. Toads, frogs, shrews, brown snakes, and ground beetles are known predators.

Status: Reported from four counties in West Virginia, but likely much more widespread than current records suggest.

Specimen: Illustrations from Grimm *et al*. (2009)

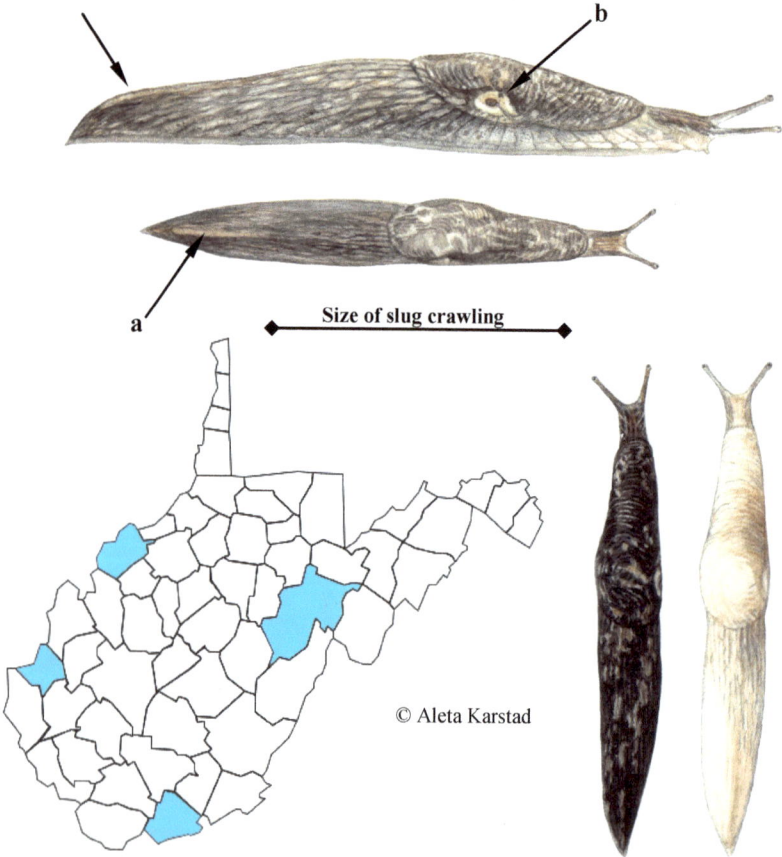

Size of slug crawling

© Aleta Karstad

Two color morphs

350

Orange-banded arion   Arionidae

Arion fasciatus (Nilsson, 1823)

Length: Adults 30-40 mm while crawling

Description: Gray overall, gray above, fading to gray-white on the sides below dark lateral bands; gray to yellowish saddle-like mantle covers only the anterior (front) part of the slugs dorsal region (back); breathing pore is in the front half of the mantle in all *Arion* (as opposed to *Deroceras* and *Limax*); reproductive pore in front of breathing pore; defense mucus clear; margins of the foot grayish-white.

Similar Species: Native slugs in West Virginia, except for *Deroceras laeve,* have mantles that nearly cover the entire body; the mantle of *D. laeve* is longer and has no lateral bands; *A. subfuscus* is larger, the reproductive pore is located behind the breathing pore, and the defense slime is yellow or orange.

Habitat: Found in mostly degraded habitats of cities and urban areas, usually living among or under stones, old wood piles and window wells; this pest species is especially problematic in gardens and can reach large numbers.

Status: Although reported from only one county in West Virginia, this species is likely more pervasive in the state.

Specimen: Illustrations from Grimm *et al.* (2009).

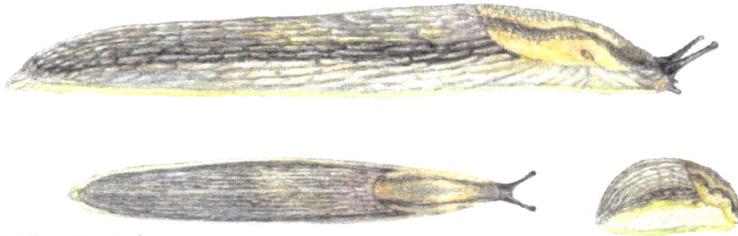

© Aleta Karstad

Size of slug crawling

Dusky Arion   Arionidae

Arion subfuscus (Draparnaud, 1805)

Length: Adults 50-60 mm while crawling

Description: Orange and brown with lateral light brownish bands; orange saddle-like mantle which covers only the anterior (front) part of the dorsal region; breathing pore is in the front half of the mantle in all *Arion* (but not in *Deroceras* and *Limax*); reproductive pore located behind the breathing pore; defense posture of slug (a); defense mucus bright yellow or orange; foot pale yellow.

Similar Species: *A. fasciatus* is smaller and has a colorless defense mucus; *L. maximus* is much larger and has a different body color and clear defense slime.

Habitat: Found in mostly degraded habitats of cities and urban areas, usually living among or under stones, old wood piles and window wells; this pest species is especially problematic in gardens and can reach substantial numbers.

Status: Even though reported from only five counties in West Virginia, this species is likely more widespread in the state, especially in highly degraded habitats and wet, upper elevation forests where it thrives.

Specimen: Illustrations from Grimm *et al.* (2009)

© Aleta Karstad

Size of slug crawling

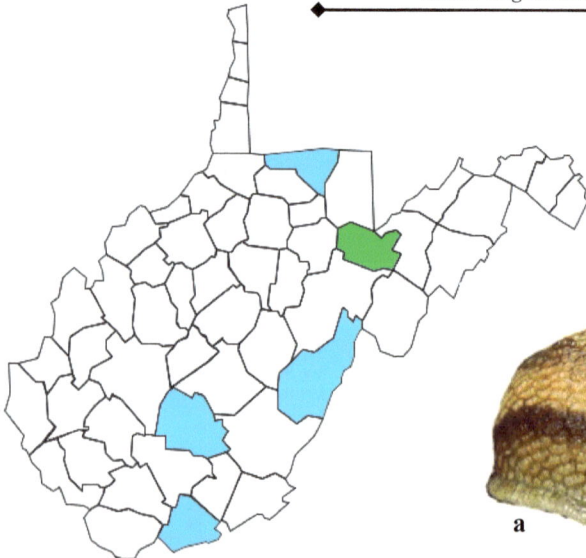

a

352

The colorful but invasive Tiger slug, *Limax maximus*

Exchanging sperm

Web image

Defensive posture

353

Tiger slug   Limacidae

Limax maximus Linnaeus 1758

Length: Adults 100-200 mm while crawling

Description: Mantle and body black spotted or mottled rather than with continuous bands; body color yellowish-gray or light brownish-gray; tentacles reddish-brown; posterior end of slug keeled, but not up to the mantle; defense mucus clear; margins of the foot and sole whitish.

Similar Species: This sizeable slug looks like no other native or exotic slug reported in West Virginia.

Habitat: Very common in degraded habitats of cities and urban areas, under railroad ties used for landscaping; the species feeds mostly on dead vegetation and fungi; other foods include roots, fruit, leafy crops and carrion (Grimm *et al.* 2009); reported to crawl onto porches to consume pet food (MacGregor pers. comm. 2010); especially problematic in gardens and can reach large numbers.

Status: Although reported from only two counties in West Virginia this species is likely more common in the state, in particular, areas around city parks and playgrounds (the species often transported to new locations via construction lumber).

Specimen: Illustrations from Grimm *et al.* (2009)

© Aleta Karstad

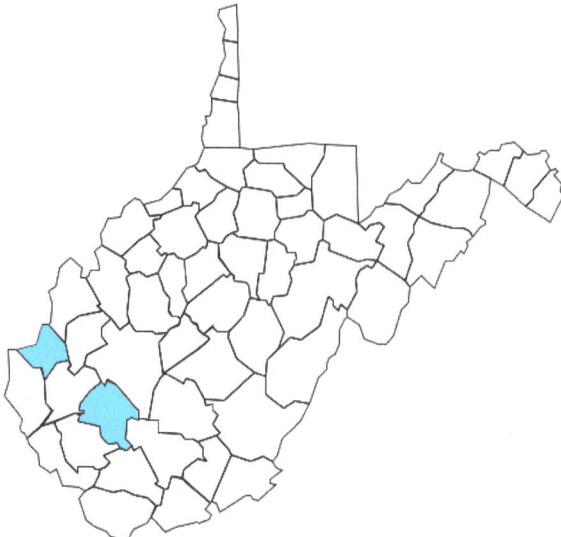

Size of slug crawling

Tree slug   Limacidae

Lehmannia marginata (Müller, 1774)

Length: Adults 60-80 mm while crawling

Description: color of mantle tan to grayish tan with a dark "smudge" down the middle of the back and conspicuous dark lateral bands; thin skinned and somewhat translucent; mucus watery.

Similar Species: *Lehmannia marginata* (not illustrated) most closely resembles *L. valentiana* (illustrated below) whose longitudinal lines on the sides are comparatively more ventral, and whose habits are more synanthropic; the two species are best distinguished, however, anatomically, being exceedingly difficult to separate based solely on external appearances.

Habitat: Found in mostly degraded habitats of cities and urban areas, in vacant lots and waste places, around meadows, and in hedge rows; in their natural habitat these slugs prefer exposed habitats on vertical surfaces such as cliffs and tree trunks, where they feed almost exclusively on lichens.

Status: Although reported from only two counties in West Virginia, this species is likely more widespread in the state, in particular, along the Ohio River Valley where flooding events would spread its footprint.

Specimen: Illustrations from Grimm *et al.* (2009)

Lehmannia valentiana

© Aleta Karstad

Size of slug crawling

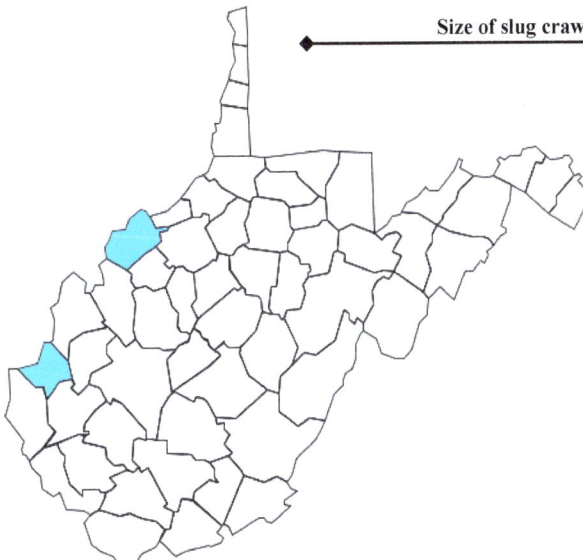

Currently Reported Exotic Slugs in WV and Other Introduced Species to North America (proportionate and life size)

Arion ater (100-200 mm)

Limax maximus (100-200 mm)

Shell of slug

Testacella haliotidea (100 mm)

Limacus flavus (75-100 mm)

Arion circumscriptus (30-40 mm)

Arion subfuscus (50-70 mm)

Arion fasciatus (30-40 mm)

Lehmannia valentiana (70 mm)

Deroceras panormitanum (30-40 mm)

Milax gagates (50-60 mm)

Arion intermedius (15-20 mm)

Deroceras reticulatum (35-50 mm)

Boettgerilla pallens (30-40 mm)

© Aleta Karstad

Calcium recycling and land snails

The skull of a bobcat doesn't go to waste as a balsam globe, *Mesodon andrewsae* (left) and a mountain tigersnail, *Anguispira jessica,* (right) glean calcium from the bone. The calcium will facilitate bodily functions, reproduction, but most importantly, shell building. Images taken on Roan Mountain, Mitchell County, North Carolina.

Part II
Land Snails of Bordering States: Kentucky, Ohio, Pennsylvania, Maryland & Virginia

This section provides detailed information on land snails that have a reasonable chance of occurring in West Virginia due to their close proximity or similar ecological requirements. If you find species that do not match gastropods reported from West Virginia, this section should cover possible candidates, unless the species is new to science or a peculiar and isolated ecological form. For each species illustrated, a West Virginia map is included with a dashed circle indicating where the species would most likely occur in the state. These potential land snail additions were extracted from several sources including Hubricht's (1985) range maps, other researchers' distribution records, and my own knowledge of these species in the adjacent states. Land snails, hereafter, are arranged on the following pages and illustrated in three basic categories, a) land snails with simple lips, not reflected, b) land snails that have reflected lips, and c) slugs. The keys begin, more or less, with the smaller species progressing to the larger taxa.

Coastal-plain tigersnail, *Anguispira fergusoni* (Bland, 1861)

Pictorial Key to Land Snails of Bordering States

A. Land Snails with simple Lips

Euconulus chersinus, Hive (3mm), pg. 361
Euconulus trochulus, Hive (2.5mm), pg. 362

Paravitrea mira, Supercoil (6.3mm), pg. 363
Paravitrea lamellidens, Supercoil (3.5mm), pg. 364
Paravitrea septadens, Supercoil (3.4mm), pg. 365
Paravitrea dentilla, Supercoil (6.8mm), pg. 366
Paravitrea placentula, Supercoil (7mm), pg. 367

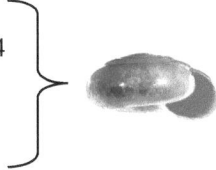

Glyphyalinia luticola, Glyph (5.7mm), pg. 369
Glyphyalinia cryptomphala, Glyph (5-6mm), pg. 370

Helicodiscus multidens, Coil (4mm), pg. 372
Helicodiscus lirellus, Coil (4mm), pg. 373
Helicodiscus diadema, Coil (4mm), pg. 374

Polygyriscus virginianus, Coil (4mm), pg. 375

Discus nigrimontanus, Disc (7mm), pg. 376

Ventridens pilsbryi, Dome (9mm), pg. 377

Anguispira fergusoni, Tigersnail (16mm), pg. 378

Mesomphix rugeli, Tigersnail (20mm), pg. 380

B. Land Snails with Reflected Lips

Vallonia parvula, Vallonia (2mm), pg. 381

Mesodon elevatus, Globe (20mm), pg. 383

Appalachina sayana kentucki, Crater (25mm), pg. 384

Triodopsis burchi, Threetooth (14mm), pg. 385
Triodopsis discoidea, Threetooth (14mm), pg. 386

***Patera* species (undetermined),** bladetooth (20mm), pg. 387

C. Native Slugs not yet recorded in West Virginia

Megapallifera wetherbyi, Mantleslug (75mm), pg. 388
Philomycus bisdosus, Mantleslug (80mm), pg. 390
Philomycus batchi, Mantleslug (80mm), pg. 392

Wild hive

Euconulidae

Euconulus chersinus (Say, 1821)

Diameter: 2.4-2.95 mm, height 2.9-3.4 mm

Description: Dome shape; lip simple; shell thin with 6-8 tightly coiled whorls; scarcely perforate; shell surface yellowish-white and rather dull; shells are translucent with a silky luster in live snails and fresh dead; transverse striae are poorly developed (microscope required); spiral striae not well defined, but always present (a very strong lens required); without teeth; body whorl slightly angular, especially in young shells.

Similar Species: This species differs from *E. fulvus* by the more elevated spire, more numerous, narrower whorls, a narrower aperture, and in fresh specimens, by the silky luster, *E. fulvus* being more glossy (Pilsbry, 1946).

Habitat: Under moist leaf litter in mixed hardwood forests.

Status: G5; although this southern species is reported from multiple counties in West Virginia, Hubricht, who spent a fair amount of time collecting in the state did not report finding this species, his closest records in Tennessee; therefore its occurrence in West Virginia remains highly questionable; area of possible occurrence in dashed circle below.

Specimen: North Carolina, Swain County, Nantahala National Forest (author's collection).

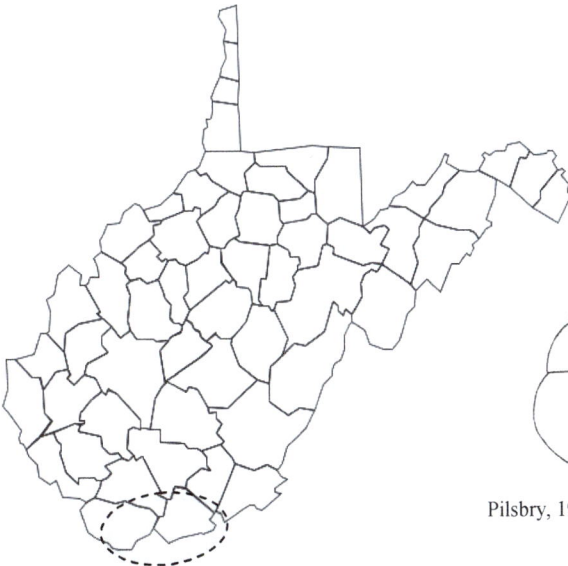

Pilsbry, 1946

Silk hive

Euconulidae

Euconulus trochulus (Reinhardt, 1883)

Diameter: 2.9 mm, height 2.5 mm

Description: Dome shape; lip simple; shell with 7 tightly coiled whorls; scarcely perforate, the narrow umbilicus is completely covered by the reflected columellar margin; pale horn with a silky luster; shell thin, translucent in live snails and fresh dead; transverse striae are poorly developed (dissecting scope required); spiral striae not well defined but always present (a strong lens required); without teeth; in mature specimens, the periphery is slightly angular or roundish (Pilsbry 1946); as with other *Euconulus,* it is important to have a series of shells for a proper ID.

Similar Species: Shell much like *E. chersinus*, but it has one whorl more in shells of similar size; *Euconulus fulvus* is more glossy and has 1 to 2 fewer whorls in adult shells.

Habitat: A species found in mixed hardwood forests on hillsides and ravines living under layers of moist leaf litter.

Status: G5; not yet recorded from West Virginia but known from counties in Virginia that border the state; area of most likely occurrence, in dashed circle below.

Specimen: Figure (a) from Kentucky, Pulaski County, Meece (FM 252087) and Figure (b) from David Kirsh and Bill Frank, at www.jaxshells.org.

a

b

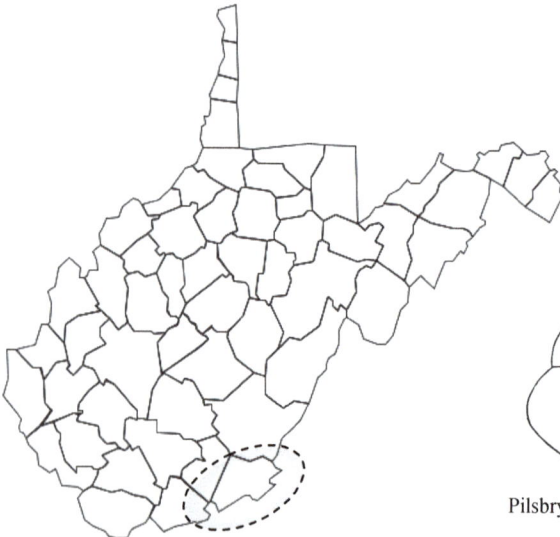

Pilsbry, 1946

Funnel supercoil

Zonitidae

Paravitrea mira Hubricht, 1975

Diameter: 6.3 mm, height 3.9 mm

Description: Depressed heliciform; lip simple; shell with 8.5 tightly coiled whorls; pale amber; glossy; umbilicate; aperture lunate; irregularly spaced, transverse striae above, becoming obsolete below; in the last whorl there are usually two rows of three, rather large teeth, sometimes connected by a low callus bridge; the immature shells containing 3 teeth in each row (a); the inner tooth of some adult shells is intermittently missing (Hubricht 1985).

Similar Species: *Paravitrea reesei* is smaller, has smaller teeth and three less whorls; *P subtilis* is smaller and its aperture contains 4-6 smaller teeth.

Habitat: Found under moist leaf litter and detritus on wooded hillsides and in ravines of mixed hardwood forests.

Status: G2; not yet recorded from West Virginia but known from counties in Virginia that border the state; area of most likely occurrence in dashed circle below.

Specimen: Kentucky, Pike County, 1.7 miles west of Meta (FM 248934).

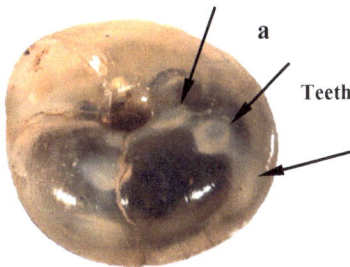

a

Teeth

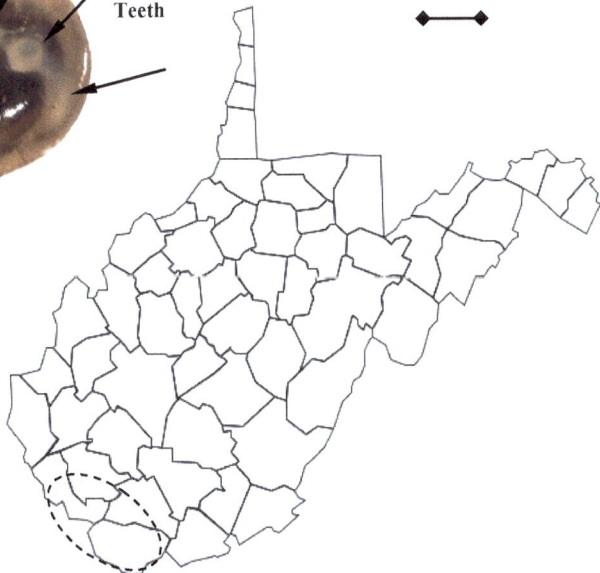

Lamellate supercoil

Paravitrea lamellidens (Pilsbry, 1898)

Diameter: 3.5-3.8 mm, height 1.6-1.9 mm
Description: Depressed heliciform; lip simple; aperture half-moon shaped; shell with 6.5 tightly coiled whorls, last whorl not expanding; perforate; dark reddish-brown to cinnamon buff; shell thin and translucent in fresh dead and live shells; transverse striae are well developed, <u>very closely and evenly spaced</u>; under a strong lens, faint spiral striae are visible on the base; three rows of obliquely radial lamellae in young shells, rarely more than one or two in adults; these lamellae can be seen through bottom of both juveniles and adults.
Similar Species: *Paravitrea multidentata* is similar in form, but is slightly smaller, typically with teeth instead of lamellae (although occasionally teeth fuse together and appear as lamellae) and has a wider umbilicus.
Habitat: Found in deep pockets of moist leaf litter, especially in rock talus located in higher elevation mixed hardwood forests.
Status: G2/S2; although this Southern Appalachian species was widely reported from West Virginia by MacMillan, all specimens examined from the Carnegie Museum have been lamellate forms of *P. multidentata,* therefore the presence of *P. lamellidens* remains highly questionable in the state.
Specimen: North Carolina, Swain County, Nantahala National Forest (author's collection).

Zonitidae

Umbilicus narrow

Lamella

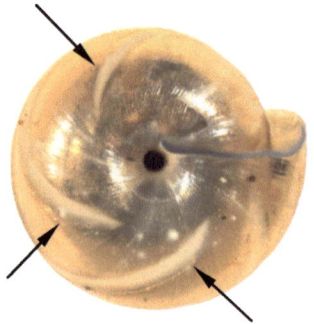

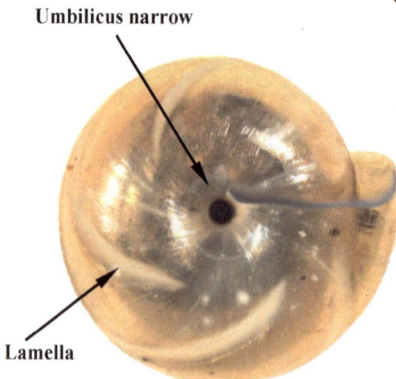

P. lamellidens

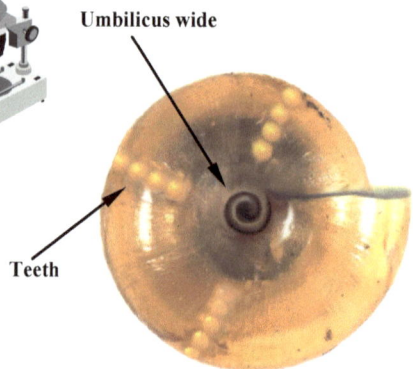

Umbilicus wide

Teeth

P. multidentata

Brown supercoil

Paravitrea septadens Hubricht, 1978

Diameter: 3.4 mm, height 1.6 mm

Description: Depressed heliciform; lip simple; shell with 6 tightly coiled whorls, the last whorl expanding slightly; shell thin, glossy and translucent in fresh and live shells; perforate and well-like; transverse striae are poorly developed and irregularly spaced; one to two radial lamellae (a) that can be seen through the bottom of fresh shells of both juveniles and usually in most adults; sometimes absent in adults.

Similar Species: Differs from *P. multidentata* in having a more rounded periphery, fewer and weaker growth lines, and lamellae instead of teeth; *P. lamellidens* is similar in size, but has a notably smaller umbilicus; *P. dentilla* is larger and has teeth not lamellae as in *P. septadens*.

Habitat: Found in the same habitat as *P. multidentata* and often found with it (Hubricht 1985); pockets of moist leaf litter located on hillsides in mixed hardwood forests.

Status: G1; a rare and range-restricted land snail, currently reported only from Dickenson and Buchanan counties, Virginia; not yet recorded from West Virginia, but known from counties in Virginia that border the state; area of most likely occurrence in West Virginia in dashed circle below.

Specimen: Virginia, Dickenson County, 1.5 miles west northwest of Bee (FM 249132).

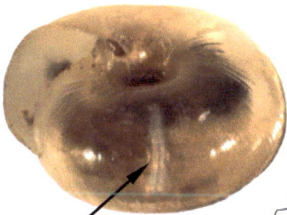

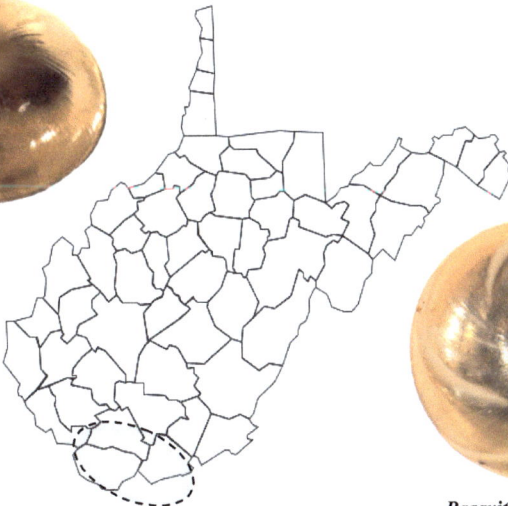

Paravitrea lamellidens

365

Comb supercoil

Zonitidae

Paravitrea dentilla Hubricht, 1978

Diameter: 6.8 mm, height 3.5 mm

Description: Depressed heliciform; lip simple; shell with 7.5-8 tightly coiled whorls, the last whorl slightly expanding; shell thin, glossy and translucent in fresh and live shells; umbilicate, well-like; nuclear whorl smooth, later whorls sculptured with numerous irregularly spaced, impressed growth lines, distinct above, but becoming obsolete below; one or more radial rows of 5 or 6 small teeth within the last whorl (a), the teeth are set close together, the bases usually touching; adults are without teeth.

Similar Species: *Paravitrea placentula* is larger and has paired teeth (in juvenile shells); young shells of *P. dentilla* might be confused with those of *P. multidentata*, but the surface of *P. dentilla* has irregularly and more widely spaced transverse striae.

Habitat: A species of river bluffs found under leaf litter and among detritus deposits on hillsides, but also upper elevation limestone talus.

Status: G1; a rare and range restricted species, currently reported in 1 county in Kentucky and 2 counties in Virginia; not yet recorded from West Virginia, but known from counties in Virginia that border the state; area of most likely occurrence in dashed circle below.

Specimen: Virginia, Washington County, 1 mi south of Damascus (FM 248919 Paratype).

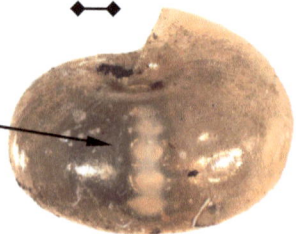

Juvenile shell 3 mm

Glossy supercoil

Zonitidae

Paravitrea placentula (Shuttleworth, 1852)

Diameter: 7.2-7.8 mm, height 3.6 mm

Description: Depressed heliciform; lip simple; shell with 5-6 tightly coiled whorls, umbilicate; the last whorl slightly expanding; corneous; shell thin, glossy and translucent in fresh and live shells; indented transverse striae are faint and irregularly spaced; two pairs of rather large teeth that are arranged close together and can be seen through the bottom of fresh dead and live shells of juveniles up to 3.5 mm in diameter (a); adults are without teeth.

Similar Species: Most similar to *Paravitrea capsella,* but larger and most importantly, contains rather large teeth in juvenile shells.

Habitat: Acidic coves, rich woods, upper elevation northern red oak and montane oak hickory forests.

Status: G3; *P. placentula* was widely reported across West Virginia by MacMillan, however no specimens examined from the Carnegie Museum labeled as *P. placentula* agree with this species, most specimens instead referring to *Paravitrea bellona,* a previously undescribed species and unknown to MacMillan in the 1940s, therefore *P. placentula* records remain doubtful for West Virginia.

Specimen: Tennessee, Blount County, White Oak Sinks, GSMNP (GSMNP collection).

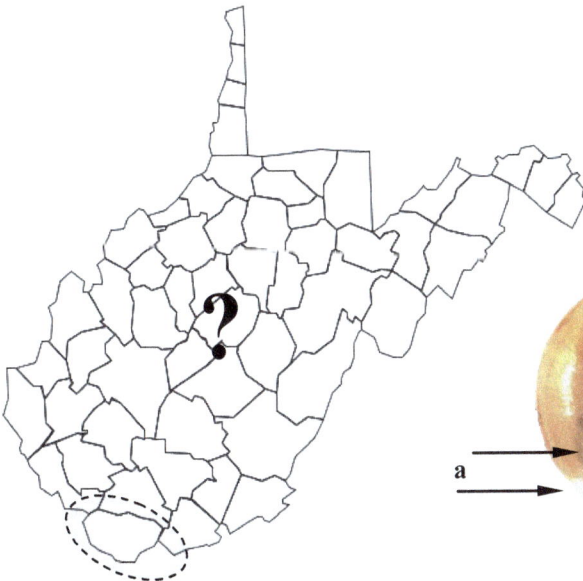

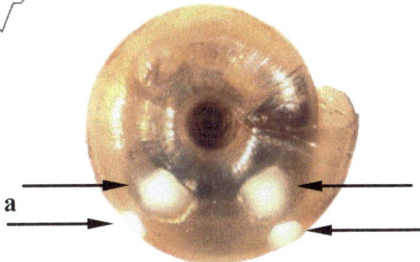

a

Young shell

367

Land snails are often seen on spider webs, easily moving across the web's sticky surface without becoming entangled. Pictured above is a juvenile *Patera panselenus,* a common gastropod of West Virginia; note the spider in the background. There is some evidence that snails are attracted to the silk nets (that trap condensed moisture) for a source of drinking water. Pictured below, the same species appears to glean water from a web. Both images from Pike County, Kentucky.

Furrowed glyph

Glyphyalinia luticola Hubricht, 1966

Zonitidae

Diameter: 5.7 mm, height 2.6 mm

Description: Depressed heliciform; lip simple; shell with 4 to 4.5 loosely coiled whorls; shell fragile, coppery (when fresh), glossy, shiny, and translucent; perforate to rimate; no teeth present; indented transverse striae are well developed, widely and nearly equally spaced, with about 20 on the last whorl; the spiral striae may be weakly defined, but are a constant feature (dissecting scope required); live animal slate-colored; deceased, and dried animal can be seen through all three views of shell.

Similar Species: This species is most similar to *Glyphyalinia indentata,* but is smaller, coppery colored, and is usually found in wetter habitats (Hubricht 1966).

Habitat: Found in a variety of moist to wet habitats; found crawling on the muddy ground in wet weather in floodplain woods; also a snail of waste ground in urban areas.

Status: G4G5; not yet recorded from West Virginia but known from counties in Virginia that border the state; area of most likely occurrence in dashed circle below.

Specimen: Louisiana, Orleans County, New Orleans near the Mississippi River between Broadway and State St. (FM 241570).

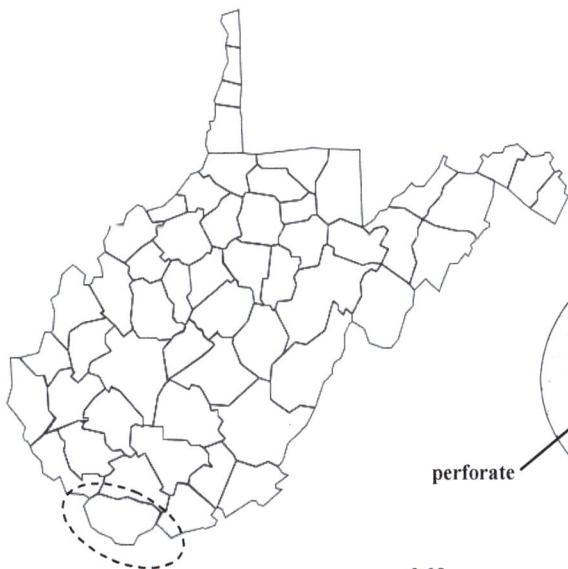

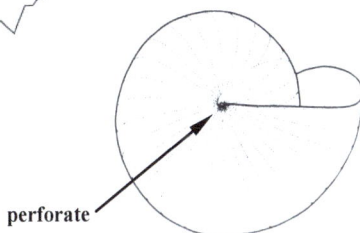

perforate

Pilsbry, 1946

Thin glyph   Zonitidae

Glyphyalinia cryptomphala (G. H. Clapp, 1915)

Diameter: 5.1-6 mm, height 2.1 mm

Description: Depressed heliciform; lip simple; shell with 5-5.5 loosely coiled whorls; imperforate, the umbilicus completely covered in all stages of growth; shell fragile, light horn to white, glossy, and semi-transparent; indented transverse striae are well developed, closely and nearly equally spaced, and continue to the base; the spiral striae are weakly defined, but a constant feature; deceased and dried animal seen through bottom view of shell.

Similar Species: *Glyphyalinia praecox* is slightly larger and is perforate, its umbilicus not completely sealed as in *G. cryptomphala*.; *G. solida* is larger by around 3 mm and has more strongly-developed indented striae.

Habitat: Often found in a variety of mixed hardwood forests under moist leaf litter along river bluffs and in ravines.

Status: G5; not yet recorded from West Virginia but known from counties in Virginia that border the state; area of most likely occurrence in dashed circle below.

Specimen: Tennessee, Coffee County, Lusk Cave, 4 miles ENE of Hillsboro (author's collection).

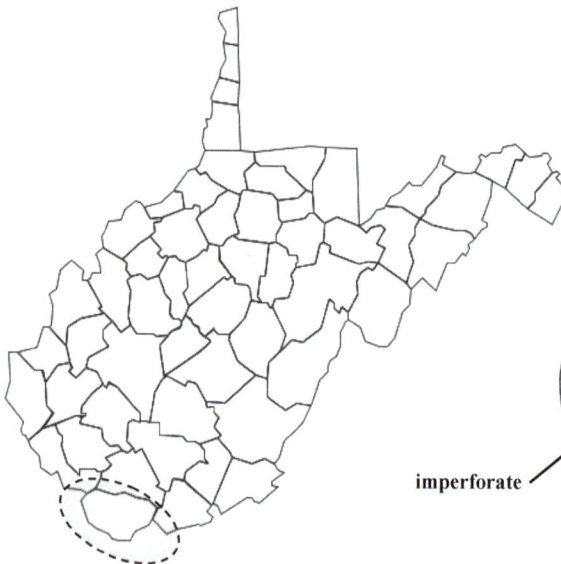

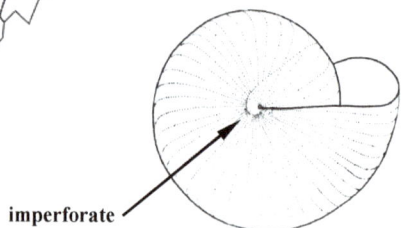

imperforate

Pilsbry, 1946

370

Twilight coil

Helicodiscus multidens Hubricht, 1962

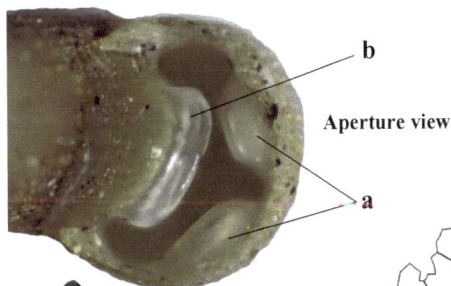

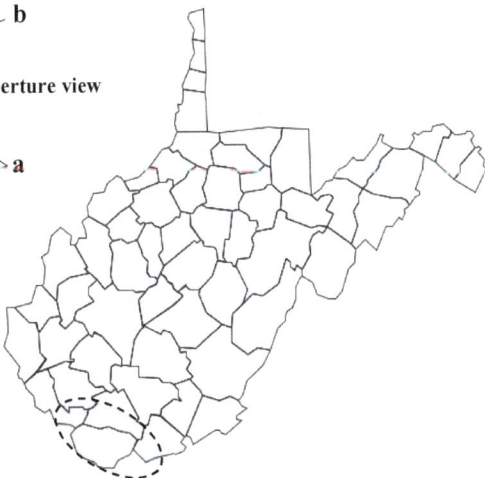

Diameter: 4.75 mm, height 1.55 mm

Description: Depressed heliciform; lip simple, aperture roundish; shell with 4.5-5 whorls; shell pale greenish-yellow, dull, opaque; widely umbilicate; in the last whorl there are usually three pairs of teeth on the outer and basal wall (a), figure shows one pair of teeth (other sets are deeper)—these teeth are radially elongate and are about twice as broad as high (unlike the teeth of other *Helicodiscus* species which are more knob-like); alternating with these paired teeth are usually three teeth on the parietal wall (b) figure shows only one (others are deeper in shell); under a strong lens the spiral striae (about 28-35 on the body whorl) are fringed (figured on opposite page) and present on all whorls, but this feature is typically lost in aging shells.

Similar Species: *Helicodiscus parallelus* and *H. notius* have knob-like teeth and are without a wide parietal tooth.

Habitat: A calciphile found under rocks and leaf litter on river bluffs and in caves.

Status: G3; not yet recorded from West Virginia, but known from counties in Virginia that border the state; area of most likely occurrence in dashed circle below.

Specimen: Tennessee, Blount County, Gregory Cave area, GSMNP (author's collection).

b

Aperture view

a

Above image illustrating the extraordinary micro-ornamentation found on the shell surface of *Helicodiscus multidens* (magnified 1000X); not an uncommon feature on other species of land snails under five millimeters. SEM image by R. Wayne Van Devender.

Rubble coil

Helicodiscus lirellus Hubricht, 1975

Diameter: 4.4 mm, height 1.89 mm

Description: Depressed heliciform-discoidal; lip simple, aperture roundish; shell with 4.5-5 whorls; shell pale greenish-yellow, dull and translucent; widely umbilicate showing all the whorls; sutures deep; within the aperture there are 2 pairs of teeth on the outer and basal walls (one set behind and deeper than first), these teeth radially elongate and distinctly separated; alternating with these are 2 teeth on the parietal wall, the parietal teeth being in front of the teeth on the outer and basal wall; these teeth are present in all stages of growth; under a strong lens sculptured with rather coarse growth wrinkles and very fine, spiral striae (not fringed), 15 -18 of these on the body whorl.

Similar Species: Most like *Helicodiscus multidens,* but has fewer and finer striae and somewhat smaller teeth; it differs from *H. diadema* in not having hairs on the striae and in having smaller teeth.

Habitat: A calciphile species found burrowing in limestone rubble at the base of a hill.

Status: G1; this globally rare land snail is currently known only from Rockbridge County, Virginia; area of very slight probability of occurrence in dashed circle below.

Specimen: Virginia, Rock Bridge County, 10 miles NW of Lexington (FM 239094 Paratype).

Aperture view illustrating the multiple alternating teeth

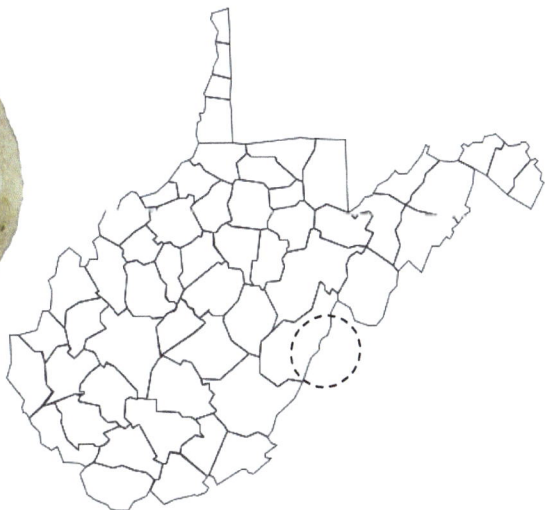

Shaggy coil

Helicodiscidae

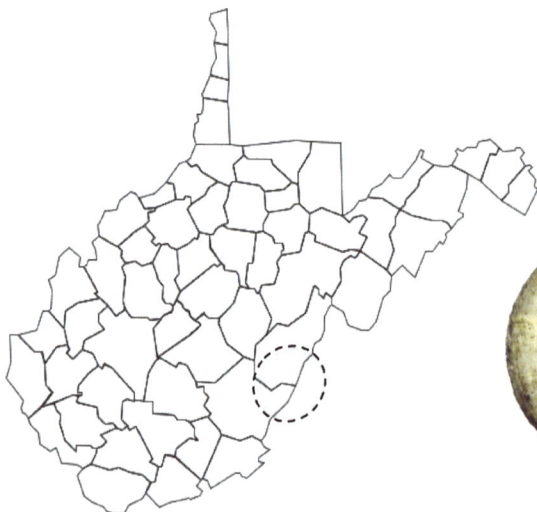

Helicodiscus diadema Grimm, 1967

Diameter: 3.2-4 mm, height 1.09– 1.34 mm

Description: Depressed heliciform-discoidal; lip simple; shell with 4.5-5 whorls; shell dull greenish-brown; widely umbilicate; sculptured with coarse growth-wrinkles and 11-13 pinched spiral threads bearing prominent, curved hairs which are most prominent on younger shells; within the last quarter whorl are 2-3 pairs of large radially elongate teeth and alternating with them, 3 parietal teeth (a); teeth on the outer and basal walls precede those on the parietal wall, are borne on a thick callous ridge and are separated by a deep, rounded sinus; the teeth on the outer wall are larger and more pointed than those on the basal wall; the cupped parietal teeth are twice as broad as high, the ends turned forward.

Similar Species: Most like *H. multidens* and *H. triodus,* but differs in having hairs.

Habitat: A calciphile found in limestone rubble at the base of hills.

Status: G1; Rare; currently known only from Rockbridge and Allegheny Counties in Virginia; area of very slight probability of occurrence in dashed circle below.

Specimen: Virginia, Allegheny County, near quarry, 8 miles NE of Covington, along US route 220. (FM 239097 Paratype).

Parietal teeth

a

Cutaway view

Virginia coil

Helicodiscidae

Polygyriscus virginianus (P.R. Burch, 1947)

Diameter: 4.2 mm, height 1.4 mm

Description: Depressed heliciform-discoidal; lip simple with a slight outward reflection; the last whorl twisting away from and leaving the preceding whorl of the shell at an angle of about 60 degrees and extending outward about 1.0 mm (a); aperture distorted; shell with 4.5 whorls; very widely umbilicate; there are extraordinary spiral fringes on young shells (see below enlarged image) mostly becoming lost as shells age; periphery roundish.

Similar Species: This species is unlike any other land snail of the region; *Helicodiscus* are the same size and body form, but without the twisted aperture.

Habitat: A calciphile that burrows deep into Elbrook limestone talus.

Status: G1; **Federally Threatened**; this globally rare land snail is currently known only from Pulaski County, Virginia, adjacent to Mercer County West Virginia.

Specimen: Virginia, Pulaski County (CM 139307).

a

Bottom view

375

Black Mountain disc

Discidae

Discus nigrimontanus (Pilsbry, 1924)

Diameter: 7.4 mm, height 2.4 mm

Description: Depressed heliciform; lip simple; aperture roundish to oval; shell with 5.25 whorls; widely umbilicate; striae are well developed and rib-like; periphery sub-angular (a); the microscopic crisscross sculpture of the embryonic whorl is weak.

Similar Species: *D. patulus* has a rounded periphery; a small parietal tooth; coarser riblets and a smaller more well-like umbilicus.

Habitat: A species of upland rocky talus slopes in mixed hardwood forests where leaf litter is sparse; common in the lower layers of deep limestone talus along the base of outcropping limestone.

Status: G4; not yet recorded from West Virginia, but known from counties in Virginia that border the state; area of most likely occurrence in dashed circle below.

Specimen: Tennessee, Fentress County, Skillman Mart Cave (GSMNP collection).

White Oaks Sinks, GSMNP,
Blount County, Tennessee

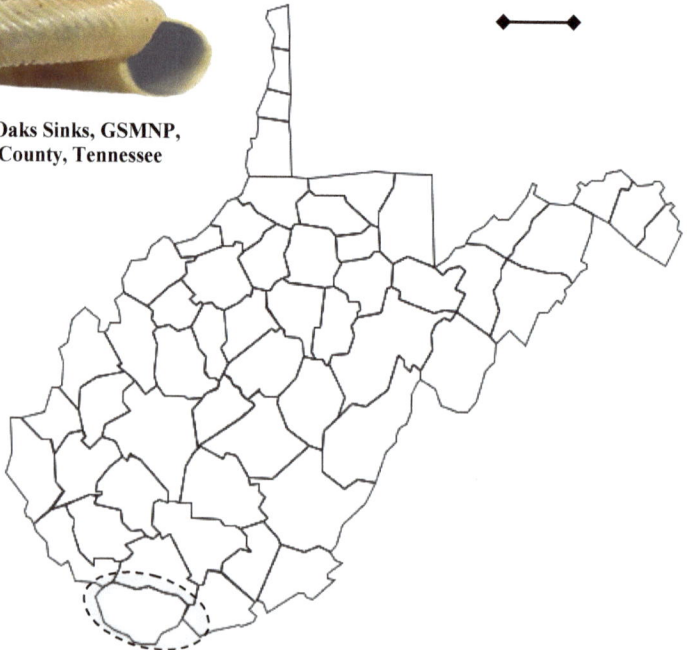

Yellow dome

Gastrodontidae

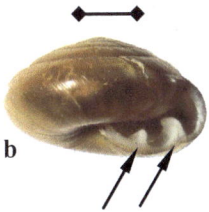

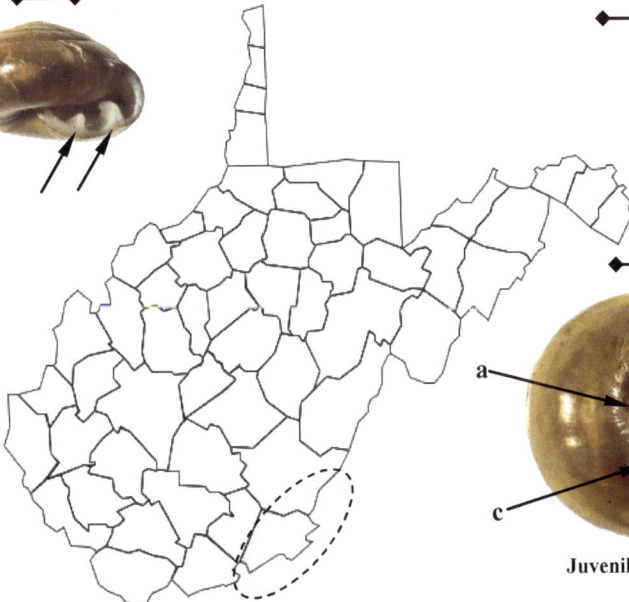

Ventridens pilsbryi Hubricht, 1964
Diameter: 8.7-9.6 mm, height 5-7 mm
Description: Depressed heliciform; lip simple; shell with 7-8 tightly coiled whorls; perforate, widest in young shells (a), becoming smaller with age, sometimes completely closed; live animal pale; shell translucent, glossy; transverse striae are distinct on top, somewhat weaker on the bottom; within the aperture of both young and adult shells there are two lamellae (b), also seen through bottom (c); member of the *V. pilsbryi* group (Hubricht 1964).
Similar Species: *V. pilsbryi* has been confused with *V. gularis,* which has a smaller, more globose shell with fewer whorls, and the live animal is nearly black.
Habitat: Found on wooded hillsides and in ravines under leaf litter and on logs, more common on limestone, but also occurs on acidic sandstone sites as well.
Status: G4; not yet recorded from West Virginia but known from counties in Virginia that border the state; area of most likely occurrence in dashed circle below.
Specimen: Tennessee, Monroe County, Cherokee NF (author's collection).

b

a

c

Juvenile shell 5 mm

Coastal-plain tigersnail

Discidae

Anguispira fergusoni (Bland, 1861)

Diameter: 15-17.5 mm, height 8-9.8 mm

Description: Depressed heliciform; lip simple; shell with 4-5 whorls; umbilicate; color features usually strong on top and sides, fading on the base; shell rather glossy for an *Anguispira*; no teeth present; transverse striae are closely spaced as in *A. alternata;* adult shells with at least 5 whorls have a count of over 80 ribs in the last whorl; last whorl rounded without a trace of peripheral angle.

Similar Species: *Anguispira mordax* has coarser ribs, having fewer than 50 ribs on the last whorl; *A. alternata* is larger with a notably wider umbilicus.

Habitat: A species of the Atlantic Coastal Plains which has moved up the floodplain of larger rivers into the Piedmont area; found around logs, hollow trees, and in the leaf litter of deciduous woods; also found in urban areas (Hubricht, 1985).

Status: G4; not yet recorded from West Virginia, but known from counties in Virginia that border the state; area of most likely occurrence in dashed circle below.

Specimen: North Carolina, Alamance County, near Dockside Dolls, beech woods (Van Devender collection).

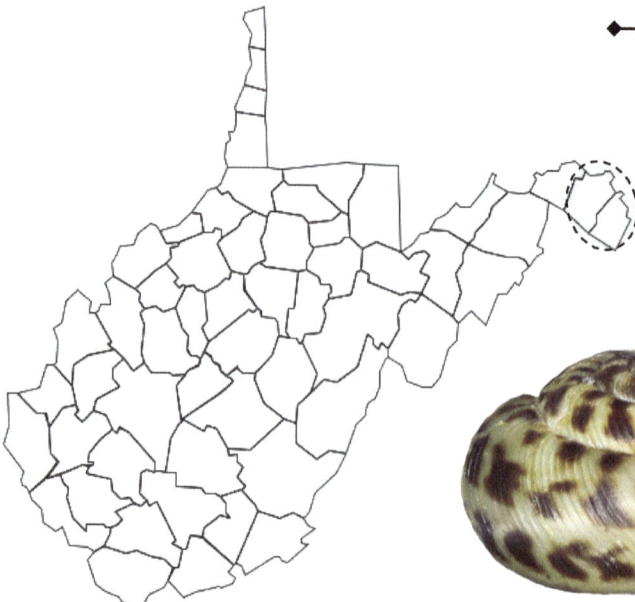

Wrinkled button

Mesomphix rugeli (W. G. Binney, 1879)

Diameter: 16-22.5 mm, height 9-12.6 mm

Description: Heliciform; lip simple; shell with 5.5-6 loosely coiled whorls; perforate; shell thin, but not fragile, greenish-horn and glossy; live animals are bluish-gray to cobalt; with or without a thin whitish callus just inside the aperture's bottom (can be seen in frontal view); transverse striae are smooth, and the shell is usually without discernible spiral rows of papillae; cannibalistic, will feed on the fresh dead of its own kind; also feeds on bird droppings (pers. obs. Roan Mountain, NC).

Similar Species: *Mesomphix perlaevis* has better developed transverse striae and spiral rows of papillae; *M. rugeli* form *oxycoccus* (Pilsbry 1946) differs from typical *M. rugeli* by having well-developed and crowded spiral papillae and a duller shell surface.

Habitat: Higher elevation mixed hardwood and spruce/fir forests under moist leaf litters; also along roads margins adjacent to above described habitat.

Status: G4; the presence of typical form *M. rugeli* in West Virginia remains questionable; of the three specimens labeled as *M. rugeli* from Carnegie, two were *M. rugeli oxycoccus* and one refers to *Mesomphix luisant*, therefore all *M. rugeli* records in WV are likely that of *M. rugeli oxycoccus*; both do, however, occur in Kentucky along Pine Mountain (Dourson 2010); area of most likely occurrence in dashed circle below.

Specimen: North Carolina, Mitchell County, Roan Mountain (author's collection).

M. rugeli form *oxycoccus*,
Carter County, Tennessee

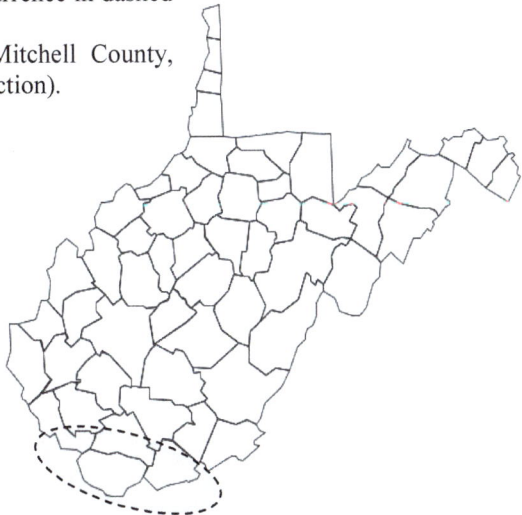

Wrinkled button, *Mesomphix rugeli*, Roan Mountain near Carvers Gap, Tennessee.

Trumpet vallonia

Valloniidae

Vallonia parvula Sterki, 1893

Diameter: 1.6-2 mm, height 1.0 mm

Description: Depressed heliciform; lip reflected, does not cross the body whorl (a & c) and strongly thickened within; shell with 3 loosely coiled whorls, transverse striae are more rib-like and paper thin (can be seen with a hand lens of 10X), the last whorl has between 30-38 ribs and is greatly expanded; widely umbilicate; horn or pale gray to nearly white; no teeth present; periphery roundish.

Similar Species: *Vallonia costata* averages fewer ribs (20-35 ribs on the last whorl) and is larger in diameter; *V. perspectiva* has a wider umbilicus, a notably thinner lip, and the aperture lip is continuous across the body whorl (b).

Habitat: A species of dry weed and grass openings.

Status: G4; not yet recorded from West Virginia but known from counties in Virginia that border the state; area of most likely occurrence in dashed circle below.

Specimen: Larry Watrous photo collection.

a

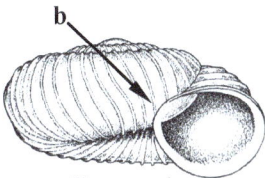

b

V. perspectiva

(Burch 1962)

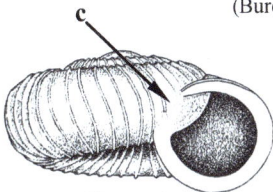

c

V. parvula

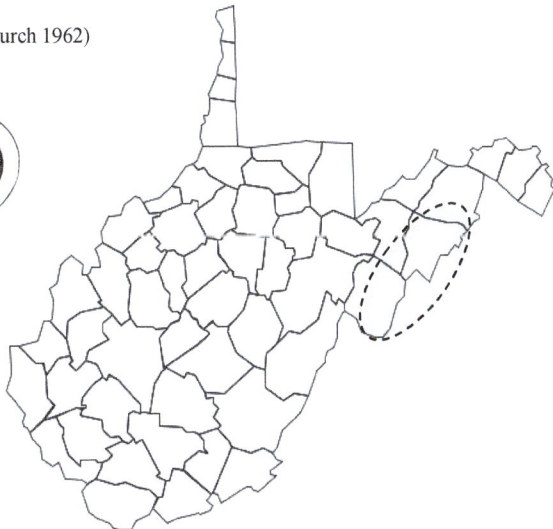

381

With perfect balance, a five millimeter wide juvenile Flat bladetooth, *Patera appressa,* crawls across a single strand of silk, demonstrating its remarkable adaptable foot. Red River Gorge, Powell County, Kentucky.

Proud globe

Polygyridae

Mesodon elevatus (Say, 1821)

Diameter: 19.8-26.3 mm, height 14-20 mm

Description: Heliciform; lip reflected and thickened; shell with 6-7 whorls; imperforate; shell solid; large parietal tooth usually present; pale yellow to light olive; adult shells without hairs; transverse and minute spiral striae are a well-developed feature and always present; shell periphery well-rounded; occasionally young and adult shells have a light band, (a).

Similar Species: *Mesodon clausus* is smaller, has a thinner shell and is without a parietal tooth; *M. zaletus* has a notably smaller parietal tooth and has a slightly more compressed shell, but this feature will vary from site to site.

Habitat: A species of limestone river buffs and rich wooded slopes in mixed hardwood forests, also found in upland sites under forest litter; the species can be especially common around the entrances of limestone caves.

Status: G5; not yet recorded from West Virginia but known from counties in Ohio, located along the Ohio River, that border the state; area of most likely occurrence in dashed circle below.

Specimen: All figures from Kentucky, Knox County, Pine Mountain State Park (author's collection).

a

A slightly lower form

Pine Mountain crater

Polygyridae

Appalachina sayana kentucki Dourson 2011

Diameter: 18-22 mm, 12-14 mm

Description: Heliciform; lip reflected; shell with 5.5 whorls; shell thin; umbilicate or rimate; usually without a parietal tooth; basal tooth rather large; pale yellow to pale olive-tan; no hairs; transverse and minute spiral striae always present; shell periphery well rounded.

Similar Species: *Appalachina sayana* (pictured below) is larger, less compact, has a wider umbilicus, a parietal tooth and its basal tooth is generally smaller than in *A. sayana kentucki.*

Habitat: An uncommon subspecies of rich upland, higher elevation mixed hardwood forests; generally found under moist leaf litter and other forest debris, becoming less common in drier sites like ridgetops and Virginia pine forests; the majority of shells found in mixed mesophytic sites.

Status: This recently described subspecies is reported from Pine Mountain, Kentucky including Pike County which borders West Virginia; area of most likely occurrence in dashed circle below.

Specimen: Kentucky, Letcher County; Bad Branch Nature Preserve (author's collection).

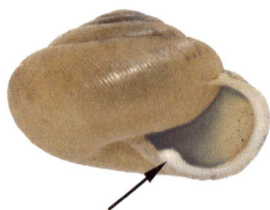

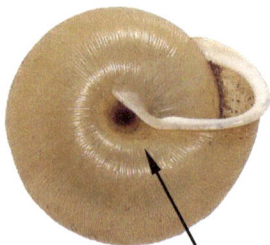

Umbilicus small

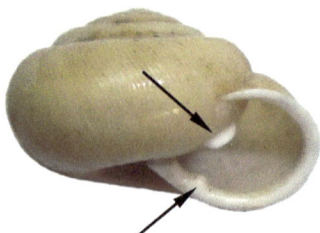

A. sayana, Yahoo Hollow, Randolph County, West Virginia

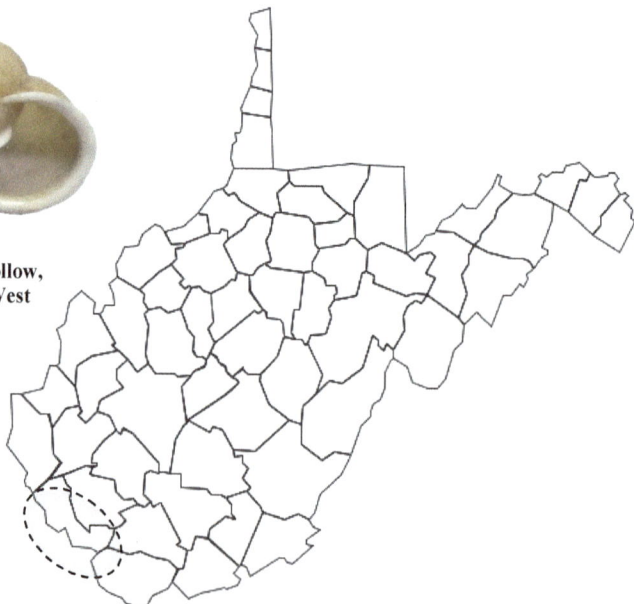

384

Pittsylvania threetooth

Polygyridae

Triodopsis burchi Hubricht, 1950

Diameter: 7.6-17.4 mm, height 4.1-7.8 mm

Description: Depressed heliciform; lip reflected; shell with 4-5.5 whorls, buffy brown and glossy; umbilicate; parietal tooth rather short but tall; palatal and basal teeth very small, the basal tooth sometimes missing altogether (a); the transverse striae are fairly well developed on the top, sides and bottom of the shell; scattered papillae are present but generally a weak feature, strongest around the umbilicus; periphery rounded.

Similar Species: Similar to *T. tennesseensis* with which it has been confused, but is smaller with smaller teeth.

Habitat: A species of the western Piedmont, which has migrated through the Roanoke Gap in the Blue Ridge Mountains into the Great Valley (Hubricht, 1950); found in upland oak forests of mountainsides, under leaf litter and logs.

Status: G3; not yet recorded from West Virginia but known from counties in Virginia that border the state; area of most likely occurrence in dashed circle below.

Specimen: All specimens from Virginia, Franklin County, upland woods (FM 266210).

a

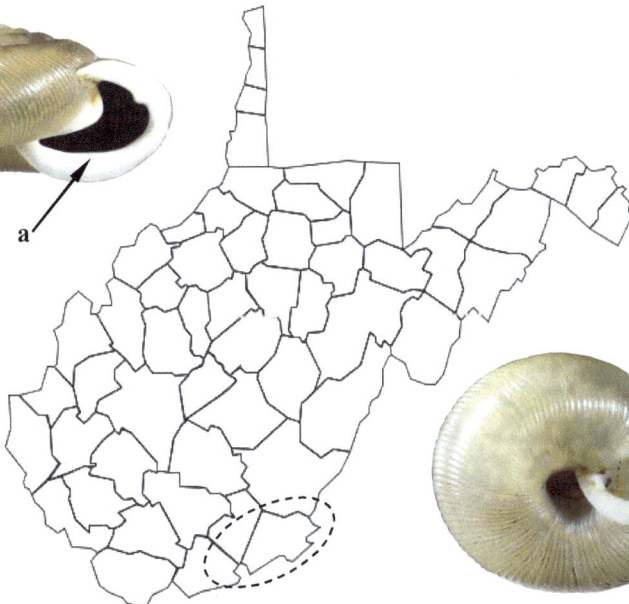

Rivercliff threetooth

Polygyridae

Triodopsis discoidea (Pilsbry, 1904)

Diameter: 16.5-19.5 mm, height 7.9-9.5 mm

Description: Depressed heliciform; lip reflected; shell with 5-5.5 whorls; umbilicate; large parietal tooth present, points above the palatal tooth; basal tooth and palatal tooth smaller; palatal tooth not situated on a buttress; dilute cream to very dilute ecru-olive; transverse striae are moderately developed on the top and sides of the shell continuing well into the umbilical region; minute papillae can usually be seen around the umbilicus.

Similar Species: The *T. tridentata* parietal tooth points at or below the palatal tooth; *T. juxtidens* is more compact in build and has a deeper transverse striae and a smaller umbilicus.

Habitat: A species found below river bluffs along the Ohio River in mixed hardwood, under leaf litter, logs and rocks.

Status: G3; not yet recorded from West Virginia but known from counties in Ohio located along the Ohio River near the border of the state; area of most likely occurrence in dashed circle below.

Specimen: Kentucky, Breckinridge County, Yellow Bank WMA (author's collection).

Wide umbilicus

Shale bladetooth

Patera species (undetermined)

Diameter: 23-24 mm, height 10 mm

Description: Depressed heliciform; lip reflected; shell with 4.5-5 whorls; imperforate; large parietal tooth present with a moderate base; basal tooth small and poorly defined (a); shell surface marked with well-developed transverse striae on top and sides, but more weakly defined on the base; papillae not particularly well developed, connected by low ridges (this feature best observed near the suture lines); shell periphery strongly angular; embryonic whorl smooth; these shell characters are described from two of the four specimens ever collected.

Similar Species: The shale bladetooth closely resembles *Patera sargentianus* (illustrated below) which inhabits limestone outcrops in northern Alabama where it is believed to be endemic; *P. appressa* is smaller and typically without an angular periphery.

Habitat: Currently known only from a highway cut through Marcellus Shale along highway 522 in Virginia.

Status: This apparently rare species was first discovered by John Cramer while searching for trilobites along a Devonian fossil site in Virginia; the shells were relatively fresh dead, indicating that the species has extant populations; not yet recorded from West Virginia but known from Fredrick County, Virginia which borders the state of West Virginia; area of most likely occurrence in dashed circle below.

Specimen: Virginia, Fredrick County, shale bank, 3.2 Miles West of Gainsboro, on west bound 522 (John Cramer collection).

angular

a

angular

Patera sargentianus, Alabama, Jackson County in town of Woodville (author's collection)

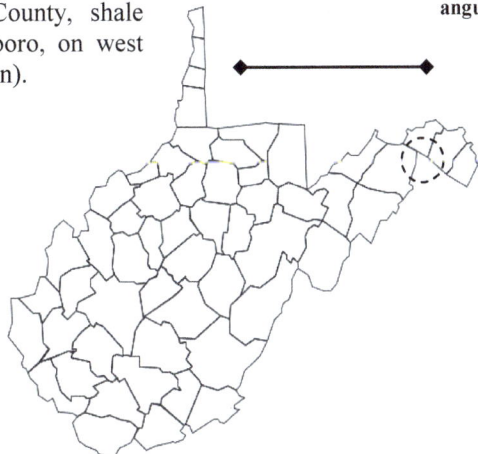

Blotchy mantleslug   Philomycidae
Megapallifera wetherbyi (W. G. Binney, 1874)
Length: Adults around 75 mm while crawling

Description; Mantle color is gray, white to cream; does not cover entire head, (as that of *Philomycus* species); mantle with a wide, uninterrupted, dorsal band running down the slug's back and broken, thinner lateral bands or spots, connected by broad, sometimes random chevron shaped bands or spots; defense mucus whitish; foot olive-gray or white.

Similar Species: *Megapallifera mutabilis* is only slightly smaller and has spots between the longitudinal bands instead of oblique connecting bands of color.

Habitat: This is a species of river bluffs and ravines. At night during warm wet weather, it can be found crawling on the trunks of smooth bark trees like the American beech (Hubricht 1985); it appears to prefer areas in which sandstone outcrops and cliffs are common (Fairbanks 1990).

Status: G2G3; this uncommon slug is reported from only five southeastern counties in Kentucky and is often found in company with *P. flexuolaris* and *P. bisdosus*; although not yet recorded from West Virginia; area of most likely occurrence in dashed circle below.

Specimen: All figures from Kentucky, Pike County; Kentucky (author's photo collection).

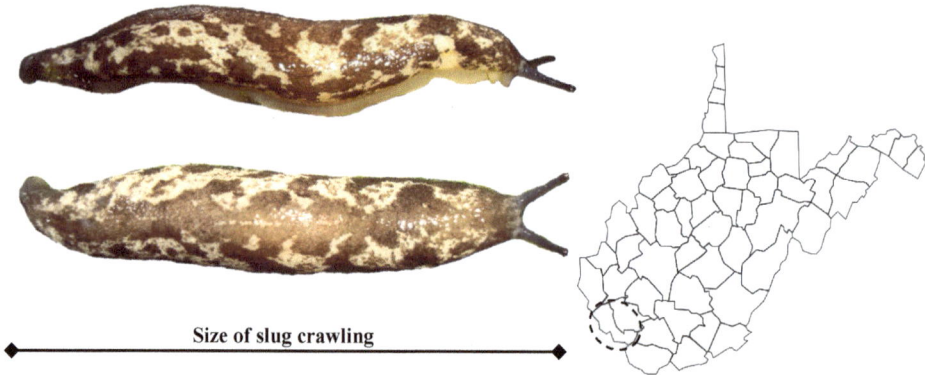

Size of slug crawling

Breaks mantleslug   Philomycidae

Philomycus bisdosus Branson, 1968
Length: Adults 50-85 mm while crawling
Description: Ground color of mantle pale-tan; starting mid-dorsally (figure a this page and figure a, page 390) a single or pair of longitudinal, brown wavy bands and on each side is a thinner, wavy band of the same color (figure b, page 390); posteriorly, the dorsal bands fuse into a single one; anteriorly all 4 bands breakup into small, brown spots; no oblique bands connecting the longitudinal bands; defense mucus whitish; margins of the foot gray (figure c, page 390); foot nearly smooth, white except for a very short black area at extreme posterior end, figure (figures d & e, page 390); head is white except for blue-black tentacles (Branson 1968);
Similar Species: Based upon external appearance and size, *P. bisdosus* is most like *P. togatus* (Fairbanks, 1989), however, the foot margin of *P. bisdosus* is gray (c), whereas the foot margin of *P. togatus* is red/orange (b); *P. bisdosus* differs from *P. venustus* by lacking the transverse oblique bands or chevrons and is a smaller slug; *P. flexuolaris* has a pale yellow defense mucus not whitish as in *P. bisdosus*.
Habitat: A species of mixed hardwood forests above the Big Sandy River.
Status: G1; a rare endemic, currently known only from Breaks Interstate Park, on the Kentucky/Virginia state border; not yet recorded from West Virginia, but known from counties in Kentucky and Virginia that border the state; area of most likely occurrence in dashed circle below.
Specimen: All images from Breaks Interstate Park, Dickerson County, Virginia; the rare white form of *P. bisdosus* illustrated was discovered under bark of a rotting log along Camp Branch, just above Russell Fork River (author's photo collection).

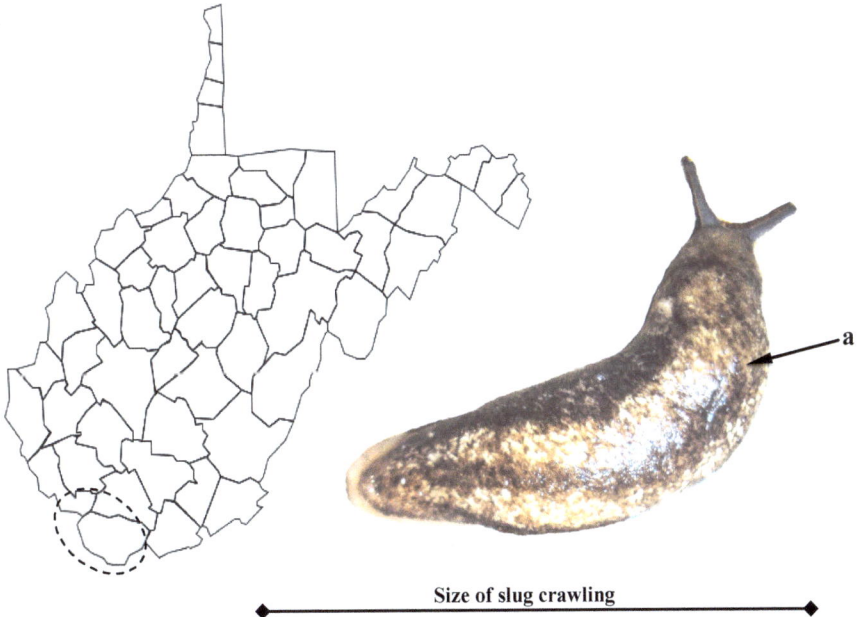

a

Size of slug crawling

Philomycus bisdosus

Kentucky mantleslug   Philomycidae

Philomycus batchi Branson, 1968

Length: Adults 50-80 mm while crawling

Description: A variable species in terms of color; mantle glistening black (figure a, page 392) with numerous indistinct black punctae (spots), but lighter individuals are not uncommon; the lower edges of the mantle are white (figure b, page 392), this feature alone appears to be the best defining external character for the species; pneumatopore surrounded by a white, irregular halo (figure c, page 392) with a faint streak of gray below it; occasionally with white blotches dorsally; the light lateral bands (if present) are more easily seen on pale specimens, figure (figure d, page 392); tail below mantle yellowish, marked on either side by 3 rows of black dashes; defense mucus pale yellow/ orange (figure e, page 392); sole of foot white to a light peach color.

Similar Species: Differs from other *Philomycus* by lacking strong longitudinal stripes, being more uniform in color, and having no distinct mantle patterns.

Habitat: Found in floodplains, hillsides, and ravines, under decaying wood and loose bark, especially beech trees.

Status: G1; not yet recorded from West Virginia but known from counties in Kentucky that border the state; area of most likely occurrence in dashed circle below.

Specimen: All figures except figure (d) from Kentucky, Fayette County; Ravens Run Nature Preserve, figure (d) from Red River Gorge, Powell County, Kentucky (author's photo collection).

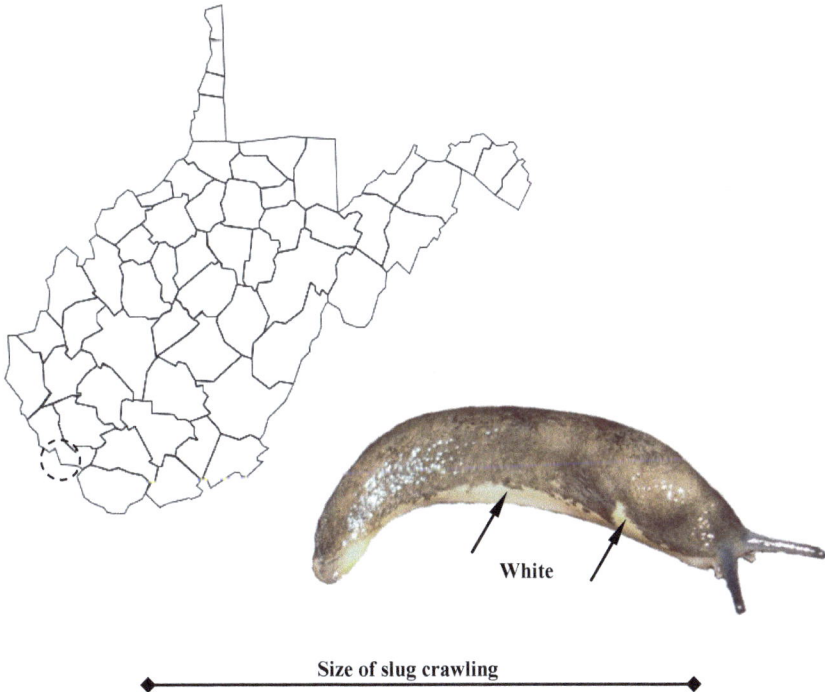

White

Size of slug crawling

Philomycus batchi

a

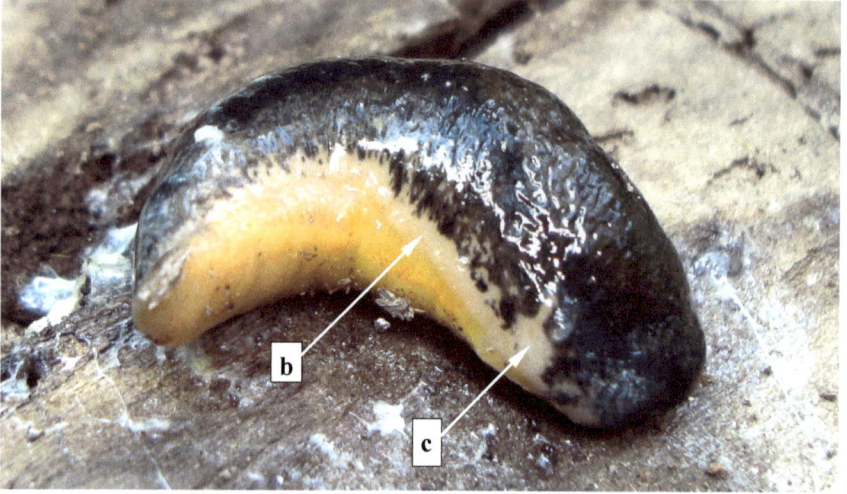

b

c

d

Two color morphs interested in mating

e

Juvenile slug

Non-Toxic Invasive Land Snail Control

Commercial slug baits and repellents contain a molluscicide that kills gastropods. Although they are widely used, they may be toxic to pets, fish, wildlife, and humans. Less toxic methods have been used by organic gardeners for years and are effective in controlling persistent slugs and shelled snails. Included here are several suggestions.

Gather individuals and place in boiling water or drench them in salt water.

Maintain a compost heap near the garden (which attracts the slugs).

Place boards, moist newspaper, cardboard, or old scraps of carpet in the garden to provide cover for the slugs during daytime hours; the slugs can then be easily harvested and euthanized.

Place cups in the garden filled with beer or grape juice to attract and drown the slugs.

Place wet dog food in piles close to infestations to attract slugs, check in the evening and simply dispose of the food and slugs.

Scatter inverted orange or grapefruit rinds among the vegetables to collect slugs which can then be harvested, mashed and composted.

Release ducks (ducks may also eat young plants) and chickens (Rhode Island Reds are best) in the garden to keep the plot clear of slugs and snails. Other birds reported to eat pesky gastropods include blackbirds, crows, jays, owls, robins, seagulls, starlings, and thrushes.

Lava rock, the spiny fruits of sweet gum trees, wood ash, oak leaves, pine needles, coffee grounds, cedar chips, Epsom salts, builders sand, nut shells, oat bran, lime, and talcum power are reported to act as barriers or repellents to slugs or are toxic to them.

Watering the soil with liquid seaweed extract has just enough of an alkaline effect that slugs detest it, however, not enough to significantly change soil pH.

Human hair, pet fur, and horsehair will entangle slugs. Ground beetles (particularly *Carabidae* beetles), box turtles, toads, frogs, lizards, salamanders, lighting bug larvae, and garter snakes are reported to eat slugs and shelled snails, so providing cover for these organisms in and around your garden will help in controlling unwanted gastropods.

Springtime is the best time of year to deal with these uninvited guests. Spring cultivation of the soil will help kill hibernating slugs and their eggs.

Removing slug slime from your hands can be accomplished with white vinegar and warm water or consider using chopsticks for picking up those slimy gastropods.

GLOSSARY

Abcission: when a plant drops one or more of its parts like leaves, fruit or seeds.

Aff.: a species whose identity is in question; means similar to a specific species, but is likely not that species.

Amorphous: shapeless.

Algific: cold-producing as in talus, a loose-rock slope affected by the movement of cold air.

Angular periphery: shell having an angular rather than a round contour (also referred to as carinate).

Angular lamella: the tooth on the parietal wall of the aperture to the right of the parietal lamella in dextral shells.

Aperture: opening or mouth of the snail shell.

Apex: the tip of a gastropod shell farthest away from its aperture.

Axis: The imaginary line around which the whorls of a coiled shell are formed.

Basal: pertaining to, situated at or forming, the base; the part of the shell furthest from the apex; the ventral portion of the aperture lip.

Basal lamella: tooth on the bottom left side of the aperture below the columellar lamella.

Basal tooth: calcareous deposit on the basal apertural lip.

Body Whorl: the last whorl of a shell, measured from the outer lip back to a point immediately above the outer lip.

Breathing pore: opening in mantle or mantle edge for passage of air into the air sac or lung cavity.

Calcareous: composed of or containing calcium carbonate.

Callus: a deposit of lime or shell material, often as a thickening near the umbilicus.

Carinate: the periphery of the shell is compressed, creating a sharper angular appearance with a ridge-like rim.

Cations: ions with a positive charge such as many minerals contained in the soil such as magnesium, aluminum and calcium.

Caudal: situated in or near the tail or posterior end.

Cf.: A species whose identity is in question; means that it is probably the species, but further examination is needed to make a definitive identification.

Columella: the internal column around which the whorls revolve; the axis of a spiral shell.

Columellar lamella: tooth on the columellar wall of the aperture.

Columellar wall: the left side of the aperture in dextral shells.

Corneous: horn-like in color.

Crypsis: the ability of one organism to avoid detection by other organism.

Depressed heliciform: shell with flattened spiral.

Dextral: spiraled to the right when the shell is held so that the apex is up and the aperture is facing the observer.

Dioecious: having both male and female individuals

Dessication: drying out.

Discoidal: having a flat spiral shell.

Embryonic whorl: the earliest whorls that are formed in the egg.

Epiphragm: a hardened mucous covering sealing the aperture that prevents desiccation during dry weather.

Endemic: native or restricted to a relatively small area such as an area or region.

Foot: the locomotory organ of mollusks; it is often modified for digging or grasping prey; the long broad ventral surface of the animal.

Globose: globular, formed like a globe, spherical.

Growth line: a line on the surface of a shell indicating a rest period during growth.

Heliciform: shell that has an elevated spire or globose form.

Hermaphrodite: possessing both male and female reproductive organs in a single individual.

Hirsute: covered with hairs.

Holarctic: a region which includes the northern parts of North America, Greenland, Europe and Asia.

Imperforate: lacking an opening on the ventral or anterior end of the shell.

Impressed: marked by a furrow.

Infra-parietal lamella: the tooth on the parietal wall to the left of the parietal lamella in dextral shells.

Lamella: a fold, "an elongated tooth", OR raised callus in the aperture of a shell.

Lip: the edge of the aperture; also called the peristome.

Loosely coiled: having few, widely expanding whorls.

Lower palatal lamella: lowermost of the two major teeth often found on the palatal wall.

Lunate: shaped like a half-moon.

Mantle: a membranous flap or outer covering of the softer parts of the mollusk; it secretes the shell.

Mouth: the opening or aperture of a gastropod shell.

Mucus: a viscid, slippery secretion, slime.

Operculate: bearing an operculum or cover to close the aperture.

Operculum: a horny or calcareous plate that serves the purpose of closing the aperture when the snail withdraws into its shell.

Outer lip: the outer edge of the aperture.

Palatal depression: indentation of the shell surface at the location of the palatal lamellae.

Palatal tooth: refers to the tooth located on the outer lip.

Palatal wall: the right side of the aperture in dextral shells.

Papillae: small calcium deposits that appear as minute bumps on the surface of the shell.

Parietal lamella: major tooth in the middle of the parietal wall of the aperture.

Parietal tooth: refers to the tooth located on inner wall of the aperture or body of the snail shell.

Pellucid: translucently clear

Penultimate whorl: the next to the last whorl in an adult shell.

Perforate: Having a minute opening in the umbilicus.

Periostracum: the chitinous external layer covering most mollusk shells.

Periphery: the part of the whorl farthest from the central axis.

Radula: a ribbon-like organ with many fine teeth used for rasping food.

Reflected: turned outward, e.g., a portion of the apertural lip of some snails' shells.

Ribs: prominent protrusions on the shell surface perpendicular to the spiral of the shell.

Rimate: the lip of the aperture slightly covers the umbilicus.

Rounded periphery: an evenly curved periphery; not angular or carinate.

Slug: a common designation for a snail without an external shell. The shell is either rudimentary and enclosed or is wanting entirely.

Senescence: biological aging.

Spiral: winding, coiling, or circling around a central axis; the form of the shell of most snails.

Spiral striae: surface features that are indented or raised on the shell surface running parallel with the whorls.

Spire: all the whorls above the aperture.

Striae: surface features that are either indented or raised in the shell surface.

Succiniform: "Shell shape" : shell that is higher than wide with a very large aperture (mouth). The spire is generally brief and the body whorl very expanded.

Suture: indentation of the shell surface where two whorls meet.

Taiga: also known as boreal forest, a northern biome characterized by coniferous forests consisting mostly of pines, spruces, and larches

Tentacles: elongated, flexible organs on the head of snails used for feeling, tasting or sensing light.

Tightly coiled: having many, narrowly expanding whorls.

Tooth: A hard, calcareous nodule or projection in or around the aperture of the shell of many snail species; usually of diagnostic value in snail taxonomy.

Transverse striae: surface features indented or raised in the shell surface running perpendicular with the whorls.

Truncate: cut off at the end; terminating abruptly; ending in a transverse line.

Umbilicate: Having an opening or cavity at the base of a gastropod shell, with the opening more than a narrow perforation.

Umbilicus: an opening in the center of the columella or axis of the shell.

Upper palatal lamella: uppermost of the two major teeth often found on the palatal wall.

Whorl: one complete turn of a gastropod shell.

Xeric: containing very little moisture; dry.

BIBLIOGRAPHY

ABBOTT, R.T. 1989. Compendium of Landshells. American Malacologists, Inc. Melbourne, Florida. 240 pp.

ANDERSON, R.C. & A.K. PRESTWOOD. 1981. Lungworms. Pp. 266-317 in W. R. Davidson, F. A. Hayes, V. F. Nettles, and F.E. Kellogg, eds., Diseases and parasites of white-tailed deer. Miscellaneous Publication. No. 7. Tall Timbers Research Station. Tallahassee, FL. 458 pp.

ARTHUR , M. A., L. M. TRITTON, & T. J. FAHEY. 1993. Dead bole mass and nutrients remaining 23 years after clear-cutting of a northern hardwood forest. Canadian Journal of Forest Reserves 23:1298-1305.

ATKINSON, J. W. 1998. Food manipulation and transport by a carnivorous land snail, *Haplotrema concavum.* Invertebrate Biology 117(2):109-113.

BONATO, V., K. G. FACURE, & W. UIEDA. 2004. Food habits of bats of subfamily Vampyrinae in Brazil. Journal of Mammalogy 708-713.

BOYCOTT, A. E. 1934. The habitats of land mollusca in Britain. Journal of Ecology 22:1-38.

BRANSON, B. A. 1973. Kentucky land mollusca: checklist, distribution, and keys for identification. Kentucky Department of Fish and Wildlife Resources. Frankfort, KY. 67 pp.

BRANSON, B. A. & D. L. BATCH. 1988. Distribution of Kentucky land snails (Mollusca: Gastropoda). Transactions of the Kentucky Academy of Science 49:101-116.

BRAUN, E. L. 1940. An ecological transect of Black Mountain, Kentucky. Ecological Monographs 10:193-241.

BURCH, J. B. 1955. Some ecological factors of the soil affecting the distribution and abundance of land snails in eastern Virginia. Nautilus 69:26-29.

BURCH, J. B. 1962. How to know the eastern land snails. Dubuque: William C. Brown Company.

BURCH, J. B. & T. A. PEARCE. 1990. Terrestrial Gastropoda. Pp. 201-309 in D. L. Dindall (ed.), Soil Biology Guide. Wiley and Sons, Inc., New Jersey

CAMERON, R. A. D. 1970. Differences in the distributions of three species Helicid snails in the limestone district of Derbyshire. Proceedings of the Royal Society of London 176:130-159.

CHASE, R. & K. C. BLANCHARD. (2006) The snail's love-dart delivers mucus to increase paternity. Proceedings of the Royal Society Biology. 273:1471-1475.

CONEY, C. C., W. A. TARPLEY, J. C. WARDEN, & J. W. NAGEL. 1982. Ecological studies of land snails in the Hiwassee River basin of Tennessee, U.S.A. Malacological Review 15:69-106.

COUNTS, C. L. III. 1982. Occurrence distribution of land snails of the family Polygyridae (Mollusca:Gastropoda:Pulmonata) in West Virginia. Brimleyana 8:145-157.

DALLINGER, R. 1993. Strategies of metal detoxification in terrestrial invertebrates. *in*: Dallinger, R. and Rainbow, P. S. (eds.). Ecotoxicology of Metals in Invertebrates. Lewis Publishers. Boca Raton, Florida. Pp. 245-289.

DALLINGER, R., & A. W. WIESER. 1984. Patterns of accumulation, distribution and liberation of Zn, Cu, Cd and Pb in different organs of the land snail Helix pomatia L. Comparative Biochemistry and Physiology-Part C: Toxicity and Pharmacology. 79 (1): 117-124.

DALLINGER, R. & A. W. WEISER. 1984a. Molecular fractionation of Zn, Cu, Cd, and Pb in the midgut gland of *Helix pomatia* L. Comparative Biochemistry and Physiology. 79:125-129.

DALLINGER, R., B. BERGER, C. GRUBER, P. HUNZIKER & S. STUZENBAUM. 2000. Metallothioneins in terrestrial invertebrates: structural aspects, biological significance, and implications for their use as biomarkers. Cellular and Molecular Biology 46:331-346.

DOUGLAS, D. A. 2011. Land snail species diversity and composition among different disturbance regimes in central and eastern Kentucky forests. (MS thesis). Eastern Kentucky University. Richmond, KY.

DOUGLAS, D. A, D. DOURSON, & R. C. CALDWELL. 2013. The land snails of White Oak Sinks, Great Smoky Mountains National Park. Southeastern Naturalist. 13(1): 166-175.

DOURSON, D. 2007. A selected land snail compilation of the Central Knobstone Escarpment on Furnace Mountain in Powell County Kentucky, USA. Journal of the Kentucky Academy of Science 68(2):119-131.

DOURSON, D. 2007. Survey Protocol for Cheat Threetooth (*Triodopsis platysayoides*) (Revised). West Virginia Department of Natural Resources, United States Fish and Wildlife Service.

DOURSON, D. 2008. The feeding behavior and diet of an endemic West Virginia land snail, *Triodopsis platysayoides*. American Malacological Bulletin. 26:153-159

DOURSON, D. 2010. Kentucky's land Snails and their ecological communities. 298 pp. Goatslug Publications, Bakersville, NC.

DOURSON, D. 2013. Land Snails of the Great Smoky Mountains National Park and Southern Appalachians. 336 pp. Goatslug Publications, Bakersville, NC.

DOURSON, D. 2012. Four new land snail species from the southern Appalachian Mountains. Journal of the North Carolina Academy of Science 128 (1):1-10.

DOURSON, D. 2011. Descriptions of three new land snails from Kentucky. Journal of the Kentucky Academy of Science . Vol. 72 (1):39-45.

DOURSON, D. & J. BEVERLY. 2009. Diversity, substrata divisions and biogeographical affinities of land snails at Bad Branch State Nature Preserve, Letcher County Kentucky, USA. Journal of the Kentucky Academy of Science. 68(2):119-131.

DOURSON, D. & M. GUMBERT. 2004. A survey of terrestrial mollusca in selected areas of the Great Smoky Mountains National Park, North Shore Road Project. Report submitted to Arcadis G & M of North Carolina, Inc. 76 pp.

DOURSON, D. & K. LANGDON. 2012. Land snails of selected rare high elevation forests and heath balds of the Great Smoky Mountains National Park. Journal of North Carolina Academy of Science: Summer 2012, Vol. 128, No. 2, pp. 27-32.

EMBERTON, K. C. 1991. The genitalic, allozymic and conchological evolution of the tribe Mesodontini (Pulmonata: Stylommatophora: Polygyridae). Malacologia 33:71-178.

EMBERTON, K. C. 1994. Polygyrid land snail phylogeny: external sperm exchange, early North American biogeography, iterative shell evolution. Biological Journal of the Linnean Society 52:241-271.

FAIRBANKS, H. L. 1998. Clarification of the taxonomic status and reproductive anatomy of *Philomycus batchi* Branson, 1968. The Nautilus 112 (1):1-5.

FOOTE, B. A. 1959. Biology and life history of the snail-killing flies belonging to the genus Sciomyza. Annals of the Entomological Society of America. 52:31-32.

FOURNIE, J. & M. CHETAIL. 1984. Calcium dynamics in land gastropods. American Zoologist 24:857-870.

GEIGER, R. 1965. The climate near the ground. Harvard University Press, Cambridge.

GETZ, L. L. 1974. Species diversity of terrestrial snails in the Great Smoky Mountains. The Nautilus 88:6-9.

GOSZ, J. R., G.E. LIKENS, & F. H. BORMANN. 1973. Nutrient release from decomposing leaf and branch litter in the Hubbard Brook Forest, New Hampshire. Ecological Monographs 43:173-191.

GRAVELAND, J. R. 1996. Avian eggshell formation in calcium-rich and calcium-poor habitats: importance of snail shells and anthropogenic calcium sources. Canadian Journal of Zoology 74:1035-1044.

GRAVELAND, J. R., & R. VAN DER WAL, J. H. 1994. Poor reproduction in forest passerines from decline of snail abundance on acidified soils. Nature 368:446-448.

GRIMM, F. W., R. G. FORSYTH, F. W. SCHUELER, & A. KARSTAD. 2009. Identifying land snails and slugs in Canada: introduced species and native genera. Canadian Food Inspection Agency. Ontario, Canada. 168 pp.

GUHA, M. M. & R. L. MITCHELL. 1966. The trace and major element composition of some deciduous trees: II. Seasonal changes. Plant and Soil 24:90-112.

399

HAMES, R. S., K. V. ROSENBERG, J. D. LOWE, & S. E. DHONDT. 2002. Adverse affects of acid rain on the (Wareborn, 1992) distribution of the woodthrush, *Hylocichla mustelina*, in North America. Proceedings of the National Academy of Science. 99 (17): 11235-11240

HICKMAN, C. P., L. S. ROBERTS & A. LARSON. 2003. Animal diversity. 3ʳ ed. McGraw-Hill, NY.

HOTOPP, K. 2002. Land snails and soil calcium in central Appalachian forests. Southeastern Naturalist 1(1): 27-44.

HOTOPP, K. 2015.A new *Triodopsis juxtidens* subspecies (Gastropoda: Pulmonata) from West Virginia, U. S. A. Zootaxa 3914(4):490-494. http://dx.doi.org/10.11646/zootaxa.3914.4.8

HOTOPP, K. & T. A. PEARCE. 2008. Land snail distributions in West Virginia. Report for contract WVFIMS 3203-2008-6305-099-025. Submitted to Michael Welch, Wildlife Resources Section, West Virginia Division of Natural Resources.

HOTOPP, K. , T. A. PEARCE & D.C.DOURSON. 2008. Land Snails of the Cheat River Canyon, West Virginia (Gastropoda: Pulmonata), Banisteria, Number31,pages 40-46.

HUBRICHT, L. 1958. New species of land snails from the eastern United States. Transactions of the Kentucky Academy of Science 19:70-76.

HUBRICHT, L. 1964. Land snails from the caves of Kentucky, Tennessee, and Alabama. Bulletin of the National Speleological Society. 26(1):33-36.

HUBRICHT, L. 1985. The distributions of the Native Land Mollusks of the Eastern United States. Fieldiana Zoology New Series, No. 24, Publication 1359: Field Museum of Natural History, Chicago, Illinois. 191 pp.

HUBRICHT L., R. S. CALDWELL, & J. G. PETRANKA. 1983. *Vitrinizonites latissimus* (Pulmonata: Zonitidae) and *Vertigo clappi* (Pupillidae) from eastern Kentucky. The Nautilus 97:20-22.

JACOT, A. P. 1935. Molluscan populations of old growth forests and re-wooded fields in the Asheville Basin of North Carolina. Ecology 16:603-605.

JENKINS, M. A., S. JOSE, & P. S. WHITE. 2007. Impacts of an exotic disease and vegetation change on foliar calcium cycling in Appalachian forests. Ecological Applications 17(3):869-881.

KALISZ, P. J. & J. E. POWELL, 2003. Effect of calcareous road dust on land snails and millipedes in acid forest soils of the Daniel Boone National Forest. Forest Ecology and Management 186:177-183.

KARLIN, E. J. 1961. Ecological relationships between vegetation in the distribution of land snails in Montana, Colorado, and New Mexico. American Midland Naturalist 65:60-66.

KELLER, H. & K. SNELL 2002. Feeding activities of slugs on Myxomycetes and macrofungi. Mycologia, 94:757-760.

KERNEY, M. P. AND R. A. D. CAMERON. 1979. Field guide to the land snails of Britain and north-west Europe.

KOENE J. M, T. S. LIEW, K. MONTAGNE-WAJER, M. SCHILTHUIZEN 2013. A syringe-like love dart injects male accessory gland products in a tropical hermaphrodite. PLoS ONE 8(7): e69968. doi:10.1371/journal.pone.0069968

LEE, J. C. 1994. The amphibians and reptiles of the Yucatán Peninsula. Cornell University Press, Ithaca /.

MACHENSTED, U. & K. MARKEL. 2001. Chapter 4: Radular structure and function. Pp. 213-236. In: The biology of terrestrial mollusks. G. M Barker, ed. CABI. New York, New York.

MACMILLAN, G. K. 1949. The land snails of West Virginia. Annals of the Carnegie Museum 31:89-238.

MARTIN, A. C., H. S. ZIM, & A. L. NELSON. 1951. American wildlife and plants: A guide to wildlife food habits. McGraw-Hill, Inc., New York.

MCHARGUE, J. S. & W. R. ROY. 1932. Mineral and nitrogen content of the leaves of some forest trees at different times in the growing season. Botanical Gazette 94:381-393.

MORITZ, C., K. S. RICHARDSON, S. FERRIER, J. STANISIC, S. E. WILLIAMS, & T. WHIFFIN. 2001. Biogeographic concordance and efficiency of taxon indicators for establishing conservation priority in a tropical rainforest biota. Proceedings of the Royal Society 268:1875-1881.

NATION, R. 2005. The influence of soil calcium on land snail diversity in the Blue Ridge Escarpment of South Carolina. Dissertation presented to Clemson University. Clemson, South Carolina.

NEKOLA, J. C. 1999. Terrestrial gastropod richness of carbonate cliff and associated habitats in the Great Lakes region of North America. Malacologia 41(1):231-252.

NEKOLA, J. C. 2003. Large-scale terrestrial gastropod community composition patterns in the Great Lakes region of North America. Diversity and Distribution. 9:55-71.

NEKOLA, J. C. and M. Barthel. 2002. Morphometric analysis of Carychium exile and Carychium exiguum in the Great Lakes region of North America. Journal of Conchology 37(5): 515.

NEKOLA, J. C. & B. F. COLES. 2010. Pupillid land snails of eastern North America. American Malacological Bulletin. 28:29-57.

NICKLAS N.L. & R. J. HOFFMAN. 1981. Apomictic parthenogenesis in a hermaphroditic terrestrial slug, Deroceras laeve (Müller). Biological Bulletin. 160:123-135.

PATTERSON, C. M. & J. B. BURCH. 1966. The chromosome cycle in the land snail Catinella vermeta (Stylommatophora: Succineidae) Malacologia. 3:309.

PEARCE, T. A. 2008. When a snail dies in the forest, how long will the shell persist? Effect of dissolution and micro-bioerosion. American Malacological Bulletin 26:111-117.

PEARCE, T. A. & A. GAERTNER. 1996. Optimal foraging and mucus-trail following in the snail-eating snail *Haplotrema concavum* (Gastropoda: Pulmonata). *Malacological Review* 29:85-99.

PETRANKA, J. G. 1982. The distribution and diversity of land snails on Big Black Mountain, Kentucky. (A thesis). University of Kentucky. Lexington, KY.

PETRANKA, J. W. 1998. Salamanders of the United States and Canada. Washington, D. C., USA: Smithsonian Institution Press.

PILSBRY, H. A. 1940. Land mollusca of North America (north of Mexico), Vol. I. Part 2. The Academy of Natural Sciences of Philadelphia Monographs.

PILSBRY, H. A. 1946. Land mollusca of North America (north of Mexico), Vol. II. Part 1. The Academy of Natural Sciences of Philadelphia Monographs.

PILSBRY, H. A. 1948. Land mollusca of North America (north of Mexico), Vol. I. Part 2. The Academy of Natural Sciences of Philadelphia Monographs.

POLLARD, E. 1975. Aspects of the ecology of *Helix pomatia* L. Journal of Animal Ecology 44:305-329.

POTTER, C. S., H. L. RAGSDALE, & C. W. BERISH. 1987. Reabsorption of foliar nutrients in a regenerating southern Appalachian forest. Oecologia 73:268-271.

RAWLS, H. & R. YATES. 1971. Fluorescence in Endodontid Snails. The Nautilus 85:17-20.

REID, F. A. 2006. Mammals of North America. Peterson Field Guide. New York: Houghton-Mifflin.

RICHTER, K. O. 1980. Evolutionary aspects of mycophagy in Ariolimax columbianus and other slugs. In: D. L. Dindal, ed., Soil Ecology as Related to Land Use Practices. Proceedings of the VII International Colloquium of Soil Biology, Washington D. C.

RICKLEFS, R. E. & K. K. MATTHEWS. 1982. Chemical characteristics of the foliage of some deciduous trees in southeastern Ontario. Canadian Journal of Botany 60:2037-2045.

RIMMER, C. C., K. P. MCFARLAND, D. C. EVERS, E. K. MILLER, Y. AUBRY, D. BUSBY & R. J. TAYLOR. 2005. Mercury concentrations in Bicknell's Thrush and other insectivorous passerines in montane forests of northeastern North America. Ecotoxicology 14:223-240.

ROODY, W. C. 2003. Mushrooms of West Virginia and the Central Appalachians. The University of Kentucky Press. Lexington, Kentucky 519 p.

ROTH, B. 1988. Identities of two Californian land mollusks described by Wesley Newcomb. Malacological Review 20:129-132.

SATHEESHKUMAR, P., A. B. KHAN, & D. SENTHILKUMAR. 2010. Marine organisms as potential supply for drug finding-a review study. Middle-East Journal of Scientific Research 5 (6): 514-519.

SCHEIFLER, R., C. SWARTZ, G. ECHEVARRIA, A. DE VAUFLEURY, P. BADOT, & J. MOREL. 2003. "Nonavailable" soil cadmium is bioavailable to snails: evidence from isotopic dilution experiments. Environmental Science Technology 37(1):81-86.

SHEARER, A. & J. W. ATKINSON. 2001. Comparative analysis of food-finding behavior of an herbivorous and a carnivorous land snail. Invertebrate Biology 120:199-205.

SLAPCINSKY, J. & B. COLES. 2004. Revision of the genus *Pilsbryna* (Gastropoda: Pulmonata: Gastrodonatidae) and comments on the taxonomic status of *Paravitrea tridens* Morrison, 1935. The Nautilus 118 (2):55-70.

TAYLOR, R. W. AND C. L. COUNTS, III. 1976. Note on some land snails from Blennerhassett Island, West Virginia. Sterkiana 63-64:11.

THABAH, A. & G. LI, Y. WANG, B. LIANG, L. HU, S. ZHANG, & G. JONES. 2007. Diet, echolocation calls, and phylogenetic affinities of the great evening bat (IA IO; Vespertilionidae): another carnivorous bat. Journal of Mammology. 88(3):728-735.

TOWNSEND, J. S., H. C. ALDRICH, L. D. WILSON, & J. R. MCCRANIE. 2007. First report of sporangia of a myxomycete (*Physarum pusillum*) on the body of a living animal, the lizard *Corytophanes cristatus'*. Mycologia. 97(2):246-248.

VANDEVENDER, A. S. & R. W. VANDEVENDER. 2003. Surveying the land snails of the southern Appalachians. Program and Abstracts of American Malacological Society: 61.

VESTERDAL. I., & K. RAULUND-RASMUSSEN 1998. Forest floor under seven tree species along a soil fertility gradient. Canadian Journal of Forest Reserves 28:1636-1647.

WALLACE, M.S., R. RAUCK, R. S. FISHER, G. CHARAPATA, D. ELLIS, & S. DISSANAYAKE. 2008. Ziconotide 98-022 Study Group. Intrathecal ziconotide for severe chronic pain: safety and tolerability results of an open-label, long-term trial. Anesthesia and Analgesia 106: 628-637.

WAREBORN, I. 1970. Environmental factors influencing the distribution of land mollusks of an oligotrophic area in southern Sweden. Oikos 2:285-291.

Index of Scientific Names

Index of Common Names

ABOUT THE AUTHOR

Dan Dourson is a biologist/naturalist/illustrator who has spent most of his adult life dedicated to the preservation, conservation, and understanding of the planet's more obscure animals. As an employee of the United States Forest Service for nearly 20 years, he worked to manage and conserve rare plants, bats, birds, reptiles, amphibians, land snails and freshwater mussels. In addition, Dan has conducted biological inventories for multiple agencies (i.e., Nature Conservancy, State Agencies and National Park Service) for numerous taxa groups and has worked for more than a decade studying the land snails of the Great Smoky Mountains National Park as part of the "All Taxa Biodiversity Inventory" through the Discover Life in America initiative. Dan has also studied land snails and other animals abroad, including Belize, Guatemala, Costa Rica and Panama and in South America, conducted land snail studies in the Tamshiyacu-Tahuayo Communal Reserve in the upper Amazon Basin of northwestern Peru.

Dan has published numerous peer review papers in prestigious journals such as the American Malacological Bulletin, The Nautilus, Southeastern Naturalist, Mesoamericana, Journal of the North Carolina Academy of Science, and others. He has described 10 new land snail species from the states of Kentucky, West Virginia, Tennessee, and North Carolina and in Central America from Belize. Dan has also authored 7 books including *"Wild Yet Tasty"* (his first), *"Kentucky's Land Snails and their Ecological Communities"*, *Biodiversity of the Maya Mountains, Belize, Central America"*, *"Land Snails of the Great Smoky Mountains National Park"*, and is currently working on *"The Land Snails of Belize Central America"*. Dan has illustrated nature books such as *"Wildflowers of Kentucky"*, *"Ginseng Dreams"*, *"Reptiles and Amphibians of Kentucky"*, *"How Snakes Work"* and multiple environmental posters used in schools around the country.

Dan remains committed to conservation work protecting the earth's most amazing and underappreciated organisms. His allegiance with the natural world is clearly reflected through his writing and simple lifestyle. He divides his time between the USA and Belize with his wife, Judy.

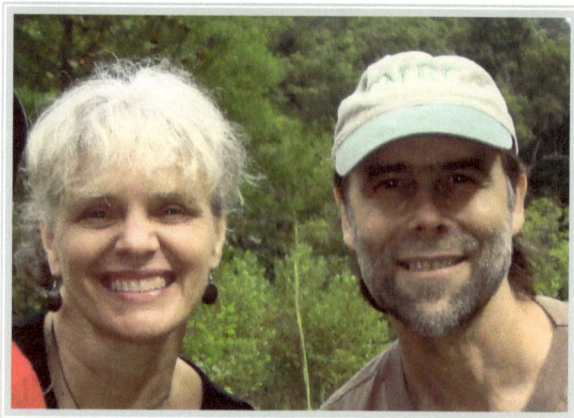

www.ingramcontent.com/pod-product-compliance
Lightning Source LLC
Chambersburg PA
CBHW041557220326
41597CB00051B/1